T0324893

METHODS IN MOLECULAR BIOLOGY

Series Editor
John M. Walker
School of Life and Medical Sciences
University of Hertfordshire
Hatfield, Hertfordshire, AL10 9AB, UK

For further volumes:
http://www.springer.com/series/7651

Programmed Cell Death

Methods and Protocols

Edited by

Hamsa Puthalakath and Christine J. Hawkins

Department of Biochemistry and Genetics, La Trobe Institute for Molecular Science,
La Trobe University, Bundoora, VIC, Australia

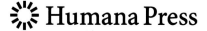

Editors
Hamsa Puthalakath
Department of Biochemistry and Genetics
La Trobe Institute for Molecular Science
La Trobe University
Bundoora, VIC, Australia

Christine J. Hawkins
Department of Biochemistry and Genetics
La Trobe Institute for Molecular Science
La Trobe University
Bundoora, VIC, Australia

Videos can also be accessed at http://link.springer.com/book/10.1007/978-1-4939-3581-9

ISSN 1064-3745   ISSN 1940-6029   (electronic)
Methods in Molecular Biology
ISBN 978-1-4939-3579-6   ISBN 978-1-4939-3581-9   (eBook)
DOI 10.1007/978-1-4939-3581-9

Library of Congress Control Number: 2016934448

Cover illustration: A high resolution image of a *Drosophila* eye in which the pro-apoptotic gene, *hid*, was expressed using an eye-specific promoter, resulting in eye ablation.

Printed on acid-free paper

This Humana Press imprint is published by Springer Nature
The registered company is Springer Science+Business Media LLC New York

Preface

The field of programmed cell death is now one of the most dynamic and fast-moving areas of research in biology. Focus of this research field encompasses the basic science of delineating the molecular mechanisms underpinning the cell death process in both mammals and other model organisms (such as *C. elegans, Drosophila,* and *S. cerevisiae*) as well as understanding its role in various human pathologies such as cancer, degenerative diseases, autoimmunity, cardiomyopathy, and sepsis. In this edition of Methods in Molecular Biology, apart from protocols specifically designed for studying programmed cell death, we also have incorporated many of the recent advances in techniques that span broader areas of biology that have been recently used or that have the potential to be incorporated into cell death research. Though the protocols are mostly described in the context of mammalian systems, we also have incorporated other systems such as plants, *Drosophila,* and yeast as tool for studying metazoan apoptosis process.

The first five chapters describe apoptosis detection techniques. These include methods to discriminate between apoptotic and autophagic cell death processes and in vivo imaging of apoptosis either by Gallium-68 labeled annexin V or by using fluorescent, activity-based probes. The latter is also a useful technique for measuring caspase activity ex vivo in resected tissues. This section also covers a method for detecting caspase activity initiation at single-cell level and peptide-based techniques for detecting or inhibiting specific caspases. The next five chapters describe methods for studying apoptosis associated with various pathologies in different organs including the lymphoid compartment (for studying sepsis), intestinal epithelium (for studying gastric abnormalities), granulocytes (for studying allergies and hypersensitivity), hepatocytes (for studying HBV infection), and cardiomyocytes (for studying cardiomyopathy and heart failure).

Lower forms of eukaryotes such as *Drosophila* have been invaluable in studying apoptosis process during development and tumorigenesis owing to their extensive use as model organism at the cutting edge of genetic research, with exciting new technologies continuing to emerge. Similarly, programmed cell death is a critical component of plant development and immunity against pathogens. Metazoan apoptotic machinery can be reconstituted in bakers' yeast *Saccharomyces cerevisiae* to study the function of candidate or established apoptotic regulators. Chapters 11–13 cover protocols and techniques for studying apoptosis in these three non-mammalian systems. Chapters 14–16 cover biochemical and biophysical methods for studying Bcl-2 family protein dynamics and protein-protein interactions during apoptosis. The last section, i.e., Chapters 17–20 includes protocols that are useful not only in apoptosis research but also in the wider areas of biological research. This section covers methods for genome editing, inducible transgenes, and proteomics.

We hope the scope of this book is sufficiently comprehensive and useful for researchers who are beginners as well as advanced in the field. We are extremely grateful to all of the colleagues who provided such high-quality contributions to this book and to the Springer Publishers for their support, especially Professor Emeritus John Walker. Finally, we are indebted to our friend and colleague Dr. Anjali Sahasrabudhe for her exceptional and tireless editorial help.

Bundoora, VIC, Australia

Hamsa Puthalakath
Christine J. Hawkins

Contents

Contributors

DAVID W. ANDREWS • *Department of Biochemistry, University of Toronto, Toronto, ON, Canada; Department of Medical Biophysics, University of Toronto, Toronto, ON, Canada; Department of Biological Sciences, Sunnybrook Research Institute, Toronto, ON, Canada*

MATTHIAS BAUWENS • *Nuclear Medicine, NUTRIM, Maastricht University Medical Center, Maastricht, Netherlands; Radiopharmacy, KU Leuven, Leuven, Belgium*

DAVID T. BLOOMER • *Department of Biochemistry and Genetics, La Trobe Institute for Molecular Science, La Trobe University, Bundoora, Melbourne, VIC, Australia*

MATTHEW BOGYO • *Department of Pathology, Stanford University School of Medicine, Stanford, CA, USA; Department of Microbiology and Immunology, Stanford University School of Medicine, Stanford, CA, USA*

LISA BOUCHIER-HAYES • *Department of Pediatrics-Hematology and Center for Cell and Gene Therapy, Baylor College of Medicine, Houston, TX, USA*

INGO L. BRAND • *Department of Biochemistry and Genetics, La Trobe Institute for Molecular Science, La Trobe University, Bundoora, Melbourne, VIC, Australia*

GABRIELA BRUMATTI • *The Walter and Eliza Hall Institute of Medical Research, Parkville, VIC, Australia; Department of Medical Biology, University of Melbourne, Parkville, VIC, Australia*

THOMAS BRUNNER • *Biochemical Pharmacology, University of Konstanz, Konstanz, Germany*

XIAOKE CHI • *Department of Biological Sciences, Sunnybrook Research Institute, Toronto, ON, Canada; Department of Chemical Biology, McMaster University, Hamilton, ON, Canada*

PETER E. CZABOTAR • *The Walter and Eliza Hall Institute of Medical Research, Parkville, VIC, Australia; Department of Medical Biology, The University of Melbourne, Melbourne, VIC, Australia*

ADRIAN N. DAUPHINEE • *Department of Biology, Dalhousie University, Halifax, NS, Canada*

EUGENIA DELGADO • *Biochemical Pharmacology, University of Konstanz, Konstanz, Germany*

GRANT DEWSON • *Walter and Eliza Hall Institute of Medical Research, Parkville, VIC, Australia; Department of Medical Biology, University of Melbourne, Melbourne, VIC, Australia*

MARCEL DOERFLINGER • *Department of Biochemistry and Genetics, La Trobe Institute for Molecular Science, La Trobe University, Melbourne, VIC, Australia*

GREGOR EBERT • *The Walter and Eliza Hall Institute of Medical Research, Parkville, VIC, Australia; Department of Medical Biology, The University of Melbourne, Parkville, VIC, Australia*

LAURA E. EDGINGTON-MITCHELL • *Drug Discovery Biology, Monash Institute of Pharmaceutical Sciences, Monash University, Parkville, VIC, Australia*

SARA R. FASSIO • *Department of Pediatrics-Hematology and Center for Cell and Gene Therapy, Baylor College of Medicine, Houston, TX, USA*

PATRICK GALLOIS • *Faculty of Life Sciences, University of Manchester, Manchester, UK*

LORENZO GALLUZZI • *Gustave Roussy Cancer Campus, Villejuif, France; Equipe 11 labellisée par la Ligue Nationale contre le Cancer, Centre de Recherche des Cordeliers, Paris, France; INSERM, U1138, Paris, France; Université Paris Descartes/Paris V, Sorbonne Paris Cité, Paris, France; Université Pierre et Marie Curie/Paris VI, Paris, France*

JASON GLAB • *Department of Biochemistry and Genetics, La Trobe Institute for Molecular Science, La Trobe University, Melbourne, VIC, Australia*

THOMAS GRABINGER • *Biochemical Pharmacology, University of Konstanz, Konstanz, Germany*

ARUNIKA HLAN GUNAWARDENA • *Department of Biology, Dalhousie University, Halifax, NS, Canada*

CHRISTINE J. HAWKINS • *Department of Biochemistry and Genetics, La Trobe Institute for Molecular Science, La Trobe University, Bundoora, Melbourne, VIC, Australia; Murdoch Children's Research Institute, Royal Children's Hospital, Parkville, VIC, Australia*

MARCO J. HEROLD • *Walter and Eliza Hall Institute of Medical Research, Parkville, VIC, Australia; Department of Medical Biology, University of Melbourne, Parkville, VIC, Australia*

ANISSA M. JABBOUR • *Murdoch Children's Research Institute, Royal Children's Hospital, Parkville, VIC, Australia; Australian Centre for Blood Diseases, Faculty of Medicine, Nursing and Health Science, Monash University, Melbourne, VIC, Australia*

JOANNA KACPRZYK • *School of Biology and Environmental Science, University College Dublin, Dublin, Ireland*

THOMAS KAUFMANN • *Institute of Pharmacology, University of Bern, Bern, Switzerland*

TANJA KITEVSKA • *Department of Biochemistry and Genetics, La Trobe Institute for Molecular Science, La Trobe University, Bundoora, Melbourne, VIC, Australia*

GUIDO KROEMER • *Equipe 11 labellisée par la Ligue Nationale contre le Cancer, Centre de Recherche des Cordeliers, Paris, France; INSERM, U1138, Paris, France; Université Paris Descartes/Paris V, Sorbonne Paris Cité, Paris, France; Université Pierre et Marie Curie/Paris VI, Paris, France; Pôle de Biologie, Hôpital Européen Georges Pompidou, AP-HP, Paris, France; Metabolomics and Cell Biology Platforms, Gustave Roussy Cancer Campus, Villejuif, France; Department of Women's and Children's Health, Karolinska University Hospital, Stockholm, Sweden*

ANDREW J. KUEH • *Walter and Eliza Hall Institute of Medical Research, Parkville, VIC, Australia; Department of Medical Biology, University of Melbourne, Parkville, VIC, Australia*

MARC KVANSAKUL • *Department of Biochemistry and Genetics, La Trobe Institute for Molecular Science, La Trobe University, Melbourne, VIC, Australia*

M. CHIARA MAIURI • *Gustave Roussy Cancer Campus, Villejuif, France; Equipe 11 labellisée par la Ligue Nationale contre le Cancer, Centre de Recherche des Cordeliers, Paris, France; INSERM, U1138, Paris, France; Université Paris Descartes/Paris V, Sorbonne Paris Cité, Paris, France; Université Pierre et Marie Curie/Paris VI, Paris, France*

GEORGE WILLIAMS MBOGO • *Department of Biochemistry and Genetics, La Trobe Institute for Molecular Science, La Trobe University, Melbourne, VIC, Australia*

PAUL F. McCABE • *School of Biology and Environmental Science, University College Dublin, Dublin, Ireland*

GAVIN P. McSTAY • *Department of Life Sciences, New York Institute of Technology, Old Westbury, NY, USA*

UELI NACHBUR • *The Walter and Eliza Hall Institute of Medical Research, Parkville, VIC, Australia; Department of Medical Biology, University of Melbourne, Parkville, VIC, Australia*

CHRISTINA NEDEVA • *Department of Biochemistry and Genetics, La Trobe Institute for Molecular Science, La Trobe University, Melbourne, VIC, Australia*

HANG NGUYEN • *Department of Biochemistry and Genetics, La Trobe Institute for Molecular Science, La Trobe University, Bundoora, Melbourne, VIC, Australia*

MELISSA J. PARSONS • *Department of Pediatrics-Hematology and Center for Cell and Gene Therapy, Baylor College of Medicine, Houston, TX, USA*

MARC PELLEGRINI • *The Walter and Eliza Hall Institute of Medical Research, Parkville, VIC, Australia; Department of Medical Biology, The University of Melbourne, Parkville, VIC, Australia*

JAMES M. PEMBERTON • *Department of Medical Biophysics, University of Toronto, Toronto, ON, Canada; Department of Biological Sciences, Sunnybrook Research Institute, Toronto, ON, Canada*

JUSTIN P. POGMORE • *Department of Biochemistry, University of Toronto, Toronto, ON, Canada; Department of Biological Sciences, Sunnybrook Research Institute, Toronto, ON, Canada*

SIMON P. PRESTON • *The Walter and Eliza Hall Institute of Medical Research, Parkville, VIC, Australia; Department of Medical Biology, The University of Melbourne, Parkville, VIC, Australia*

HAMSA PUTHALAKATH • *Department of Biochemistry and Genetics, La Trobe Institute for Molecular Science, La Trobe University, Melbourne, VIC, Australia*

RAMONA REINHART • *Institute of Pharmacology, University of Bern, Bern, Switzerland*

BORIS RELJIĆ • *The Walter and Eliza Hall Institute of Medical Research, Parkville, VIC, Australia; Department of Medical Biology, The University of Melbourne, Parkville, VIC, Australia*

HYUNG DON RYOO • *Department of Cell Biology, New York University School of Medicine, New York, NY, USA*

VALENTINA SICA • *Gustave Roussy Cancer Campus, Villejuif, France; Equipe 11 labellisée par la Ligue Nationale contre le Cancer, Centre de Recherche des Cordeliers, Paris, France; INSERM, U1138, Paris, France; Faculté de Medicine, Université Paris Saclay/Paris XI, Le Kremlin-Bicêtre, France; Université Paris Descartes/Paris V, Sorbonne Paris Cité, Paris, France*

DAVID A. STROUD • *Department of Biochemistry and Molecular Biology, Monash Biomedicine Discovery Institute, Monash University, Clayton, Melbourne, VIC, Australia*

DEEPIKA VASUDEVAN • *Department of Cell Biology, New York University School of Medicine, New York, NY, USA*

ANDREW I. WEBB • *Systems Biology and Personalized Medicine Division, The Walter and Eliza Hall Institute, Parkville, VIC, Australia*

SIMONE WICKI • *Institute of Pharmacology, University of Bern, Bern, Switzerland*

Chapter 1

Detection of Apoptotic Versus Autophagic Cell Death by Flow Cytometry

Valentina Sica, M. Chiara Maiuri, Guido Kroemer, and Lorenzo Galluzzi

Abstract

Different modes of regulated cell death (RCD) can be initiated by distinct molecular machineries and their morphological manifestations can be difficult to discriminate. Moreover, cells responding to stress often activate an adaptive response centered around autophagy, and whether such a response is cytoprotective or cytotoxic cannot be predicted based on morphological parameters only. Molecular definitions are therefore important to understand various RCD subroutines from a mechanistic perspective. In vitro, various forms of RCD including apoptosis and autophagic cell death can be easily discriminated from each other with assays that involve chemical or pharmacological interventions targeting key components of either pathway. Here, we detail a straightforward method to discriminate apoptosis from autophagic cell death by flow cytometry, based on the broad-spectrum caspase inhibitor Z-VAD-fmk and the genetic inhibition of ATG5.

Key words Autophagy, Immunogenic cell death, Mitochondrial outer membrane permeabilization, Necrosis, Mitochondrial permeability transition, Necroptosis

1 Introduction

Cell death can be accidental, meaning that its course cannot be altered, or regulated, which means that it can be inhibited or at least retarded by specific pharmacological or genetic interventions [1–3]. Thus, at odds with its accidental counterpart, regulated cell death (RCD) is precipitated by the activation of a genetically encoded molecular machinery, which generally occurs once adaptive responses to stress fail [4–6]. As a notable exception to this tendency, programmed cell death (PCD) constitutes a peculiar case of RCD that is activated in a completely physiological manner, in the context of post-embryonic development or adult tissue homeostasis [1, 7].

During the past three decades, several systems for the classification of RCD have been proposed, based on morphological, biochemical, or functional features [8–14]. It soon became clear

Hamsa Puthalakath and Christine J. Hawkins (eds.), *Programmed Cell Death: Methods and Protocols*, Methods in Molecular Biology, vol. 1419, DOI 10.1007/978-1-4939-3581-9_1, © Springer Science+Business Media New York 2016

that defining RCD instances based on their morphology is rather inappropriate, because it suffers from a considerable degree of operator-dependency, and it does not convey any mechanistic information [15, 16]. Nowadays, a molecular classification of RCD, based on objectively quantifiable biochemical parameters, is favored [17]. Thus, apoptosis is currently defined as caspase-3-dependent variant of RCD, while autophagic cell death is defined as a form of RCD that mechanistically impinges on the molecular machinery for autophagy [17, 18]. However, detecting caspase-3 activation or any biochemical manifestations of autophagy in the course of RCD is not sufficient for defining it as apoptotic or autophagic. Indeed, caspase-3 activation occurs in several apoptosis-unrelated settings [19]. Along similar lines, the adaptive response of eukaryotic cells to stress often (if not always) involves an autophagic component, which generally supports (rather than compromises) the reestablishment of homeostasis and cell survival [20–22].

These observations imply that functional assays are required to properly identify apoptotic and autophagic instances of RCD, as well as other forms or RCD including necroptosis [23, 24]. While implementing such functional assays in vivo may be complicated, they can be carried out in vitro in a relatively straightforward manner, by appropriately combining (1) the detection of reliable indicators of cell death, and (2) the use of pharmacological or genetic interventions that inhibit caspase-3 or essential components of the autophagic machinery.

It is widely accepted that plasma membrane permeabilization (PMP) constitutes the most reliable (if not the only) marker of dead cells, at least in vitro [1]. Indeed, while several other biochemical processes can accompany (and be mechanistically involved in) RCD, most (if not all) of them: (1) are not universally associated with it; and (2) are not always irreversible. For instance, caspase-3 is activated not only during the terminal phases of apoptosis, but also in a reversible manner in the course of erythroid differentiation (to which it provides a critical contribution) [25]. Additional events that generally accompany RCD, such as mitochondrial outer membrane permeabilization (MOMP) and phosphatidylserine (PS) exposure, can be monitored to obtain kinetic insights into the process, but are inappropriate as sole biomarkers of cell death [5, 26]. In vitro, PMP, MOMP, and PS exposure can be conveniently measured by flow cytometry, after co-staining living cells with the exclusion dye propidium iodide (PI) and either the mitochondrial transmembrane potential ($\Delta\psi_m$)-sensitive fluorophore 3,3'-dihexyloxacarbocyanine iodide ($DiOC_6(3)$), or a fluorescent variant of the PS-binding protein annexin A5 (ANXA5, best known as AnnV) [27].

Several pharmacological agents have been developed to inhibit components of the apoptotic or autophagic machinery, and are now commercially available. For instance, the non-cleavable caspase substrate *N*-benzyloxycarbonyl-Val-Ala-Asp(O-Me) flu-oromethylketone (Z-VAD-fmk) is commonly employed as an inhibitor of apoptosis [28, 29], while the phosphoinositide-3-ki-nase (PI3K) inhibitor 3-methyladenine (3-MA) can be used to block autophagic responses, which often rely on the PI3K-dependent synthesis of phosphatidylinositol-3-phosphate [30, 31]. Still, many of these compounds have specificity issues, warranting the use of targeted genetic tools, including RNA interference and the gene knockout technology. General recommendations for the implementation of appropriate assays involving RNA interference or knockout cells go beyond the scope of this chapter and can be found in the literature [32–34].

Here, we provide a detailed description of a simple, cytofluo-rometric assay for the discrimination of apoptotic and autophagic cell death in vitro, based on the simultaneous detection of PMP and MOMP exposure in human cancer cells responding to a lethal stimulus in normal conditions, in the presence of Z-VAD-fmk, or upon the small interfering RNA (siRNA)-mediated downregula-tion of ATG5. With the appropriate variations, this protocol is suitable for the identification of apoptosis and autophagic cell death in most, if not all, cultured mammalian cells.

2 Materials

2.1 Disposables and Equipment

1. 6- and 12-well plates for cell culture.
2. 75 cm^2 flasks for cell culture.
3. 5 mL, 12×75 mm FACS tubes.
4. 1.5 mL microcentrifuge tubes.
5. 15 and 50 mL conical centrifuge tubes.
6. Cytofluorometer: FACScan or FACSVantage (BD, San Jose, USA) or equivalent, equipped with an argon ion laser emitting at 488 nm and controlled by the operational/analytical software CellQuest™ Pro (BD) or equivalent (*see* **Note 1**).

2.2 Cell Maintenance

1. Complete growth medium for human osteosarcoma U2OS cells: Dulbecco modified Eagle's medium (DMEM) contain-ing 3.0 g/L D-glucose, 1.5 mM L-glutamine, supplemented with 100 mM 2-[4-(2-hydroxyethyl)piperazin-1-yl]ethanesul-fonic acid (HEPES) buffer and 10 % fetal bovine serum (FBS) (*see* **Note 2**).
2. Phosphate buffered saline (PBS, 1×): 137 mM NaCl, 2.7 mM KCl, 4.3 mM Na_2HPO_4, 1.4 mM KH_2PO_4 in deionized water (dH_2O), adjust pH to 7.4 with 2 N NaOH.
3. Trypsin–EDTA: 0.25 % trypsin–0.38 g/L (1 mM) EDTA×4 Na^+ in Hank's balanced salt solution (*see* **Note 3**).

2.3 RNA Interference

1. siUNR (sense 5′-GCCGGUAUGCCGGUUAAGUdTdT-3′), 100 µM stock solution in dH$_2$O, stored at –20 °C (*see* **Notes 4 and 5**).

2. siATG5 (sense 5′-UUUCUUCUUAGGCCAAAGGdTdT-3′), 100 µM, stock solution in dH$_2$O, stored at –20 °C (*see* **Notes 4 and 6**).

3. Transfection reagent: HiPerFect® or equivalent (*see* **Note 7**).

4. Transfection medium: Opti-MEM® with Glutamax™ and phenol red (*see* **Note 8**).

2.4 Pharmacological Treatments and DiOC6(3)/PI Co-staining

1. *N*-benzyloxycarbonyl-Val-Ala-Asp (O-Me) fluoromethylketone (Z-VAD-fmk): 20 mM stock solution in dimethylsulfoxide (DMSO), stored at –20 °C (*see* **Note 9**).

2. Staurosporine (STS): 2 mM stock solution in DMSO, stored at –20 °C (*see* **Note 10**).

3. 3,3′-dihexyloxacarbocyanine iodide (DiOC$_6$(3)): 40 µM stock solution in 100 % ethanol,, stored at –20 °C under protection from light (*see* **Notes 11 and 12**).

4. Propidium iodide (PI): 1 mg/mL stock solution in dH$_2$O, stored at 4 °C under protection from light (*see* **Notes 13 and 14**).

3 Methods

3.1 Cell Maintenance

1. U2OS cells are routinely maintained in complete growth medium within 75 cm^2 flasks, in standard culture conditions (37 °C, 5 % CO$_2$) (*see* **Note 15**).

2. When the culture reach 70–80 % confluence (*see* **Note 16**), discard the supernatant by aspiration, wash gently adherent cells with pre-warmed PBS (*see* **Note 17**), and incubate them with ~3 mL 0.25 % (w/v) Trypsin–EDTA solution for 1–3 min at 37 °C (*see* **Notes 19**).

3. As soon as cells are detached (*see* **Note 20**), add complete growth medium to the cell suspension (*see* **Note 21**).

4. Maintenance cell cultures can be generated by transferring aliquots of the cell suspension to new 75 cm^2 flasks, and propagated as described in **steps 1–3** in Subheading 3.1 (*see* **Notes 15–23**).

5. For RNA interference, seed 2.0×10^5 U2OS cells in 6-well plates, in 2 mL growth medium per well (*see* **Note 24**), and proceed to Subheading 3.2.

6. For pharmacological treatments, seed 1.5×10^5 U2OS cells in 12-well plates, in 1 mL growth medium per well (*see* **Note 25**), and proceed to Subheading 3.3.

3.2 RNA Interference

1. When cells reach a confluence of 40–60 % (*see* **Note 26**), dilute 50 pmol siRNA (final concentration in wells = 25 nM) in 62.5 µL Opti-MEM® (solution A), and 7.5 µL HiPerFect® in 55 µL Opti-MEM® (solution B), and allow both solutions to stand at RT for 5–10 min (*see* **Note 27**).

2. Mix solution A and B gently, and incubate at RT additional for 15–20 min, to allow for the formation of HiPerFect®:siRNA transfection complexes (transfection solution) (*see* **Notes 28** and **29**).

3. Replace growth medium with 1.875 mL complete growth medium.

4. Add 125 µL of the transfection solution to each well (*see* **Note 30**), and incubate plates under standard culture conditions (37 °C, 5 % CO_2).

5. 4–24 h later (*see* **Notes 31** and **32**), detach transfected cells (500 µL trypsin–EDTA per well), seed them in 12-well plates (0.8×10^5 cells in 1 mL growth medium per well) (*see* **Note 25**), and proceed to Subheading 3.3.

3.3 Pharmacological Treatments and DiOC6(3)/PI Co-staining

1. 24 h after seeding non-transfected U2OS cells, as described in **step 6** in Subheading 3.1 (*see* **Note 33**), gently remove supernatant and substitute with 1 mL complete culture medium alone (or containing an equivalent amount of solvent, negative control condition), or supplemented with 1 µM STS (or the cell death inducer of choice), 50 µM Z-VAD-fmk (additional control condition), or 1 µM STS + 50 µM Z-VAD-fmk.

2. Alternatively, 24 h after seeding transfected U2OS cells, as described in **step 3** in Subheading 3.2 (*see* **Note 33**), gently remove supernatant and substitute with 1 mL complete culture medium alone (or containing an equivalent amount of solvent, negative control condition), or supplemented with 1 µM STS (or the cell death inducer of choice).

3. When the stimulation period is over, collect culture supernatants in 5 mL FACS tubes (*see* **Notes 34** and **35**) and detach adherent cells with ~0.5 mL trypsin–EDTA, following a wash with ~0.5 mL pre-warmed PBS (*see also* **steps 2** and **3** in Subheading 3.1, and **Notes 17–19**).

4. Following complete detachment (*see* **Note 20**), add 1 mL complete growth medium to each well (*see* **Note 21**), and transfer cells to the FACS tube containing the corresponding supernatant.

5. Spin down cell suspensions at $300 \times g$, RT, for 5 min.

6. Discard supernatants, resuspend cells in 200–400 µL of staining solution (40 nM $DiOC_6(3)$ in complete growth medium) (*see* **Notes 36–38**), and incubate them for 20–30 min in the dark at 37 °C (5 % CO_2) (*see* **Note 39**).

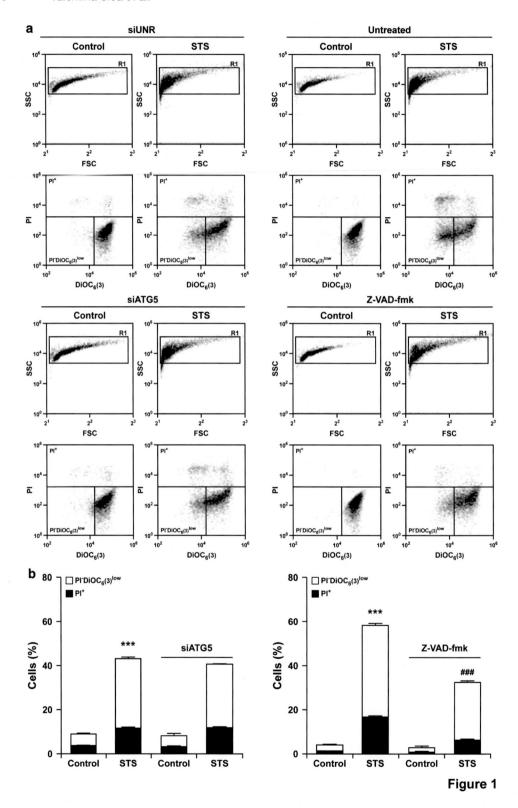

Figure 1

7. Add PI to each sample at a final concentration of 1 μg/mL, and incubate samples for additional 2–5 min under protection from light (*see* **Notes 40** and **41**).

8. Analyze samples on a conventional flow cytometer allowing for the simultaneous assessment of light scattering parameters (forward and side scatter, FSC and SSC) and fluorescence in two separate channels (e.g., green and red) (*see* **Notes 42–45**).

9. *Bona fide* apoptotic cell death is retarded in the presence of Z-VAD-fmk, but normally accelerated upon the pharmacological or genetic inhibition of core components of the machinery for autophagy (*see* Fig. 1). Conversely, bona fide autophagic cell death is insensitive to Z-VAD-fmk, but can be retarded by the pharmacological or genetic inhibition of autophagy (not shown).

4   Notes

1. The manufacturer recommends to periodically check flow rate, laser alignment and fluorescence stability, to ensure technical reliability from the instrument. Moreover, it is recommended to align/calibrate the cytometer with standard beads for flow cytometry, as per manufacturer's recommendations, prior to each experimental session.

2. Recommended for U2OS cells by the American Type Culture Collection (ATCC, Manassas, VA, USA).

3. Under optimal storage conditions (–20 °C, protected from light), trypsin–EDTA is stable for at least 18 months. Repeated freeze-thawing should be avoided by storing the reagent in aliquots of 2–10 mL. Once thawed, the trypsin–EDTA is stable at 4 °C for approximately 2 weeks.

4. According to the manufacturer, lyophilized siRNA are stable for at least 3 years if stored appropriately (–20 °C, protected from light). Under appropriate storage conditions reconstituted siRNA stock solution are stable for at least 6 months. We recommend storing the reagent in small aliquots (5–20 μL), to avoid repeated freeze-thawing.

Fig. 1 Apoptotic cell death induced by staurosporine. Human osteosarcoma U2OS cells were transfected with a control siRNA (siUNR) or with an ATG5-targeting siRNA (siATG5) for 48 h, as detailed in Subheading 3.2, and then left untreated or treated with 1 μM staurosporine (STS) for additional 24 h. Alternatively, non-transfected U2OS cells were maintained in control conditions or treated with 1 μM STS, alone or in combination with Fig50 μM Z-VAD-fmk, as detailed in Subheading 3.3. Twenty-four later cells were processed for the cytofluorometric quantification of plasma membrane permeabilization and mitochondrial transmembrane potential dissipation as detailed in Subheading 3.3. In panel (**a**), representative *dot plots* are reported. In panel (**b**), quantitative data are reported (means ± SD, $n = 2$ parallel samples, ***$p < 0.001$, as compared to untreated or siUNR-transfected cells; ###$p < 0.001$, as compared to cells treated with STS only; two-sided, unpaired Student's *t* test)

5. siUNR is a custom siRNA unrelated to the human and murine genome [35]. Other commercial or noncommercial siRNAs can be employed to generate appropriate negative control conditions for the siRNA-mediated downregulation of ATG5.

6. Core components of the autophagic machinery other than ATG5, including ATG7 and beclin 1 (BECN1), can be targeted instead, or in parallel. Ideally, an instance of RCD should be tagged as autophagic if (1) it can be delayed by the genetic inhibition of at least two distinct components of the core machinery for autophagy; or (2) it can be delayed by the knockdown/knockout of one component of the core autophagic machinery as well as by chemical inhibitors of autophagy [31].

7. Several liposomal transfection reagents commercially available can substitute for HiPerFect®, including Oligofectamine™ (Gibco®-Life Technologies) and DharmaFECT™ (GE Dharmacon, Lafayette, CO, USA). Each of these reagents attains maximal efficacy with a specific transfection protocol.

8. For most cell types, Opti-MEM® can be replaced by FBS-, antibiotic-, HEPES-, and sodium pyruvate-free growth medium.

9. Under appropriate storage conditions (–20 °C, sealed and protected from light), undissolved Z-VAD-fmk is stable for at least 2 years. If stored at –20 °C and under protection from light, stock solutions are stable for at least 1 year. It is recommended to avoid repeated freeze-thawing by storing the reconstituted product in small aliquots (10–50 μL).

10. According to the manufacturer, STS stock solutions are stable for at least 6 months, if stored at –20 °C and protected from light.

11. If stored at –20 °C and protected from light, $DiOC_6(3)$ stock solution is stable for at least 12 months. Unnecessary exposure to light should be avoided to prevent photobleaching.

12. $DiOC_6(3)$ exhibits excitation/emission peaks at 482/504 nm, respectively.

13. Undissolved PI is stable for at least 12 months under standard storage conditions (at room temperature, and protected from light). PI stock solution is stable for at least 6 months, if stored at 4 °C and protected from light.

14. PI exhibits excitation/emission peaks at: (1) 482/504 nm, respectively, in aqueous solution; and (2) 535/617 nm, respectively, when bound to DNA.

15. The choice of the support for maintenance cultures (i.e., 25 cm², 75 cm², or 175 cm² flasks) depends on the amount of cells needed for experimental determination and other factors (e.g., limited space within incubators). As an indication, a 75 cm² flask of U2OS cells at 60–70 % confluence contains approximately $4–5 \times 10^6$ cells.

16. Both under-confluence and over-confluence in maintenance cultures should be avoided, as the former may be associated with a considerable genetic drift in the cell population, and the latter may impose a metabolic burden that affects cell viability.

17. This step ensures the removal of residual traces of FBS, which inactivates trypsin. Washing should not be protracted or harsh to avoid a sizeable loss of cells, especially for cell types that per se are relatively prone to detachment like human colorectal carcinoma HCT 116 cells.

18. TrypLE™ Express can substitute for trypsin–EDTA. As compared to trypsin–EDTA, TrypLE™ Express exhibits improved stability at 4 °C and RT, and does not require inactivation.

19. Optimal detachment time may vary from <1 min to several minutes, depending on cell type and culture conditions. As an indication most cancer cell lines are properly detached in 1–3 min. Over-trypsinization should be avoided, as it can result in cellular damage and/or phenotypic alterations.

20. Detachment can be verified by visual inspection.

21. Addition of complete growth medium at this step ensures the FBS-dependent inactivation of residual trypsin activity.

22. The ATCC recommends to subculture U2OS cells at a ratio of 1:3–1:6. We observed that U2OS cells can be safely subcultured at a ratio 1:8 without noticeable shifts in phenotype and behavioral traits.

23. As a general recommendation, immortalized cells should be kept in the exponential growth phase, and be maintained in culture for a limited, predetermined number of passages. This calls for a relatively large stock of cryopreserved cells.

24. The amount of cells required to generate cultures that are suitable for transfection 24 h later vary quite considerably with cell type and culture conditions.

25. The amount of cells required to generate cultures that are suitable for pharmacological treatment 24 h later vary quite considerably with cell type and culture conditions.

26. In our experience, transfection efficacy drops remarkably when confluence >50 %.

27. These conditions are appropriate for transfecting cells in 1 well of a 6-well plate. They can be readily scaled up to transfect several wells with the same transfection solution.

28. The transfection solution may appear cloudy.

29. Transfection is carried out entirely at RT under a common safety cabinet. However, it is a good practice to maintain siRNA stock solutions and HiPerFect® in ice bath (and to return them to storage conditions immediately after use).

30. Transfection complexes have a very high affinity for the plasma membrane. They should therefore be added to wells dropwise and evenly (covering the whole surface of the growth medium), in order to avoid intra-well variations in transfection efficiency and potential toxicity.

31. Transfection is generally rapid (<4 h), but cells can be allowed to reach 70–80 % confluence in 6-well plates prior to sub-culturing.

32. Ideally, cells should be ready for treatment (*see* also **Note 25**) in 12-well plates as soon as the siRNA-mediated downregulation of ATG5 achieves maximal efficacy. ATG5 and most other proteins are optimally downregulated by siRNAs 48 h after transfection. However, checking transfection efficacy and kinetics by quantitative real-time PCR or immunoblotting in preliminary experiments is strongly recommended.

33. Particularly sensitive cells may require prolonged adaptation times to resume sufficient proliferation.

34. Confluence and general status should be verified on light microscopy before collection.

35. Some forms or RCD are accompanied by the detachment of cells from the substrate. It is therefore important to collect supernatants (unless the experiment is designed as to include only viable cells) to avoid a considerable underestimation of cell death.

36. Fluorochrome-containing solutions should be thoroughly mixed before use to eliminate precipitates. Moreover, they should be shielded from light to minimize photobleaching.

37. At low concentrations (20–40 nM), $DiOC_6(3)$ rapidly accumulates within energized mitochondria in a virtually non-saturable fashion. To label all samples from the same experiment in a homogenous manner, we recommend to: (1) carefully remove supernatants (by aspiration and inversion of FACS tubes on paper); (2) employ a unique staining solution for all samples; (3) repeatedly and thoroughly mix the staining solution throughout the experiment; (4) employ an equal volume of staining solution for all samples; and (5) take particular care at avoiding cell clumps during labeling.

38. At high concentrations (>100 nM), $DiOC_6(3)$ exhibit remarkable self-quenching, which renders it inappropriate to measure $\Delta\psi_m$. Moreover, at high doses, $DiOC_6(3)$ may label other intracellular compartments, including the endoplasmic reticulum.

39. Prolonged incubation with $DiOC_6(3)$ (>40 min) may be toxic for some cell types, while an excessively short labeling time (<20 min) may result in non-homogenous or incomplete staining. Preliminary experiments should address the toxicity of $DiOC_6(3)$ in the experimental setting of choice.

40. As an alternative, PI can be added directly to the staining solution, at the final concentration of 0.5–1 μg/mL (*see* **step 6**, Subheading 3.3). Preliminary experiments to assess the toxicity of PI are recommended in this case.

41. FSC and SSC reflect cell size and the so-called "refractive index," respectively. The refractive index depends on various parameters, including cell shape and granularity.

42. We generally employ channel FL1 for the detection of $DiOC_6(3)$ and channel FL3 for the detection of PI.

43. If >30 samples must be analyzed, we suggest to carry out staining and acquisition on <24 samples at a time, to homogenize the exposure of cells to $DiOC_6(3)$ and PI.

44. Two-color analyses are relatively straightforward and can be carried out on the flow cytometer proprietary software. Alternatively, several software packages for the analysis of cytofluorometric data are available online.

45. To provide adequate statistical power, we recommend to acquire and analyze at least 10,000 events exhibiting normal FSC and SSC values per sample.

Acknowledgements

We are indebted to Dr. Ilio Vitale (Università di Roma "Tor Vergata") for careful reading of the manuscript. Authors are supported by the Ligue contre le Cancer (équipe labellisée); Agence National de la Recherche (ANR); Association pour la recherche sur le cancer (ARC); Cancéropôle Ile-de-France; AXA Chair for Longevity Research; Institut National du Cancer (INCa); Fondation Bettencourt-Schueller; Fondation de France; Fondation pour la Recherche Médicale (FRM); the European Commission (ArtForce); the European Research Council (ERC); the LabEx Immuno-Oncology; the SIRIC Stratified Oncology Cell DNA Repair and Tumor Immune Elimination (SOCRATE); the SIRIC Cancer Research and Personalized Medicine (CARPEM); and the Paris Alliance of Cancer Research Institutes (PACRI).

References

1. Galluzzi L, Bravo-San Pedro JM, Vitale I, Aaronson SA, Abrams JM, Adam D, Alnemri ES, Altucci L, Andrews D, Annicchiarico-Petruzzelli M, Baehrecke EH, Bazan NG, Bertrand MJ, Bianchi K, Blagosklonny MV, Blomgren K, Borner C, Bredesen DE, Brenner C, Campanella M, Candi E, Cecconi F, Chan FK, Chandel NS, Cheng EH, Chipuk JE, Cidlowski JA, Ciechanover A, Dawson TM, Dawson VL, De Laurenzi V, De Maria R, Debatin KM, Di Daniele N, Dixit VM, Dynlacht BD, El-Deiry WS, Fimia GM, Flavell RA, Fulda S, Garrido C, Gougeon ML, Green DR, Gronemeyer H, Hajnoczky G, Hardwick JM, Hengartner MO, Ichijo H, Joseph B, Jost PJ, Kaufmann T, Kepp O, Klionsky DJ, Knight RA, Kumar S, Lemasters JJ, Levine B, Linkermann A, Lipton SA, Lockshin RA,

Lopez-Otin C, Lugli E, Madeo F, Malorni W, Marine JC, Martin SJ, Martinou JC, Medema JP, Meier P, Melino S, Mizushima N, Moll U, Munoz-Pinedo C, Nunez G, Oberst A, Panaretakis T, Penninger JM, Peter ME, Piacentini M, Pinton P, Prehn JH, Puthalakath H, Rabinovich GA, Ravichandran KS, Rizzuto R, Rodrigues CM, Rubinsztein DC, Rudel T, Shi Y, Simon HU, Stockwell BR, Szabadkai G, Tait SW, Tang HL, Tavernarakis N, Tsujimoto Y, Vanden Berghe T, Vandenabeele P, Villunger A, Wagner EF, Walczak H, White E, Wood WG, Yuan J, Zakeri Z, Zhivotovsky B, Melino G, Kroemer G (2015) Essential versus accessory aspects of cell death: recommendations of the NCCD 2015. Cell Death Differ 22(1):58–73. doi:10.1038/cdd.2014.137

2. Vanden Berghe T, Linkermann A, Jouan-Lanhouet S, Walczak H, Vandenabeele P (2014) Regulated necrosis: the expanding network of non-apoptotic cell death pathways. Nat Rev Mol Cell Biol 15(2):135–147. doi:10.1038/nrm3737

3. Linkermann A, Green DR (2014) Necroptosis. N Engl J Med 370(5):455–465. doi:10.1056/NEJMra1310050

4. Sica V, Galluzzi L, Bravo-San Pedro JM, Izzo V, Maiuri MC, Kroemer G (2015) Organelle-specific initiation of autophagy. Mol Cell 59(4):522–539. doi:10.1016/j.molcel.2015.07.021

5. Tait SW, Green DR (2010) Mitochondria and cell death: outer membrane permeabilization and beyond. Nat Rev Mol Cell Biol 11(9):621–632. doi:10.1038/nrm2952

6. Galluzzi L, Bravo-San Pedro JM, Kroemer G (2014) Organelle-specific initiation of cell death. Nat Cell Biol 16(8):728–736. doi:10.1038/ncb3005

7. Fuchs Y, Steller H (2011) Programmed cell death in animal development and disease. Cell 147(4):742–758. doi:10.1016/j.cell.2011.10.033

8. Galluzzi L, Maiuri MC, Vitale I, Zischka H, Castedo M, Zitvogel L, Kroemer G (2007) Cell death modalities: classification and pathophysiological implications. Cell Death Differ 14(7):1237–1243. doi:10.1038/sj.cdd.4402148

9. Garg AD, Martin S, Golab J, Agostinis P (2014) Danger signalling during cancer cell death: origins, plasticity and regulation. Cell Death Differ 21(1):26–38. doi:10.1038/cdd.2013.48

10. Krysko DV, Garg AD, Kaczmarek A, Krysko O, Agostinis P, Vandenabeele P (2012) Immunogenic cell death and DAMPs in cancer therapy. Nat Rev Cancer 12(12):860–875. doi:10.1038/nrc3380

11. Kroemer G, Galluzzi L, Kepp O, Zitvogel L (2013) Immunogenic cell death in cancer therapy. Annu Rev Immunol 31:51–72. doi:10.1146/annurev-immunol-032712-100008

12. Kaczmarek A, Vandenabeele P, Krysko DV (2013) Necroptosis: the release of damage-associated molecular patterns and its physiological relevance. Immunity 38(2):209–223. doi:10.1016/j.immuni.2013.02.003

13. Kepp O, Senovilla L, Vitale I, Vacchelli E, Adjemian S, Agostinis P, Apetoh L, Aranda F, Barnaba V, Bloy N, Bracci L, Breckpot K, Brough D, Buque A, Castro MG, Cirone M, Colombo MI, Cremer I, Demaria S, Dini L, Eliopoulos AG, Faggioni A, Formenti SC, Fucikova J, Gabriele L, Gaipl US, Galon J, Garg A, Ghiringhelli F, Giese NA, Guo ZS, Hemminki A, Herrmann M, Hodge JW, Holdenrieder S, Honeychurch J, Hu HM, Huang X, Illidge TM, Kono K, Korbelik M, Krysko DV, Loi S, Lowenstein PR, Lugli E, Ma Y, Madeo F, Manfredi AA, Martins I, Mavilio D, Menger L, Merendino N, Michaud M, Mignot G, Mossman KL, Multhoff G, Oehler R, Palombo F, Panaretakis T, Pol J, Proietti E, Ricci JE, Riganti C, Rovere-Querini P, Rubartelli A, Sistigu A, Smyth MJ, Sonnemann J, Spisek R, Stagg J, Sukkurwala AQ, Tartour E, Thorburn A, Thorne SH, Vandenabeele P, Velotti F, Workenhe ST, Yang H, Zong WX, Zitvogel L, Kroemer G, Galluzzi L (2014) Consensus guidelines for the detection of immunogenic cell death. Oncoimmunol 3(9):e955691. doi:10.4161/21624011.2014.955691

14. Linkermann A, Stockwell BR, Krautwald S, Anders HJ (2014) Regulated cell death and inflammation: an auto-amplification loop causes organ failure. Nat Rev Immunol 14(11):759–767. doi:10.1038/nri3743

15. Kroemer G, Galluzzi L, Vandenabeele P, Abrams J, Alnemri ES, Baehrecke EH, Blagosklonny MV, El-Deiry WS, Golstein P, Green DR, Hengartner M, Knight RA, Kumar S, Lipton SA, Malorni W, Nunez G, Peter ME, Tschopp J, Yuan J, Piacentini M, Zhivotovsky B, Melino G (2009) Classification of cell death: recommendations of the Nomenclature Committee on Cell Death 2009. Cell Death Differ 16(1):3–11. doi:10.1038/cdd.2008.150

16. Taylor RC, Cullen SP, Martin SJ (2008) Apoptosis: controlled demolition at the cellular level. Nat Rev Mol Cell Biol 9(3):231–241. doi:10.1038/nrm2312

17. Galluzzi L, Vitale I, Abrams JM, Alnemri ES, Baehrecke EH, Blagosklonny MV, Dawson TM, Dawson VL, El-Deiry WS, Fulda S, Gottlieb E, Green DR, Hengartner MO, Kepp O, Knight RA, Kumar S, Lipton SA, Lu X,

Madeo F, Malorni W, Mehlen P, Nunez G, Peter ME, Piacentini M, Rubinsztein DC, Shi Y, Simon HU, Vandenabeele P, White E, Yuan J, Zhivotovsky B, Melino G, Kroemer G (2012) Molecular definitions of cell death subroutines: recommendations of the Nomenclature Committee on Cell Death 2012. Cell Death Differ 19(1):107–120. doi:10.1038/cdd.2011.96

18. Wirawan E, Vanden Berghe T, Lippens S, Agostinis P, Vandenabeele P (2012) Autophagy: for better or for worse. Cell Res 22(1):43–61. doi:10.1038/cr.2011.152

19. Galluzzi L, Kepp O, Trojel-Hansen C, Kroemer G (2012) Non-apoptotic functions of apoptosis-regulatory proteins. EMBO Rep 13(4):322–330. doi:10.1038/embor.2012.19

20. Mizushima N, Levine B, Cuervo AM, Klionsky DJ (2008) Autophagy fights disease through cellular self-digestion. Nature 451(7182):1069–1075. doi:10.1038/nature06639

21. Galluzzi L, Pietrocola F, Bravo-San Pedro JM, Amaravadi RK, Baehrecke EH, Cecconi F, Codogno P, Debnath J, Gewirtz DA, Karantza V, Kimmelman A, Kumar S, Levine B, Maiuri MC, Martin SJ, Penninger J, Piacentini M, Rubinsztein DC, Simon HU, Simonsen A, Thorburn AM, Velasco G, Ryan KM, Kroemer G (2015) Autophagy in malignant transformation and cancer progression. EMBO J 34(7):856–880. doi:10.15252/embj.201490784

22. Green DR, Levine B (2014) To be or not to be? How selective autophagy and cell death govern cell fate. Cell 157(1):65–75. doi:10.1016/j.cell.2014.02.049

23. Vanden Berghe T, Grootjans S, Goossens V, Dondelinger Y, Krysko DV, Takahashi N, Vandenabeele P (2013) Determination of apoptotic and necrotic cell death in vitro and in vivo. Methods 61(2):117–129. doi:10.1016/j.ymeth.2013.02.011

24. Jouan-Lanhouet S, Riquet F, Duprez L, Vanden Berghe T, Takahashi N, Vandenabeele P (2014) Necroptosis, in vivo detection in experimental disease models. Semin Cell Dev Biol 35:2–13. doi:10.1016/j.semcdb.2014.08.010

25. Zermati Y, Garrido C, Amsellem S, Fishelson S, Bouscary D, Valensi F, Varet B, Solary E, Hermine O (2001) Caspase activation is required for terminal erythroid differentiation. J Exp Med 193(2):247–254

26. Kepp O, Galluzzi L, Lipinski M, Yuan J, Kroemer G (2011) Cell death assays for drug discovery. Nat Rev Drug Discov 10(3):221–237. doi:10.1038/nrd3373

27. Galluzzi L, Aaronson SA, Abrams J, Alnemri ES, Andrews DW, Baehrecke EH, Bazan NG, Blagosklonny MV, Blomgren K, Borner C, Bredesen DE, Brenner C, Castedo M, Cidlowski JA, Ciechanover A, Cohen GM, De Laurenzi V, De Maria R, Deshmukh M, Dynlacht BD, El-Deiry WS, Flavell RA, Fulda S, Garrido C, Golstein P, Gougeon ML, Green DR, Gronemeyer H, Hajnoczky G, Hardwick JM, Hengartner MO, Ichijo H, Jaattela M, Kepp O, Kimchi A, Klionsky DJ, Knight RA, Kornbluth S, Kumar S, Levine B, Lipton SA, Lugli E, Madeo F, Malomi W, Marine JC, Martin SJ, Medema JP, Mehlen P, Melino G, Moll UM, Morselli E, Nagata S, Nicholson DW, Nicotera P, Nunez G, Oren M, Penninger J, Pervaiz S, Peter ME, Piacentini M, Prehn JH, Puthalakath H, Rabinovich GA, Rizzuto R, Rodrigues CM, Rubinsztein DC, Rudel T, Scorrano L, Simon HU, Steller H, Tschopp J, Tsujimoto Y, Vandenabeele P, Vitale I, Vousden KH, Youle RJ, Yuan J, Zhivotovsky B, Kroemer G (2009) Guidelines for the use and interpretation of assays for monitoring cell death in higher eukaryotes. Cell Death Differ 16(8):1093–1107. doi:10.1038/cdd.2009.44

28. Cain K, Inayat-Hussain SH, Couet C, Cohen GM (1996) A cleavage-site-directed inhibitor of interleukin-1 beta-converting enzyme-like proteases inhibits apoptosis in primary cultures of rat hepatocytes. Biochem J 314(Pt 1):27–32

29. Slee EA, Zhu H, Chow SC, MacFarlane M, Nicholson DW, Cohen GM (1996) Benzyloxycarbonyl-Val-Ala-Asp (OMe) fluoromethylketone (Z-VAD.FMK) inhibits apoptosis by blocking the processing of CPP32. Biochem J 315(Pt 1):21–24

30. Seglen PO, Gordon PB (1982) 3-Methyladenine: specific inhibitor of autophagic/lysosomal protein degradation in isolated rat hepatocytes. Proc Natl Acad Sci U S A 79(6):1889–1892

31. Klionsky DJ, Abdalla FC, Abeliovich H, Abraham RT, Acevedo-Arozena A, Adeli K, Agholme L, Agnello M, Agostinis P, Aguirre-Ghiso JA, Ahn HJ, Ait-Mohamed O, Ait-Si-Ali S, Akematsu T, Akira S, Al-Younes HM, Al-Zeer MA, Albert ML, Albin RL, Alegre-Abarrategui J, Aleo MF, Alirezaei M, Almasan A, Almonte-Becerril M, Amano A, Amaravadi R, Amarnath S, Amer AO, Andrieu-Abadie N, Anantharam V, Ann DK, Anoopkumar-Dukie S, Aoki H, Apostolova N, Arancia G, Aris JP, Asanuma K, Asare NY, Ashida H, Askanas V, Askew DS, Auberger P, Baba M, Backues SK, Baehrecke EH, Bahr BA, Bai XY, Bailly Y, Baiocchi R, Baldini G, Balduini W, Ballabio A, Bamber BA, Bampton ET, Banhegyi G, Bartholomew CR, Bassham DC, Bast RC Jr, Batoko H, Bay BH, Beau I, Bechet DM, Begley TJ, Behl C, Behrends C, Bekri S, Bellaire B, Bendall LJ, Benetti L, Berliocchi L,

Bernardi H, Bernassola F, Besteiro S, Bhatia-Kissova I, Bi X, Biard-Piechaczyk M, Blum JS, Boise LH, Bonaldo P, Boone DL, Bornhauser BC, Bortoluci KR, Bossis I, Bost F, Bourquin JP, Boya P, Boyer-Guittaut M, Bozhkov PV, Brady NR, Brancolini C, Brech A, Brenman JE, Brennand A, Bresnick EH, Brest P, Bridges D, Bristol ML, Brookes PS, Brown EJ, Brumell JH, Brunetti-Pierri N, Brunk UT, Bulman DE, Bultman SJ, Bultynck G, Burbulla LF, Bursch W, Butchar JP, Buzgariu W, Bydlowski SP, Cadwell K, Cahova M, Cai D, Cai J, Cai Q, Calabretta B, Calvo-Garrido J, Camougrand N, Campanella M, Campos-Salinas J, Candi E, Cao L, Caplan AB, Carding SR, Cardoso SM, Carew JS, Carlin CR, Carmignac V, Carneiro LA, Carra S, Caruso RA, Casari G, Casas C, Castino R, Cebollero E, Cecconi F, Celli J, Chaachouay H, Chae HJ, Chai CY, Chan DC, Chan EY, Chang RC, Che CM, Chen CC, Chen GC, Chen GQ, Chen M, Chen Q, Chen SS, Chen W, Chen X, Chen YG, Chen Y, Chen YJ, Chen Z, Cheng A, Cheng CH, Cheng Y, Cheong H, Cheong JH, Cherry S, Chess-Williams R, Cheung ZH, Chevet E, Chiang HL, Chiarelli R, Chiba T, Chin LS, Chiou SH, Chisari FV, Cho CH, Cho DH, Choi AM, Choi D, Choi KS, Choi ME, Chouaib S, Choubey D, Choubey V, Chu CT, Chuang TH, Chueh SH, Chun T, Chwae YJ, Chye ML, Ciarcia R, Ciriolo MR, Clague MJ, Clark RS, Clarke PG, Clarke R, Codogno P, Coller HA, Colombo MI, Comincini S, Condello M, Condorelli F, Cookson MR, Coombs GH, Coppens I, Corbalan R, Cossart P, Costelli P, Costes S, Coto-Montes A, Couve E, Coxon FP, Cregg JM, Crespo JL, Cronje MJ, Cuervo AM, Cullen JJ, Czaja MJ, D'Amelio M, Darfeuille-Michaud A, Davids LM, Davies FE, De Felici M, de Groot JF, de Haan CA, De Martino L, De Milito A, De Tata V, Debnath J, Degterev A, Dehay B, Delbridge LM, Demarchi F, Deng YZ, Dengjel J, Dent P, Denton D, Deretic V, Desai SD, Devenish RJ, Di Gioacchino M, Di Paolo G, Di Pietro C, Diaz-Araya G, Diaz-Laviada I, Diaz-Meco MT, Diaz-Nido J, Dikic I, Dinesh-Kumar SP, Ding WX, Distelhorst CW, Diwan A, Djavaheri-Mergny M, Dokudovskaya S, Dong Z, Dorsey FC, Dosenko V, Dowling JJ, Doxsey S, Dreux M, Drew ME, Duan Q, Duchosal MA, Duff K, Dugail I, Durbeej M, Duszenko M, Edelstein CL, Edinger AL, Egea G, Eichinger L, Eissa NT, Ekmekcioglu S, El-Deiry WS, Elazar Z, Elgendy M, Ellerby LM, Eng KE, Engelbrecht AM, Engelender S, Erenpreisa J, Escalante R, Esclatine A, Eskelinen EL, Espert L, Espina V, Fan H, Fan J, Fan QW, Fan Z, Fang S, Fang Y, Fanto M, Fanzani A, Farkas T, Farre JC, Faure M, Fechheimer M, Feng CG, Feng J, Feng Q, Feng Y, Fesus L, Feuer R, Figueiredo-Pereira ME, Fimia GM, Fingar DC, Finkbeiner S, Finkel T, Finley KD, Fiorito F, Fisher EA, Fisher PB, Flajolet M, Florez-McClure ML, Florio S, Fon EA, Fornai F, Fortunato F, Fotedar R, Fowler DH, Fox HS, Franco R, Frankel LB, Fransen M, Fuentes JM, Fueyo J, Fujii J, Fujisaki K, Fujita E, Fukuda M, Furukawa RH, Gaestel M, Gailly P, Gajewska M, Galliot B, Galy V, Ganesh S, Ganetzky B, Ganley IG, Gao FB, Gao GF, Gao J, Garcia L, Garcia-Manero G, Garcia-Marcos M, Garmyn M, Gartel AL, Gatti E, Gautel M, Gawriluk TR, Gegg ME, Geng J, Germain M, Gestwicki JE, Gewirtz DA, Ghavami S, Ghosh P, Giammarioli AM, Giatromanolaki AN, Gibson SB, Gilkerson RW, Ginger ML, Ginsberg HN, Golab J, Goligorsky MS, Golstein P, Gomez-Manzano C, Goncu E, Gongora C, Gonzalez CD, Gonzalez R, Gonzalez-Estevez C, Gonzalez-Polo RA, Gonzalez-Rey E, Gorbunov NV, Gorski S, Goruppi S, Gottlieb RA, Gozuacik D, Granato GE, Grant GD, Green KN, Gregorc A, Gros F, Grose C, Grunt TW, Gual P, Guan JL, Guan KL, Guichard SM, Gukovskaya AS, Gukovsky I, Gunst J, Gustafsson AB, Halayko AJ, Hale AN, Halonen SK, Hamasaki M, Han F, Han T, Hancock MK, Hansen M, Harada H, Harada M, Hardt SE, Harper JW, Harris AL, Harris J, Harris SD, Hashimoto M, Haspel JA, Hayashi S, Hazelhurst LA, He C, He YW, Hebert MJ, Heidenreich KA, Helfrich MH, Helgason GV, Henske EP, Herman B, Herman PK, Hetz C, Hilfiker S, Hill JA, Hocking LJ, Hofman P, Hofmann TG, Hohfeld J, Holyoake TL, Hong MH, Hood DA, Hotamisligil GS, Houwerzijl EJ, Hoyer-Hansen M, Hu B, Hu CA, Hu HM, Hua Y, Huang C, Huang J, Huang S, Huang WP, Huber TB, Huh WK, Hung TH, Hupp TR, Hur GM, Hurley JB, Hussain SN, Hussey PJ, Hwang JJ, Hwang S, Ichihara A, Ilkhanizadeh S, Inoki K, Into T, Iovane V, Iovanna JL, Ip NY, Isaka Y, Ishida H, Isidoro C, Isobe K, Iwasaki A, Izquierdo M, Izumi Y, Jaakkola PM, Jaattela M, Jackson GR, Jackson WT, Janji B, Jendrach M, Jeon JH, Jeung EB, Jiang H, Jiang JX, Jiang M, Jiang Q, Jiang X, Jimenez A, Jin M, Jin S, Joe CO, Johansen T, Johnson DE, Johnson GV, Jones NL, Joseph B, Joseph SK, Joubert AM, Juhasz G, Juillerat-Jeanneret L, Jung CH, Jung YK, Kaarniranta K, Kaasik A, Kabuta T, Kadowaki M, Kagedal K, Kamada Y, Kaminskyy VO, Kampinga HH, Kanamori H, Kang C, Kang KB, Kang KI, Kang R, Kang YA, Kanki T, Kanneganti TD, Kanno H, Kanthasamy AG, Kanthasamy A,

Karantza V, Kaushal GP, Kaushik S, Kawazoe Y, Ke PY, Kehrl JH, Kelekar A, Kerkhoff C, Kessel DH, Khalil H, Kiel JA, Kiger AA, Kihara A, Kim DR, Kim DH, Kim EK, Kim HR, Kim JS, Kim JH, Kim JC, Kim JK, Kim PK, Kim SW, Kim YS, Kim Y, Kimchi A, Kimmelman AC, King JS, Kinsella TJ, Kirkin V, Kirshenbaum LA, Kitamoto K, Kitazato K, Klein L, Klimecki WT, Klucken J, Knecht E, Ko BC, Koch JC, Koga H, Koh JY, Koh YH, Koike M, Komatsu M, Kominami E, Kong HJ, Kong WJ, Korolchuk VI, Kotake Y, Koukourakis MI, Kouri Flores JB, Kovacs AL, Kraft C, Krainc D, Kramer H, Kretz-Remy C, Krichevsky AM, Kroemer G, Kruger R, Krut O, Ktistakis NT, Kuan CY, Kucharczyk R, Kumar A, Kumar R, Kumar S, Kundu M, Kung HJ, Kurz T, Kwon HJ, La Spada AR, Lafont F, Lamark T, Landry J, Lane JD, Lapaquette P, Laporte JF, Laszlo L, Lavandero S, Lavoie JN, Layfield R, Lazo PA, Le W, Le Cam L, Ledbetter DJ, Lee AJ, Lee BW, Lee GM, Lee J, Lee JH, Lee M, Lee MS, Lee SH, Leeuwenburgh C, Legembre P, Legouis R, Lehmann M, Lei HY, Lei QY, Leib DA, Leiro J, Lemasters JJ, Lemoine A, Lesniak MS, Lev D, Levenson VV, Levine B, Levy E, Li F, Li JL, Li L, Li S, Li W, Li XJ, Li YB, Li YP, Liang C, Liang Q, Liao YF, Liberski PP, Lieberman A, Lim HJ, Lim KL, Lim K, Lin CF, Lin FC, Lin J, Lin JD, Lin K, Lin WW, Lin WC, Lin YL, Linden R, Lingor P, Lippincott-Schwartz J, Lisanti MP, Liton PB, Liu B, Liu CF, Liu K, Liu L, Liu QA, Liu W, Liu YC, Liu Y, Lockshin RA, Lok CN, Lonial S, Loos B, Lopez-Berestein G, Lopez-Otin C, Lossi L, Lotze MT, Low P, Lu B, Lu Z, Luciano F, Lukacs NW, Lund AH, Lynch-Day MA, Ma Y, Macian F, MacKeigan JP, Macleod KF, Madeo F, Maiuri L, Maiuri MC, Malagoli D, Malicdan MC, Malorni W, Man N, Mandelkow EM, Manon S, Manov I, Mao K, Mao X, Mao Z, Marambaud P, Marazziti D, Marcel YL, Marchbank K, Marchetti P, Marciniak SJ, Marcondes M, Mardi M, Marfe G, Marino G, Markaki M, Marten MR, Martin SJ, Martinand-Mari C, Martinet W, Martinez-Vicente M, Masini M, Matarrese P, Matsuo S, Matteoni R, Mayer A, Mazure NM, McConkey DJ, McConnell MJ, McDermott C, McDonald C, McInerney GM, McKenna SL, McLaughlin B, McLean PJ, McMaster CR, McQuibban GA, Meijer AJ, Meisler MH, Melendez A, Melia TJ, Melino G, Mena MA, Menendez JA, Menna-Barreto RF, Menon MB, Menzies FM, Mercer CA, Merighi A, Merry DE, Meschini S, Meyer CG, Meyer TF, Miao CY, Miao JY, Michels PA, Michiels C, Mijaljica D, Milojkovic A, Minucci S, Miracco C, Miranti CK, Mitroulis I, Miyazawa K, Mizushima N, Mograbi B, Mohseni S, Molero X, Mollereau B, Mollinedo F, Momoi T, Monastyrska I, Monick MM, Monteiro MJ, Moore MN, Mora R, Moreau K, Moreira PI, Moriyasu Y, Moscat J, Mostowy S, Mottram JC, Motyl T, Moussa CE, Muller S, Munger K, Munz C, Murphy LO, Murphy ME, Musaro A, Mysorekar I, Nagata E, Nagata K, Nahimana A, Nair U, Nakagawa T, Nakahira K, Nakano H, Nakatogawa H, Nanjundan M, Naqvi NI, Narendra DP, Narita M, Navarro M, Nawrocki ST, Nazarko TY, Nemchenko A, Netea MG, Neufeld TP, Ney PA, Nezis IP, Nguyen HP, Nie D, Nishino I, Nislow C, Nixon RA, Noda T, Noegel AA, Nogalska A, Noguchi S, Notterpek L, Novak I, Nozaki T, Nukina N, Nurnberger T, Nyfeler B, Obara K, Oberley TD, Oddo S, Ogawa M, Ohashi T, Okamoto K, Oleinick NL, Oliver FJ, Olsen LJ, Olsson S, Opota O, Osborne TF, Ostrander GK, Otsu K, Ou JH, Ouimet M, Overholtzer M, Ozpolat B, Paganetti P, Pagnini U, Pallet N, Palmer GE, Palumbo C, Pan T, Panaretakis T, Pandey UB, Papackova Z, Papassideri I, Paris I, Park J, Park OK, Parys JB, Parzych KR, Patschan S, Patterson C, Pattingre S, Pawelek JM, Peng J, Perlmutter DH, Perrotta I, Perry G, Pervaiz S, Peter M, Peters GJ, Petersen M, Petrovski G, Phang JM, Piacentini M, Pierre P, Pierrefite-Carle V, Pierron G, Pinkas-Kramarski R, Piras A, Piri N, Platanias LC, Poggeler S, Poirot M, Poletti A, Pous C, Pozuelo-Rubio M, Praetorius-Ibba M, Prasad A, Prescott M, Priault M, Produit-Zengaffinen N, Progulske-Fox A, Proikas-Cezanne T, Przedborski S, Przyklenk K, Puertollano R, Puyal J, Qian SB, Qin L, Qin ZH, Quaggin SE, Raben N, Rabinowich H, Rabkin SW, Rahman I, Rami A, Ramm G, Randall G, Randow F, Rao VA, Rathmell JC, Ravikumar B, Ray SK, Reed BH, Reed JC, Reggiori F, Regnier-Vigouroux A, Reichert AS, Reiners JJ Jr, Reiter RJ, Ren J, Revuelta JL, Rhodes CJ, Ritis K, Rizzo E, Robbins J, Roberge M, Roca H, Roccheri MC, Rocchi S, Rodemann HP, Rodriguez de Cordoba S, Rohrer B, Roninson IB, Rosen K, Rost-Roszkowska MM, Rouis M, Rouschop KM, Rovetta F, Rubin BP, Rubinsztein DC, Ruckdeschel K, Rucker EB 3rd, Rudich A, Rudolf E, Ruiz-Opazo N, Russo R, Rusten TE, Ryan KM, Ryter SW, Sabatini DM, Sadoshima J, Saha T, Saitoh T, Sakagami H, Sakai Y, Salekdeh GH, Salomoni P, Salvaterra PM, Salvesen G, Salvioli R, Sanchez AM, Sanchez-Alcazar JA, Sanchez-Prieto R, Sandri M, Sankar U, Sansanwal P, Santambrogio L, Saran S, Sarkar S, Sarwal M, Sasakawa C, Sasnauskiene A, Sass M, Sato K, Sato M,

Schapira AH, Scharl M, Schatzl HM, Scheper W, Schiaffino S, Schneider C, Schneider ME, Schneider-Stock R, Schoenlein PV, Schorderet DF, Schuller C, Schwartz GK, Scorrano L, Sealy L, Seglen PO, Segura-Aguilar J, Seiliez I, Seleverstov O, Sell C, Seo JB, Separovic D, Setaluri V, Setoguchi T, Settembre C, Shacka JJ, Shanmugam M, Shapiro IM, Shaulian E, Shaw RJ, Shelhamer JH, Shen HM, Shen WC, Sheng ZH, Shi Y, Shibuya K, Shidoji Y, Shieh JJ, Shih CM, Shimada Y, Shimizu S, Shintani T, Shirihai OS, Shore GC, Sibirny AA, Sidhu SB, Sikorska B, Silva-Zacarin EC, Simmons A, Simon AK, Simon HU, Simone C, Simonsen A, Sinclair DA, Singh R, Sinha D, Sinicrope FA, Sirko A, Siu PM, Sivridis E, Skop V, Skulachev VP, Slack RS, Smaili SS, Smith DR, Soengas MS, Soldati T, Song X, Sood AK, Soong TW, Sotgia F, Spector SA, Spies CD, Springer W, Srinivasula SM, Stefanis L, Steffan JS, Stendel R, Stenmark H, Stephanou A, Stern ST, Sternberg C, Stork B, Stralfors P, Subauste CS, Sui X, Sulzer D, Sun J, Sun SY, Sun ZJ, Sung JJ, Suzuki K, Suzuki T, Swanson MS, Swanton C, Sweeney ST, Sy LK, Szabadkai G, Tabas I, Taegtmeyer H, Tafani M, Takacs-Vellai K, Takano Y, Takegawa K, Takemura G, Takeshita F, Talbot NJ, Tan KS, Tanaka K, Tang D, Tanida I, Tannous BA, Tavernarakis N, Taylor GS, Taylor GA, Taylor JP, Terada LS, Terman A, Tettamanti G, Thevissen K, Thompson CB, Thorburn A, Thumm M, Tian F, Tian Y, Tocchini-Valentini G, Tolkovsky AM, Tomino Y, Tonges L, Tooze SA, Tournier C, Tower J, Towns R, Trajkovic V, Travassos LH, Tsai TF, Tschan MP, Tsubata T, Tsung A, Turk B, Turner LS, Tyagi SC, Uchiyama Y, Ueno T, Umekawa M, Umemiya-Shirafuji R, Unni VK, Vaccaro MI, Valente EM, Van den Berghe G, van der Klei IJ, van Doorn W, van Dyk LF, van Egmond M, van Grunsven LA, Vandenabeele P, Vandenberghe WP, Vanhorebeek I, Vaquero EC, Velasco G, Vellai T, Vicencio JM, Vierstra RD, Vila M, Vindis C, Viola G, Viscomi MT, Voitsekhovskaja OV, von Haefen C, Votruba M, Wada K, Wade-Martins R, Walker CL, Walsh CM, Walter J, Wan XB, Wang A, Wang C, Wang D, Wang F, Wang G, Wang H, Wang HG, Wang HD, Wang J, Wang K, Wang M, Wang RC, Wang X, Wang YJ, Wang Y, Wang Z, Wang ZC, Wansink DG, Ward DM, Watada H, Waters SL, Webster P, Wei L, Weihl CC, Weiss WA, Welford SM, Wen LP, Whitehouse CA, Whitton JL, Whitworth AJ, Wileman T, Wiley JW, Wilkinson S, Willbold D, Williams RL, Williamson PR, Wouters BG, Wu C, Wu DC, Wu WK, Wyttenbach A, Xavier RJ, Xi Z, Xia P, Xiao G, Xie Z, Xu DZ, Xu J, Xu L, Xu X, Yamamoto A, Yamashina S, Yamashita M, Yan X, Yanagida M, Yang DS, Yang E, Yang JM, Yang SY, Yang W, Yang WY, Yang Z, Yao MC, Yao TP, Yeganeh B, Yen WL, Yin JJ, Yin XM, Yoo OJ, Yoon G, Yoon SY, Yorimitsu T, Yoshikawa Y, Yoshimori T, Yoshimoto K, You HJ, Youle RJ, Younes A, Yu L, Yu SW, Yu WH, Yuan ZM, Yue Z, Yun CH, Yuzaki M, Zabirnyk O, Silva-Zacarin E, Zacks D, Zacksenhaus E, Zaffaroni N, Zakeri Z, Zeh HJ 3rd, Zeitlin SO, Zhang H, Zhang HL, Zhang J, Zhang JP, Zhang L, Zhang MY, Zhang XD, Zhao M, Zhao YF, Zhao Y, Zhao ZJ, Zheng X, Zhivotovsky B, Zhong Q, Zhou CZ, Zhou C, Zhu WG, Zhu XF, Zhu X, Zhu Y, Zoladek T, Zong WX, Zorzano A, Zschocke J, Zuckerbraun B (2012) Guidelines for the use and interpretation of assays for monitoring autophagy. Autophagy 8(4):445–544

32. Fellmann C, Lowe SW (2014) Stable RNA interference rules for silencing. Nat Cell Biol 16(1):10–18. doi:10.1038/ncb2895

33. Poulin GB (2011) A guide to using RNAi and other nucleotide-based technologies. Brief Funct Genomics 10(4):173–174. doi:10.1093/bfgp/elr025

34. Shalem O, Sanjana NE, Zhang F (2015) High-throughput functional genomics using CRISPR-Cas9. Nat Rev Genet 16(5):299–311. doi:10.1038/nrg3899

35. de La Motte RT, Galluzzi L, Olaussen KA, Zermati Y, Tasdemir E, Robert T, Ripoche H, Lazar V, Dessen P, Harper F, Pierron G, Pinna G, Araujo N, Harel-Belan A, Armand JP, Wong TW, Soria JC, Kroemer G (2007) A novel epidermal growth factor receptor inhibitor promotes apoptosis in non-small cell lung cancer cells resistant to erlotinib. Cancer Res 67(13):6253–6262. doi:10.1158/0008-5472.CAN-07-0538

Chapter 2

In Vivo Apoptosis Imaging Using Site-Specifically ^{68}Ga-Labeled Annexin V

Matthias Bauwens

Abstract

Noninvasive molecular imaging, using positron emission tomography (PET), is an important technique to visualize metabolic processes in vivo. It also allows to visualize the process of apoptosis, by using radiolabeled compounds such as Annexin V, that bind to extracellular phosphatidylserine (PS). This chapter describes the radiosynthesis of ^{68}Ga-labeled Annexin V and how to noninvasively image apoptosis in vivo.

Key words Annexin V, ^{68}Ga, PET, Noninvasive imaging

1 Introduction

Apoptosis plays a role in a large number of diseases, such as neuro-degenerative diseases, ischemic damage, autoimmune disorders, and many types of cancer. Timely assessment of in vivo apoptotic cell death in a particular tissue, for example in cancer tissue upon chemotherapeutic treatment, is of crucial importance to optimize treatment strategies, thereby improving patient survival and welfare. Classical techniques, such as computed tomography (CT) and magnetic resonance imaging (MRI), often require several months of treatment before a (relatively large-scale) anatomical effect can be seen [1]. Molecular imaging, on the other hand, can be performed within several days after the first treatment, showing a change in tissue metabolic rate. An even faster possibility is to visualize (apoptotic) cell death: this may allow to assess tumor cell death within one day [2, 3]. The general concept of molecular imaging (of apoptosis) requires radiolabeled compounds, also known as tracers, that can specifically bind to apoptotic cells, and with reasonable pharmacokinetics and biodistribution. Several radiolabeled apoptotic-targeting compounds have been developed for this purpose, with moderate success. The most well-known class of compounds is derived from Annexin V, targeting externalized

Hamsa Puthalakath and Christine J. Hawkins (eds.), *Programmed Cell Death: Methods and Protocols*, Methods in Molecular Biology, vol. 1419, DOI 10.1007/978-1-4939-3581-9_2, © Springer Science+Business Media New York 2016

phosphatidylserine (PS), but other radiopharmaceuticals such as zinc dipicolylamine (also targeting PS), the Aposense family (targeting gamma-carboxyglutamic-acid (Gla)-domain proteins and intermembrane pH differences) and several others have also been described [4–8]. Numerous clinical trials with ^{99m}Tc-Annexin V have been performed, but widespread clinical application has not been achieved for various reasons [9–12]. Over the years, site-specifically radiolabeled analogues of Annexin V have been developed, where the radioisotope is specifically placed on a position of the Annexin, which is outside of the binding region, further improving the potential of radiolabeled annexin V [13, 14].

It is important to note that imaging apoptosis in vivo requires a relatively large degree of apoptosis in a tissue. In general, untreated tumors have a "background" level of apoptosis of 1–10 % (*see* Refs. 15 and 16 for examples), so visualizing a therapeutic effect requires a substantial short-term induction of apoptosis. This implies that drugs with long-lasting effects may not show noticeable efficacy when analyzed by imaging apoptosis at any particular time. Another confounding factor is necrosis: necrotic tissue (either from direct necrosis or through necroptosis) is very prevalent in cancer tissue, and also allows Annexin V to bind to "externalized" PS on exposed cell membranes.

This chapter focuses on the production and in vivo applicability of site-specifically ^{68}Ga-labeled Annexin V. The application is described both in an anti-Fas antibody mouse model and in a tumor bearing mouse model. The anti-Fas mouse model is a very straightforward method to induce fast and massive hepatic apoptosis (upto 70 % within 3 h), which can easily be visualized [17]. However, accumulation of pharmaceuticals in the liver (even those that do not induce any apoptosis) could lead to high level of false positives in this model. A tumor model, although more labor and time intensive, is better suited to fully analyze the properties of a radiopharmaceutical targeting apoptosis, as it mimics the human situation more closely. In this chapter we describe how to in vivo visualize apoptosis in a Burkitt's lymphoma in mice, using established chemotherapy and radiotherapy. Of course, the technique also applies to any new chemotherapeutic apoptosis-inducing agent under investigation.

2   Materials

2.1   Equipment

1. Radiosynthesis module: a synthesis module, with sufficient shielding and suitable for handling ^{68}Ga, should be available. Manual synthesis, although technically possible, results in high radiation dose to the extremities and should be avoided. We used a module that was developed in-house, but commercial modules are available from several suppliers (*see* **Note 1**).

2. High Pressure Liquid Chromatography (HPLC): quality control is adequate by using an HPLC system with a UV detector at 254 nm and a 3-in. radiometric NaI(Tl) detector.

3. microPET: Focus 220 microPET device (Siemens). Alternatively, this may be another standalone device, or a combination with other imaging modalities (PET/CT, PET/MRI). Ideally, the microPET is equipped with monitoring apparatus for mouse health (temperature, respiratory rate).

4. MRI: small animal Bruker Biospec MR scanner (Bruker Biospin), operating at 9.4 T and using 3D Turbo RARE for image acquisition. The acquisition parameters for the 3D Turbo RARE experiment are as follows: a $256 \times 96 \times 96$ data matrix is acquired covering a field of view of $8 \times 3 \times 3$ cm, resulting in an isotropic resolution of 312 μm; repetition time: 900 ms; effective echo time: 42 ms, RARE factor: 10; four dummy scans and one average. The total scan time is approximately 15 min.

5. Image analysis software: suitable software for analyzing μPET and μMRI images should be available. We use PMOD, but other programs are also suitable (AMIDE, ASIPRO, etc.).

6. Gamma-counter: automated NaI(Tl) gamma counter (we use Wallac 1480 Wizard 3″, Perkin-Elmer).

7. Cryomicrotome: suitable for cutting 10–50 μm slices of tissues at −20 to −30 °C.

8. Autoradiography device, including phosphor screens with a high spatial resolution (50 μm or better): Cyclone plus phosphor imager (Perkin-Elmer).

9. 10-kDa molecular weight cutoff (MWCO) filter unit, placed in a suitable centrifuge: a MWCO filter unit allows to filter any unlabelled ^{68}Ga from ^{68}Ga-labeled Annexin V (*see* **Note 1**).

10. pH paper and pH meter (*see* **Note 2**).

2.2 Solutions

All chemicals should be of analytical or pharmaceutical grade, unless otherwise mentioned. All solutions should be prepared freshly, stored at room temperature unless otherwise mentioned and used the same day.

1. ^{68}Ge/^{68}Ga generator: a ^{68}Ge/^{68}Ga generator of sufficient quality should be available. There are several suppliers at the moment, with minor differences in their output. The HCl-concentration required to elute ^{68}Ga from these generators differs in-between companies (ranging from 0.05 to 0.6 M), which will impact the concentration of buffer required in subsequent steps in order to reach a certain pH. In our case, ^{68}Ga was eluted with 0.6 M HCl.

2. Cys2-Annexin V: Obtained by site directed mutagenesis of the cDNA of human AnxA5, expressed in *E. coli* and purified to homogeneity (purity >95 %). More detailed production can be

found in Refs. 13 and 14 (*see* **Note 3**). 700 µg Cys2-AnxA5 corresponds to about 20 nmol.

3. Metal-free water: any water coming into contact with unbound ^{68}Ga should be as much as possible free of iron. This can be achieved by using doubly distilled or deionized water (milli-Q water), or by purchasing high-purity metal-free water (preferred) (*see* **Note 4**).

4. DOTA-maleimide (Macrocyclics, CheMatech, or similar companies in high quality): The DOTA functions as a chelator to trap the ^{68}Ga, while maleimide will bind to a free sulfur function (as in Cystein). Typically, we dissolve 0.8 mg in 1 ml of metal-free water and divide this into batches of 20 µl, which can be stored at −20 °C for upto 6 months.

5. 0.5 M sodium acetate buffer solution (pH 5.5): dissolve 4.1 g sodium acetate in 100 ml of metal-free water.

6. Hepes buffer: 25 mM Hepes, 40 mM NaCl: dissolve 0.595 g Hepes and 0.233 g NaCl in 100 ml milli-Q water.

7. 10 mM dithiothreitol solution (DTT): dissolve 15.4 mg in 10 ml milliQ water in a fumehood.

8. 2.5M TRIS buffer: dissolve 3.03 g tris(hydroxymethyl)aminomethane (TRIS) in 10 ml milli-Q water.

9. Purified hamster anti-Fas mAb (Dose: 0.2 µg of anti-Fas mAb per gram mouse, to be injected i.v.). Store at 4 °C (*see* **Note 5**).

2.3 Animals

1. The mouse model is highly dependent on the type of research a scientific center is focusing on, but should be a well-established model and free of (mouse and human) pathogens. In our case, we work with NMRI mice or Severe Combined Immune Deficient (SCID)-mouse (C.B-17/Icr scid/scid) (Harlan) [18]. Maintenance of SCID mice is done in specific pathogen-free rooms with a high degree of protection, and according to national and international legislation.

2. Tumor-bearing mice: we use a human B-lymphoblast cell line Daudi, derived from a Burkitt's lymphoma. Upon inoculating 5.10^5 cells (in 200 µl PBS) subcutaneously near each shoulder, it takes 5–6 weeks for an approximately 1 ml size tumor is reached (*see* **Note 6**). These tumors have a background degree of apoptosis of 1–5 %, which can be increased manifold by adequate chemotherapy or radiation therapy.

3 Methods

3.1 Radiosynthesis

1. Production of ^{68}Ga-DOTA-maleimide.

   • Adjust the pH of the ^{68}Ga eluate to pH 4 ± 0.5 by addition of 350 µl of a 3 M sodium acetate buffer solution.

- Add 20 nmol DOTA-maleimide (20 µl of a 1 mg per ml solution) (this can be increased to 25 nmol if yields are unsatisfactory).

- Heat mixture at 90 °C for 8 min and subsequently cool to room temperature.

- Adjust pH to 7–7.5 by addition of 150 µl of a 2.5 M TRIS buffer (pH 11.2).

2. Production of ^{68}Ga-DOTA-maleimide-Annexin V (^{68}Ga-AnxV).

- Add 250 µl DTT solution to a solution of 700 µg Cys2-AnxA5 in 250 µl Hepes buffer to reduce any intermolecu-lardisulfide bridges.

- Incubate for 2 h at 37 °C and then apply to a 10-kDa MWCO filter unit (500 µl capacity). Centrifuge at 12,000×g for 15 min.

- Wash the protein retained on the filter five times by centrifugation with 200 µl of a Hepes buffer (5 min centrifugation) to remove the reductant and subsequently recover and transfer into 200 µl Hepes buffer. About 80 % of the protein can be recovered, the remaining 20 % is lost due to stickiness to the filter unit (*see* **Note 7**).

- Add the reduced Cys-AnxA5 to the ^{68}Ga-DOTA-maleimide and heat for 15 min at 37 °C.

- Purify by applying the entire sample onto a 10-kDa MWCO filter unit (20 ml capacity) and centrifuging for 5 min at 3220×g. After washing with 2 ml Hepes buffer and again centrifuging once, the purified ^{68}Ga-Cys*-AnxA5 can be recovered in 200–300 µl of Hepes buffer. Losses due to adsorption to the filter unit are typically about 10 % (*see* **Note 7**).

3.2 PET Imaging

1. Anti-Fas mouse model.

- Sedate NMRI mice with isoflurane anesthesia (2.5 % for induction and 1.0–1.5 % in O_2 for maintenance during scanning).

- Inject mice in a tail vein with purified hamster anti-Fas mAb (0.2 µg/g of anti-Fas mAb, dissolved in 200 µl PBS) (treated group) or with 200 µl PBS (control group). Allow the mice to wake up and roam freely in an individual cage for 90 min, with access to water but not food.

- Sedate the mice again using isoflurane, and inject the mice via the tail vein with 7–15 MBq of ^{68}Ga-Cys2-Anx (*see* **Note 8**).

- Acquire dynamic images of the tracer distribution with the small animal PET camera for 60 min, keeping the liver in the center of the field of view. We use 12 times 5 s, 6 times 10 s, 6 times 30 s, 5 times 1 min, and 10 times 5 min as acquisition time frames. MRI imaging is optional, but not required.

- Sacrifice the animals using an overdose of pentobarbital or other suitable means. Take a 100 μl blood sample and dissect and weigh organs of interest. Measure the activity in an automated gamma counter to calculate the percentage of injected dose in each organ.

2. Burkitt's lymphoma mouse model.

- Inject cyclophosphamide (125 mg/kg) (treated group) or PBS (control group) intraperitoneally 1 day before μPET scanning (no sedation needed) [19].

- Irradiate the tumors of mice in the treated group upto 10 Gy per tumor 4–6 h before μPET scanning (sedation using isoflurane) [20].

- At the start of the μPET imaging experiment, sedate the mice with isoflurane anesthesia (2.5 % for induction and 1.0–1.5 % in O_2 for maintenance during scanning).

- Inject the mice via the tail vein with 7–15 MBq of ^{68}Ga-Cys2-Anx.

- Acquire dynamic images of the tracer distribution with the small animal PET camera for 60 min, keeping the tumor in the center of the field of view.

- Immediately after the PET scan, perform the MRI scan. Make sure to not change the position of the animals by using a dedicated transportable mouse bed.

- Sacrifice the animals using an overdose of pentobarbital or other suitable means. Take a 100 μl blood sample and dissect and weigh the tumor and organs of interest. Quickly measure the activity of the tumor in an automated gamma counter, then measure the other organs to calculate the percentage of injected dose in the tumor and each organ.

- After measuring the radioactivity in the tumor, quickly freeze the tumor using isopentane in a bath of dry ice or by snap freezing (*see* **Note 9**). Cut sample into 10–50 μm slices at –25 °C, process slices for autoradiography, TUNEL, or H and E staining according to manufacturers' guidelines (*see* **Note 10**).

3. Image analysis.

- In vivo mages are analyzed using PMOD software, by fusing PET and MRI images using manual coregistration. MRI images are used for organ and tumor delineation, while superimposed PET images are used to interpret tracer accumulation (*see* Fig. 1).

- Autoradiography images are matched to and compared with TUNEL and H and E staining. Heterogeneous uptake is very frequent (*see* Fig. 2).

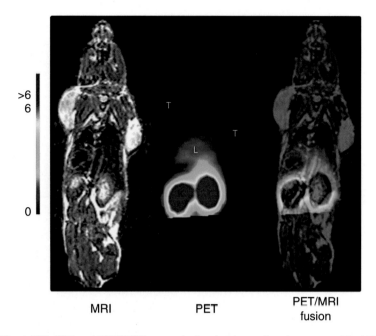

Fig. 1 MRI, PET, and PET/MRI image of a treated tumor-bearing mouse. The kidneys (**K**) are clearly visible on both MRI and PET images, as is the liver (**L**). The tumors cannot be distinguished on the PET image, but are clearly visible on the MRI image allowing delineation of the tumors. The scale indicates the intensity of the radioactive signal and is expressed as an SUV value (Standardized Uptake Value, radioactivity in Bq per volume in the region of interest, divided by the total amount of injected dose and multiplied by the animal weight)

4   Notes

1. Radiation protection is not to be underestimated in case of ^{68}Ga. For example, a dose of 100 MBq, at a distance of 30 cm (for example with an unshielded source on a bench), yields a dose of 0.2 mSv per hour. Knowing that a ^{68}Ga-generator can typically deliver upto 1500 MBq, the legal limit of 20 mSv per year is easily reached if inadequate protection is applied. Fully shielding ^{68}Ga (stopping more than 99 % of radiation) requires more than 5 cm of lead, which is difficult to achieve in practice around large equipment such as centrifuges.

2. Although a pH meter is more accurate, pH paper is preferred when measuring the pH of radioactive samples, as pH meters consistently become contaminated with radioactivity, rendering them unfit for other work.

3. The purity of the Annexin V preparation should be as high as possible, both in terms of protein purity and in absence of low molecular weight impurities such as DTT. For example, an equimolar presence of cysteine or DTT in comparison to the

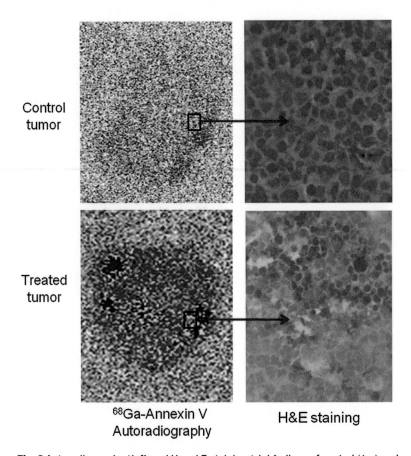

Control
tumor

Treated
tumor

^{68}Ga-Annexin V   H&E staining
Autoradiography

Fig. 2 Autoradiography (*left*) and H and E staining (*right*) slices of control (*top*) and treated (*bottom*) tumors. Note the heterogeneous uptake of ^{68}Ga-Annexin V, indicating heterogeneity of the degree of apoptosis within the tumor (confirmed by staining)

protein, may not show on routine protein purity analysis, but will dramatically reduce the yield of the radiosynthesis. When in doubt, perform an additional dialysis or centrifugal filtration over a MWCO filter prior to radiosynthesis.

4. Iron behaves very similar to gallium in many ways, including binding to chelators such as DOTA. Considering the low amount of ^{68}Ga (10^9 Bq corresponds to about 10 pmol), iron contaminations even from using metal spatula can be very disruptive to the radiochemical yield.

5. While storage of anti-Fas mAb should be done at 4 °C, care should be taken to warm up the antibody to at least room temperature (and preferably 37 °C) prior to i.v. injection for optimal animal comfort.

6. The location of the tumor is important: while tumors generally develop better when inoculated subcutaneously in the abdomen

region, such a location renders imaging more difficult as the clearance pathway of radiopharmaceuticals, either renally or hepatically, would be in to close proximity to the tumor. Alternative locations, such as near the shoulders or in the femur region, are therefore preferred. Additionally, care should be taken to use tumors with only minor amounts of necrosis: necrotic tissue is not well perfused (hampering uptake of the ^{68}Ga-Anx) and necrotic cells also present exposed PS, which is indistinguishable from extracellular PS as seen in apoptotic cells.

7. The stickiness to the filter can be reduced by washing the filter with PBS prior to administration of the compound.

8. The injection of the radiopharmaceutical should be done via a catheter, to allow fast and easy administration (reducing radiation dose to the technician) and to reduce the amount of paravenous injection.

9. Avoid placing the tumor directly in liquid nitrogen. In our experience, this results in brittle tissue, which is difficult to cut in the microtome.

10. Thicker slices allow for more radioactivity in the slice, enhancing autoradiography data, while thinner slices allow for better staining. Instead of going for an average thickness, it is best to alternatively cut 10 and 50 μm slices for each purpose.

Acknowledgements

This work was financially supported by the European Union through the grant Euregional PACT II by the Interreg IV program of Grensregio Vlaanderen-Nederland (IVA-VLANED-1.20) and the Center of Excellence "MoSAIC" of the K.U. Leuven.

References

1. Therasse P, Arbuck SG, Eisenhauer EA, Wanders J, Kaplan RS, Rubinstein L, Verweij J, Van Glabbeke M, van Oosterom AT, Christian MC, Gwyther SG (2000) New guidelines to evaluate the response to treatment in solid tumors. European Organization for Research and Treatment of Cancer, National Cancer Institute of the United States. J Natl Cancer Inst 92(3):205–216

2. Neves AA, Brindle KM (2006) Assessing responses to cancer therapy using molecular imaging. Biochim Biophys Acta 1766(2): 242–261

3. De Saint-Hubert M, Prinsen K, Mortelmans L, Verbruggen A, Mottaghy FM (2009) Molecular imaging of cell death. Methods 48(2):178–187

4. Koulov AV, Stucker KA, Lakshmi C, Robinson JP, Smith BD (2003) Detection of apoptotic cells using a synthetic fluorescent sensor for membrane surfaces that contain phosphatidylserine. Cell Death Differ 10(12):1357–1359

5. Cohen A, Ziv I, Aloya T, Levin G, Kidron D, Grimberg H, Reshef A, Shirvan A (2007) Monitoring of chemotherapy-induced cell death in melanoma tumors by N, N′-Didansyl-L-cystine. Technol Cancer Res Treat 6(3):221–234

6. Aloya R, Shirvan A, Grimberg H, Reshef A, Levin G, Kidron D, Cohen A, Ziv I (2006) Molecular imaging of cell death *in vivo* by a

novel small molecule probe. Apoptosis 11(12):2089–2101

7. Zeng W, Wang X, Xu P, Liu G, Eden HS, Chen X (2015) Molecular imaging of apoptosis: from micro to macro. Theranostics 5(6):559–582

8. Ogawa K, Aoki M (2014) Radiolabeled apoptosis imaging agents for early detection of response to therapy. Sci World J 2014:732603

9. Yang TJ, Haimovitz-Friedman A, Verheij M (2012) Anticancer therapy and apoptosis imaging. Exp Oncol 34(3):269–276

10. Blankenberg FG, Katsikis PD, Tait JF, Davis RE, Naumovski L, Ohtsuki K, Kopiwoda S, Abrams MJ, Strauss HW (1999) Imaging of apoptosis (programmed cell death) with 99mTc annexin V. J Nucl Med 40(1):184–191

11. Hofstra L, Liem IH, Dumont EA, Boersma HH, van Heerde WL, Doevendans PA, De Muinck E, Wellens HJ, Kemerink GJ, Reutelingsperger CP, Heidendal GA (2000) Visualisation of cell death *in vivo* in patients with acute myocardial infarction. Lancet 356(9225):209–212

12. Vangestel C, Peeters M, Mees G, Oltenfreiter R, Boersma HH, Elsinga PH, Reutelingsperger C, Van Damme N, De Spiegeleer B, Van de Wiele C (2011) *In vivo* imaging of apoptosis in oncology: an update. Mol Imaging 10(5): 340–358

13. De Saint-Hubert M, Mottaghy FM, Vunckx K, Nuyts J, Fonge H, Prinsen K, Stroobants S, Mortelmans L, Deckers N, Hofstra L, Reutelingsperger CP, Verbruggen A, Rattat D (2010) Site-specific labeling of 'second generation' annexin V with 99mTc(CO)3 for improved imaging of apoptosis *in vivo*. Bioorg Med Chem 18(3):1356–1363

14. Bauwens M, De Saint-Hubert M, Devos E, Deckers N, Reutelingsperger C, Mortelmans L, Himmelreich U, Mottaghy FM, Verbruggen A (2011) Site-specific 68Ga-labeled Annexin A5 as a PET imaging agent for apoptosis. Nucl Med Biol 38(3):381–392

15. Sakaguchi Y, Stephens LC, Makino M, Kaneko T, Strebel FR, Danhauser LL, Jenkins GN, Bull JM (1995) Apoptosis in tumors and normal tissues induced by whole body hyperthermia in rats. Cancer Res 55(22):5459–5464

16. Tan S, Peng X, Peng W, Zhao Y, Wei Y (2015) Enhancement of oxaliplatin-induced cell apoptosis and tumor suppression by 3-methyladenine in colon cancer. Oncol Lett 9(5):2056–2062

17. Feldmann G, Lamboley C, Moreau A, Bringuier A (1998) Fas-mediated apoptosis of hepatic cells. Biomed Pharmacother 52(9):378–385

18. Ghetie MA, Richardson J, Tucker T, Jones D, Uhr JW, Vitetta ES (1990) Disseminated or localized growth of a human B-cell tumor (Daudi) in SCID mice. Int J Cancer 45(3):481–485

19. Takei T, Kuge Y, Zhao S, Sato M, Strauss HW, Blankenberg FG, Tait JF, Tamaki N (2004) Time course of apoptotic tumor response after a single dose of chemotherapy: comparison with 99mTc-annexin V uptake and histologic findings in an experimental model. J Nucl Med 45(12):2083–2087

20. Mirkovic N, Meyn RE, Hunter NR, Milas L (1994) Radiation-induced apoptosis in a murine lymphoma *in vivo*. Radiother Oncol 33(1):11–16

Chapter 3

Detection of Active Caspases During Apoptosis Using Fluorescent Activity-Based Probes

Laura E. Edgington-Mitchell and Matthew Bogyo

Abstract

Activity-based probes (ABPs) are reactive small molecules that covalently bind to active enzymes. When tagged with a fluorophore, ABPs serve as powerful tools to investigate enzymatic activity across a wide variety of applications. In this chapter, we provide detailed methods for using fluorescent ABPs to detect the activity of caspases during the onset of apoptosis in vitro. We describe how these probes can be used to biochemically profile caspase activity in vitro using fluorescent SDS-PAGE as well as their application to imaging protease activity in live animals and tissues.

Key words Caspases, Proteases, Apoptosis, Activity-based probes, Fluorescent SDS-PAGE, Imaging

1 Introduction

Apoptosis is a highly regulated form of cell death that is mediated by a family of cysteine proteases called caspases [1, 2]. The extrinsic cell death pathway is initiated at the cell surface through ligation of death receptors. This leads to the dimerization and activation of the initiator caspase-8. Alternatively, the intrinsic death pathway is stimulated in response to intracellular signals such as DNA damage or mitochondrial stress. This results in formation of the apoptosome and subsequent activation of caspase-9, another initiator caspase. Once active, caspase-8 and -9 can then cleave and activate the downstream executioner caspases (-3, -6, and -7), which are common to both pathways. These proteases can then hydrolyze peptide bonds after aspartic acid residues of numerous other proteins. These cleavages lead to inactivation of substrates and initiate signaling pathways that promote the controlled disassembly of the cell, and ultimately, death.

Because caspases are synthesized as inactive enzymes, measurement of their total expression using antibody-based methods does not always reflect their activity levels [3–6]. Further to this point, caspases are also subject to inhibition by IAPs (inhibitors of

Hamsa Puthalakath and Christine J. Hawkins (eds.), *Programmed Cell Death: Methods and Protocols*, Methods in Molecular Biology, vol. 1419, DOI 10.1007/978-1-4939-3581-9_3, © Springer Science+Business Media New York 2016

apoptosis) once activated. To study the function of proteases in normal cellular processes and disease states, tools are necessary to directly assess their activity.

Fluorogenic substrates are one type of tool that can be used to measure caspase activity [3]. These molecules contain a peptide sequence that is based on the preferred cleavage site of each caspase and a fluorophore. When an active caspase cleaves the substrate, a detectable shift in the fluorescent properties of the fluorophore occurs. Fluorescence emission therefore correlates with levels of active enzyme. Substrate-based probes are commercially available and are marketed as being specific for individual caspase family members. In reality, these probes are not truly specific given the overlap in specificity for multiple caspases and even other protease families. A particular substrate may be cleaved more efficiently by one caspase over the others; however, the others are likely able to cleave it as well. As a result, measuring the activity of an individual caspase is often difficult with this type of probe. Consequently, many reports in the literature that have used fluorogenic probes to ascribe functions to one caspase over another are difficult to interpret.

An alternative approach to measuring caspase activity utilizes activity-based probes (ABPs) [3, 5, 6]. These are small molecules that bind irreversibly to active caspases. Like the fluorogenic substrates, ABPs contain a peptide sequence based on the preferred cleavage site for caspases (including a P1 aspartate residue) and a fluorophore (*see* Fig. 1a, b). The key feature of an ABP, however, is a reactive functional group called a warhead that covalently modifies the enzyme in an activity-dependent manner. A common warhead for caspase probes (and other cysteine proteases) is the acyloxymethyl ketone, which binds selectively to the active-site cysteine residue.

The binding of an ABP to the caspase is strong enough to survive the denaturing conditions of SDS-PAGE, and fluorescence emitted by the fluorophore can be detected by scanning the gel with a flatbed laser scanner. This unique feature of ABPs enables identification of the specific caspases that are active within a particular sample at a given time. ABPs can also be administered in vivo, allowing for in situ determination of caspase activity by imaging the accumulation of fluorescence in live mice or excised tissues. Probe-labeled caspases can then be analyzed biochemically by fluorescent SDS-PAGE, allowing for identification of the exact enzymes that lead to fluorescence accumulation in vivo.

Two examples of ABPs for caspases are AB50 and LE22 [7, 8]. Both these probes contain an acyloxymethyl ketone warhead and are tagged with a Cy5 fluorophore. They differ slightly in their peptide specificity region and their target caspases. AB50 contains a Glu-Pro-Asp sequence and binds to active caspase-3 and -7 (*see* Fig. 1c). The peptide sequence in LE22 is Val-Glu-Ile-Asp and it labels active

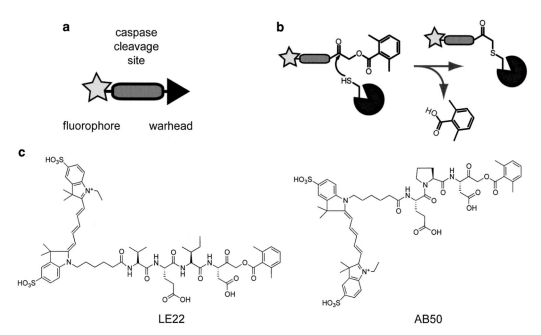

Fig. 1 Activity-based probes for caspases. (**a**) Schematic of the key features of an activity-based probe. (**b**) Mechanism of a cysteine protease binding to an ABP with an acyloxymethyl ketone warhead. (**c**) Structures of two fluorescent activity-based probes for caspases, AB50 and LE22. Figure adapted from Ref. 3

caspase-6 in addition to -3 and -7 (*see* Fig. 1c). Both these probes are also able to label the initiator caspases-8 and -9; however, given that the levels of these enzymes are dramatically lower than the executioner caspases, they are not often detected in whole cell extracts. LE22 and AB50 suffer from slight cross-reactivity with another cysteine protease called legumain; however, this is not usually a problem since all probe-labeled targets can be tracked biochemically.

We previously utilized AB50 and LE22 to detect caspase activity in COLO205 colon cancer cells after inducing apoptosis with an anti-death receptor 5 antibody (anti-DR5) [7, 8]. Both probes labeled caspases in a dose-dependent manner (*see* Fig. 2a). As outlined above, their specificities were slightly different, with AB50 labeling active caspase-3/-7 and LE22 labeling -3/-6/-7. Band identity was confirmed by pretreatment with the caspase-3/-7-specific inhibitor AB13 (*see* Fig. 2a) and by immunoprecipitation with caspase-specific antibodies (*see* Fig. 2b). We also utilized these probes to image caspase activation in tumor-bearing mice treated with anti-DR5 chemotherapy. Upon induction of apoptosis, LE22 fluorescence accumulated in the tumor, as demonstrated by whole animal imaging (*see* Fig. 2c) and ex vivo imaging (*see* Fig. 2d). The increase in fluorescence between vehicle- and drug-treated mice corresponded with an increase in caspase labeling as shown by fluorescent SDS-PAGE analysis (*see* Fig. 2d).

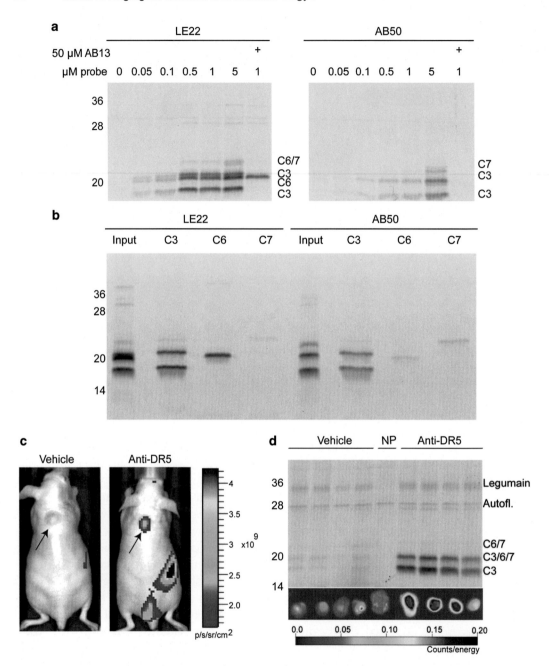

Fig. 2 Detection of caspase activity with ABPs in vitro and in vivo. (**a**) Dose curve of activity-dependent caspase labeling in apoptotic cells. COLO205 colon cancer cells were treated with anti-DR5 antibody for 4 h to induce apoptosis. For the last 30 min of treatment, the ABPs LE22 or AB50 were added at the indicated dose to detect caspase activity, followed by fluorescent SDS-PAGE. Where indicated, AB13, a caspase-3/-7-selective inhibitor was added prior to ABP addition. (**b**) Confirmation of the identity of the caspases labeled in (**a**) by immunoprecipitation with cleaved caspase antibodies. (**c**) In vivo detection of caspase activity. Nude mice bearing subcutaneous COLO205 tumors (*red arrows*) were treated with anti-DR5 antibody or vehicle for 11 h, followed by injection of the ABP LE22. After 1 h, mice were imaged for Cy5 fluorescence using an IVIS100 whole animal imaging machine. (**d**) Tumors were extracted from the mice in (**c**), imaged ex vivo, and then lysed for fluorescent SDS-PAGE analysis. Panels (**a–d**) reproduced with permission from Ref. 9

The following protocols describe the application of covalent ABPs to detect caspase activity in vitro and in vivo [9]. We indicate appropriate conditions for using LE22 and AB50 to detect caspase activity; however, these proteases can be adapted for any covalent ABP that targets any protease. Optimization of timing, dosage, and buffer conditions may be required.

2   Materials

2.1   Components for Labeling Caspases with Activity-Based Probes

1. Cells (chosen based on model).
2. 6-well plates.
3. Cell scrapers.
4. Complete medium for cells (varies by cell type).
5. Caspase activation agent (typically an apoptosis stimulant; chosen based on model).
6. Cold Phosphate buffered saline (PBS).
7. Hypotonic Lysis Buffer: 50 mM PIPES, pH 7.4, 10 mM KCl, 5 mM $MgCl_2$, 2 mM EDTA, 1 % NP-40, 4 mM DTT (add DTT fresh when ready to use). Store at 4 °C.
8. Bicinchoninic acid protein quantitation (BCA) kit or equivalent.
9. Fluorescent activity-based probe: 100× stock solution in dimethyl sulfoxide (DMSO). (This concentration will vary by probe. For AB50 and LE22, 100× = 100 μM.)
10. 4× sample buffer: 40 % glycerol (v/v), 200 mM Tris–HCl, pH 6.8, 8 % SDS (w/v), 0.04 % bromophenol blue (w/v), 5 % beta-mercaptoethanol (v/v). Store aliquots at −20 °C or at room temperature for 1–2 weeks.

2.2   Components for Fluorescent SDS-PAGE

1. SDS-PAGE gel (15 % acrylamide is ideal for caspase separation).
2. Gel running apparatus.
3. Laemmli Running Buffer: 25 mM Tris base, 192 mM glycine, 0.1 % SDS. The pH should be ~8.3 with no adjustment required.
4. Fluorescent protein ladder (such as Precision Plus Protein™ Dual Color Standards from BioRad).
5. Flatbed laser scanner with appropriate filters, for example Typhoon from GE Healthcare or Odyssey from LiCOR.

2.3   Immunoprecipitation Components

1. Probe-labeled lysate.
2. Protein A/G agarose beads.
3. Caspase-specific antibodies.

4. Immunoprecipitation buffer (IP buffer): 1× PBS, pH 7.4, 0.5 % NP-40 (v/v), 1 mM EDTA. Store at room temperature for 1 year.

5. 0.9 % sodium chloride in water.

6. 4× sample buffer (*see* **step 10** of Subheading 2.1).

7. SDS-PAGE components (*see* Subheading 2.2).

2.4 Components for In Vivo Detection of Caspases

1. Mice (chosen based on model).

2. Caspase activation agent (typically an apoptosis stimulant; chosen based on model).

3. Fluorescent activity-based probe, for example, AB50 or LE22.

4. Dimethyl sulfoxide (DMSO).

5. *P*hosphate *b*uffered *s*aline (PBS).

6. Isoflurane to anesthetize the mice.

7. Non-fluorescent (alfalfa-free) chow.

8. Beard trimmer.

9. Nair hair removal lotion or equivalent.

10. Insulin syringes.

11. Restrainer for tail vein injections.

12. Anesthesia vaporizer.

13. Fluorescent mouse imager such as the IVIS or FMT or BioFLECT or equivalent.

14. Dissection tools.

15. Muscle lysis buffer: 1× PBS, pH 7.4, 1 % Triton X-100 (v/v), 0.1 % SDS (w,v), 0.5 % sodium deoxycholate (w,v). Store at 4 °C.

16. BCA kit or equivalent.

17. 4× sample buffer (*see* **step 10** of Subheading 2.1).

18. SDS-PAGE components (*see* Subheading 2.2).

3 Methods

3.1 Labeling of Active Caspases in Cell Lysates

This protocol describes the most basic assay for activity-based labeling of caspases in a complex proteome. Labeling cell lysates will provide a snapshot of the caspase activity at the time the cells are harvested. The timing and degree of caspase activation will depend on the method used to stimulate apoptosis. If this information is not readily available in the literature for the cell line used, a time course and/or dose curve may be performed to determine optimal conditions for caspase activation. This assay detects caspases at neutral pH, making it less likely to detect common off-targets of caspase ABPs such as legumain or cathepsins, which are active in the acidic conditions of the lysosome.

1. Seed cells in 6-well dishes at 80 % confluency (depending on the size of the cells, this will be about $1–1.5 \times 10^6$ cells).

2. The following day, refresh media and stimulate apoptosis/caspase activation by preferred method (*see* **Notes 1** and **2**).

3. Harvest the cells at the optimal time point. In cell death models, some of the cells may have already detached and will be floating in the media. Using cell scrapers or the broad end of a P200 pipette tip, dislodge the remaining cells.

4. Transfer the cells in media to a microfuge tube and spin using a tabletop centrifuge at $1000 \times g$ for 1–2 min.

5. Aspirate media with care, so as not to disturb the cell pellet and then add 1 mL cold PBS to the cells. Pipette up and down 2–3 times to wash the pellet. Keep tubes on ice from this point to preserve caspase activity.

6. Centrifuge as in **step 4** above and aspirate PBS. Then add 50 µL of hypotonic lysis buffer, resuspend pellet, and incubate on ice for 10 min to lyse the cells.

7. At this point, the sample may be snap frozen in liquid nitrogen and stored at –20 °C for later use, or be advanced to the next step.

8. Spin at $20,000 \times g$ in a tabletop centrifuge at 4 °C for ~15 min and then transfer supernatant to a fresh microfuge tube.

9. Quantify total protein concentration using a BCA kit or equivalent (*see* **Note 3**).

10. Aliquot 50–100 µg lysate into a fresh tube, bringing the volume to 20 µl with hypotonic lysis buffer.

11. Add 0.2 µl of a 100× DMSO stock of the activity-based probe. For AB50 or LE22, use 100 µM stock solution (*see* **Notes 4–6**).

12. Incubate the sample at 37 °C. Labeling time may vary by ABP. For LE22 and AB50, 30 min is the optimal labeling time.

13. Add 6.7 µl 4× sample buffer to quench the reaction. Samples may be snap frozen in liquid nitrogen and stored at –20 °C or directly analyzed by SDS-PAGE (*see* Subheading 3.3).

3.2 Labeling of Active Caspases in Intact/Dying Cells

This protocol depends on the ability of the ABP to freely penetrate cells and enter the cytoplasm as in the case of caspases. AB50 and LE22 are both permeable. One of the major caveats of caspase probes is their varied propensity to label the lysosomal proteases (cathepsins and/or legumain) at acidic pH due to similarities in structure and mechanism of catalysis [7, 8, 10]. If the probe enters the cell by endocytosis or macropinocytosis, it may pass through the lysosome before reaching the cytosol, labeling these off-targets in addition to caspases.

1. Seed cells in 6-well dishes to 80 % confluence (depending on the size of the cells, this will be about $1-1.5 \times 10^6$ cells).

2. The following day, aspirate medium and add 1 mL fresh medium. Stimulate apoptosis/caspase activation by preferred method (*see* **Note 7**).

3. The time at which the probe is added to the cells will vary by experiment and probe. To obtain a more cumulative picture of the caspases that become activated over time, the probe can be added as soon as the apoptosis stimulus is added and incubated throughout the course of the experiment (*see* **Note 8**). To obtain a snapshot of the particular caspases that are active at a given time point after stimulation, add the probe during the desired window. To do so, add 1 μl of a 1000× DMSO stock solution of ABP directly to the media and mix well. (Use 1 μl of 1 mM stock to make a final concentration of 1 μM. The final DMSO concentration should be no more than 0.2 %.) (*see* **Notes 9** and **10**).

4. Harvest the cells at the optimal time point after probe addition (shorter incubation for snapshots, longer for cumulative pictures) by detaching them from the dish. This can be achieved using cell scrapers or the broad end of a P200 pipette tip. Collect the cells and media and centrifuge at $1000 \times g$ for 1–2 min. Then carefully aspirate the media and wash once with cold PBS, making sure to resuspend the pellet. Remove the PBS.

5. Resuspend the pellet in 50 μl hypotonic lysis buffer and incubate on ice for ten minutes. At this point, the sample can be snap frozen in liquid nitrogen and stored at –20 °C, or the protocol may be continued.

6. Centrifuge at $20,000 \times g$ in a bench top centrifuge at 4 °C for 15 min. Transfer the supernatant to a fresh microfuge tube. Reserve 5 μl for protein quantification by BCA or equivalent. To the remaining 45 μl, add 15 μl 4× sample buffer. The sample may then be frozen at –20 °C, or boiled and loaded onto a gel according to Subheading 3.3.

3.3 Detection of Probe-Labeled Caspases by Fluorescent SDS-PAGE

This protocol enables detection of the caspases that are labeled with probe in Subheadings 3.1, 3.2, 3.4, and 3.5.

1. Prepare 15 % acrylamide gel according to standard protocols [11].

2. Boil protein samples at 95 °C for 5–10 min and load samples, along with a fluorescent protein ladder on the gel.

3. Run gel according to standard protocol [11]. Stop just after the dye front runs off (*see* **Note 11**).

4. To visualize the active caspases, scan the gel on a Typhoon or Odyssey scanner (or equivalent) at the wavelength appropriate for the fluorophore used (*see* **Notes 12–16**).

3.4 Immuno-precipitations of Probe-Labeled Lysates to Confirm ABP Targets

Once ABP-labeled species have been visualized by SDS-PAGE, it may be necessary to confirm their identity. This can usually be achieved by pre-treating samples with selective inhibitors to compete for probe labeling; however, in the case of caspases, truly selective inhibitors do not exist. They tend to react broadly with members of the caspase family as well as lysosomal cysteine proteases like legumain and cathepsins. For this reason, immunoprecipitation may be the ideal way to validate probe-labeled species. If enough material remains, the sample used for SDS-PAGE analysis may be used directly in the immunoprecipitation assay.

1. Prepare probe-labeled lysate according to Subheadings 3.1, 3.2 or 3.5.

2. After boiling in sample buffer, reserve 50 μg of probe-labeled protein lysate to load as an input sample for SDS PAGE analysis. Store at −20 °C.

3. For each immunoprecipitation, aliquot 100^+ μg probe-labeled lysate into microfuge tubes (*see* **Note 17**).

4. Add 500 μl IP buffer and 5–10 μl antibody and incubate on ice for 10 min (*see* **Note 18**).

5. Aliquot a 40-μl slurry of protein A/G agarose beads to a new microfuge tube. Add 500 μl IP buffer to wash beads and centrifuge at high speed on a tabletop centrifuge for 30 s. Remove buffer, add 50 μl fresh IP buffer, and transfer beads to the tube containing the protein + antibody. Rock tube overnight at 4 °C.

6. The next morning, transfer supernatant to a 2-mL microfuge tube. To precipitate and concentrate the proteins in the supernatant, add enough acetone to fill the tube and freeze at −80 °C for at least 2 h. Then spin the tube at high speed at 4 °C for 15 min. Aspirate the acetone, taking great care not to disturb the protein pellet. Allow the pellet to dry completely before dissolving it in 30 μl of 1× sample buffer.

7. While the supernatant is precipitating, wash the agarose beads four times with IP buffer and then once with 0.9 % sodium chloride. After the last wash, remove all of the remaining supernatant (*see* **Note 19**). Then add 30 μl 2× sample buffer and boil the beads to elute the immunoprecipitated proteins.

8. Analyze immunoprecipitation by SDS-PAGE as described in Subheading 3.3. Load the input sample first, followed by the pull down and the supernatant (*see* **Notes 20** and **21**).

3.5 In Vivo Detection of Caspase Activity

Protocol details for in vivo experiments will largely depend on the mouse model and probes used, but these are guidelines to follow that will help increase the success rate of noninvasive imaging experiments. Since light does not penetrate tissue particularly well, the most successful noninvasive imaging experiments will involve superficial tissues, for example, subcutaneous tumors. Deeper tissues may need to be excised and imaged ex vivo. Models involving at least a threefold induction of protease activity will have the highest chance of obtaining enough contrast for imaging. If whole animal fluorescence imaging is to be performed, optimal results will be obtained if mice are fed alfalfa-free chow for 4–7 days prior to imaging (*see* **Note 22**). Ensure that all animal experiments are approved by the appropriate Institutional Animal Care and Use Committee (IACUC) and that government regulations regarding the care and use of laboratory animals are followed.

1. Thin the hair using a small beard trimmer and use hair removal lotion to dissolve any remaining hair. Make sure to completely wash off the lotion, since extended exposure can burn the mice (*see* **Note 23**).

2. Induce caspase activity by administering the appropriate stimulant (*see* **Notes 24** and **25**). In parallel, treat negative control mice with the vehicle in which the drug was administered.

3. At the peak of caspase activation, administer the activity-based probe intravenously by tail vein injection. The probe will then circulate throughout the mouse and bind to active caspases. Free, unbound probe will clear from the body primarily through the renal system (*see* **Note 26**). It is also a good idea to include a "no-probe" control to distinguish probe fluorescence from autofluorescence.

4. Anesthetize the mouse using isoflurane, and image it using the appropriate filters on a whole animal-imaging machine (IVIS, FMT, or BioFLECT) (*see* **Note 27**).

5. Kill the mouse and remove the tissue of interest. Place the tissue on a sheet of black paper and image it ex vivo using the same filters on an IVIS or FMT machine.

6. Next, a biochemical analysis of the labeled caspases should be performed. At this stage, the tissue may be snap frozen or lysed directly.

7. Place the tissue in a microfuge tube. Add an appropriate volume of muscle lysis buffer and homogenize the tissue (*see* **Notes 28** and **29**). Spin the homogenates at $20,000 \times g$ in a benchtop centrifuge for 15 min, and then transfer the supernatant to a fresh tube.

8. Determine the protein concentration using a BCA kit or equivalent and solubilize the protein with 4× sample buffer.

9. Analyze the probe-labeled lysate (50–100 μg total protein) by fluorescent SDS-PAGE described in Subheading 3.3.

4 Notes

1. The timing will vary depending on the cell line and stimulant used.

2. "No-treatment" controls are also important to include.

3. Remember to account for DTT in the buffer by making protein standards in hypotonic lysis buffer at the same dilution as the samples.

4. For LE22 or AB50, the optimal concentration for lysate labeling is 1 μM. Probe concentration varies by molecule; a dose curve will aid in optimizing if this information is not already available.

5. It is highly recommended to run a "no-probe" control in parallel to account for any background/autofluorescent bands that may appear in the gel.

6. A protease inhibitor may be used to validate the selectivity of the probe. Add it *before* the probe at the recommended concentration. The length of the pre-incubation time will vary by inhibitor.

7. "No-treatment" controls are also important to include.

8. If the time course is longer than ~4 h, fresh probe may need to be added, depending on probe stability and the concentration used.

9. 1 μM is optimal for LE22 and AB50.

10. Various protease inhibitors may be used to validate the selectivity of the ABP. The inhibitor should be added *before* the addition of the probe. The pre-incubation time and dose will vary by inhibitor.

11. Ensure that the caspases (16–21 KDa) do not run off the gel.

12. For LE22 and AB50, the Cy5 filter set should be used.

13. The gel can be left in the glass plates during scanning. The scan depth must be changed from "platen" to +3 mm. If the gel is removed from the plates for the scan, rinse it with water to avoid tears and do not touch the gel with ungloved fingers (this causes autofluorescence). Use the platen setting.

14. The photomultiplier tube (PMT/sensitivity) setting should be adjusted such that the scan is not overexposed or may be increased to boost signal.

15. Mature, active caspases should appear in the 16–21 kDa range.

16. At this point, the proteins on the gel may be transferred to a membrane and blotted according to standard western protocols. This is a good way to compare protease activity with total expression.

17. This amount may vary depending on the intensity of the labeling. If weak, use more.

18. This antibody should be carefully chosen such that it binds selectively to the protease of interest. It is critical that the antibody be raised towards the subunit to which the probe binds. In the case of caspases, the probes bind to the active site cysteine, which is found within the *large subunit*.

19. An insulin syringe is helpful to remove the liquid without aspirating the beads.

20. Sometimes it is helpful to leave a blank lane in between each sample.

21. Bands in the pull down lane confirm that the labeled protease corresponds to the target of the immunoprecipitating antibody. When immunoprecipitation is complete, this band should be absent from the "supernatant" sample.

22. Normal chow causes autofluorescence in the abdominal region.

23. This step is only required for whole animal imaging experiments, as fluorescent light cannot not pass through the hair.

24. Stimulant, dosage, and the time it takes to produce active caspases are dependent on the model and must be optimized.

25. For LE22 or AB50, inject 20 nmol of probe (10 % DMSO/PBS, 100 μl volume).

26. For LE22 and AB50, maximal caspase labeling occurs after 1 h.

27. For LE22 and AB50, use the Cy5 filter.

28. Generally, add 10 μl buffer for every 1 mg of tissue.

29. Sonication is a good way to disrupt the tissue; however, Dounce homogenizers, bead beaters, or other means of mechanical disruption may be effective, depending on the tissue.

References

1. Li J, Yuan J (2008) Caspases in apoptosis and beyond. Oncogene 27(48):6194–6206

2. McIlwain DR, Berger T, Mak TW (2015) Caspase functions in cell death and disease. Cold Spring Harbor Perspect Biol 7(4)

3. Edgington LE, Verdoes M, Bogyo M (2011) Functional imaging of proteases: recent advances in the design and application of substrate-based and activity-based probes. Curr Opin Chem Biol 15(6):798–805

4. Deu E, Verdoes M, Bogyo M (2012) New approaches for dissecting protease functions to improve probe development and drug discovery. Nat Struct Mol Biol 19(1):9–16

5. Sanman LE, Bogyo M (2014) Activity-based profiling of proteases. Annu Rev Biochem 83:249–273

6. Serim S, Haedke U, Verhelst SH (2012) Activity-based probes for the study of proteases: recent advances and developments. ChemMedChem 7(7):1146–1159

7. Edgington LE, Berger AB, Blum G, Albrow VE, Paulick MG, Lineberry N, Bogyo M (2009) Noninvasive optical imaging of apoptosis by caspase-targeted activity-based probes. Nat Med 15(8):967–973

8. Edgington LE, van Raam BJ, Verdoes M, Wierschem C, Salvesen GS, Bogyo M (2012) An optimized activity-based probe for the study of caspase-6 activation. Chem Biol 19(3):340–352

9. Edgington LE, Bogyo M (2013) In vivo imaging and biochemical characterization of protease function using fluorescent activity-based probes. Curr Protoc Chem Biol 5(1):25–44

10. Edgington LE, Verdoes M, Ortega A, Withana NP, Lee J, Syed S, Bachmann MH, Blum G, Bogyo M (2013) Functional imaging of legumain in cancer using a new quenched activity-based probe. J Am Chem Soc 135(1):174–182

11. SR Gallagher (2012) One-dimensional SDS gel electrophoresis of proteins. Curr Protoc Mol Biol. Chapter 10:Unit 10 12A

Chapter 4

Detection of Initiator Caspase Induced Proximity in Single Cells by Caspase Bimolecular Fluorescence Complementation

Melissa J. Parsons, Sara R. Fassio, and Lisa Bouchier-Hayes

Abstract

The caspase family of proteases includes key regulators of apoptosis and inflammation. The caspases can be divided into two groups, the initiator caspases and the executioner caspases. Initiator caspases include caspase-2, caspase-8, and caspase-9 and are activated by proximity-induced dimerization upon recruitment to large molecular weight protein complexes called activation platforms. This protocol describes an imaging-based technique called caspase *Bi*molecular *Fl*uorescence *C*omplementation (BiFC) that measures induced proximity of initiator caspases. This method uses nonfluorescent fragments of the fluorescent protein Venus fused to initiator caspase monomers. When the caspase is recruited to its activation platform, the resulting induced proximity of the caspase monomers facilitates refolding of the Venus fragments into the full molecule, reconstituting its fluorescence. Thus, the assembly of initiator caspase activation platforms can be followed in single cells in real time. Induced proximity is the most apical step in the activation of initiator caspases, and therefore, caspase BiFC is a robust and specific method to measure initiator caspase activation.

Key words Apoptosis, BiFC, Caspases, Confocal, Fluorescence, Imaging, Microscopy, Venus

1 Introduction

The caspases are a family of *cy*steine-dependent *asp*artate-directed prote*ases* that are responsible for the initiation and execution of apoptosis. The caspases are classified as either initiator caspases, including caspase-2, -8, -9, and -10, or executioner caspases, such as caspase-3, -6, and -7 [1]. Initiator caspases are so-called because they are the first to be activated in a pathway. The predominant function of active initiator caspases in apoptotic pathways is to activate downstream executioner caspases. Active executioner caspases then proteolytically cleave numerous regulatory and structural proteins within the cell usually via cleavage at specific aspartate residues [2]. A third group of caspases exists that includes caspase-1, -4, -5, -11, and -12. These caspases are similar in structure to

Hamsa Puthalakath and Christine J. Hawkins (eds.), *Programmed Cell Death: Methods and Protocols*, Methods in Molecular Biology, vol. 1419, DOI 10.1007/978-1-4939-3581-9_4, © Springer Science+Business Media New York 2016

the initiator caspases but are not involved in the regulation of apoptosis. Members of this group are referred to as inflammatory caspases and have key roles in the regulation of innate immunity.

Initiator caspases are activated by proximity-induced dimerization following recruitment to large molecular weight complexes known as activation platforms (Fig. 1). For example, ligation of death receptors such as TNFR1, TRAIL, or Fas by their cognate ligands leads to the formation of the activation platform for the initiator caspase-8. This activation platform is a multimeric signaling complex called the DISC (*d*eath-*i*nducing *s*ignaling *c*omplex) [3]. The DISC consists of proteins that oligomerize via homotypic protein–protein interaction domains. The main component linking death receptor ligation to caspase-8 activation is a protein called FADD [4]. FADD contains two protein–protein interaction motifs: a *d*eath *d*omain (DD) and a *d*eath *e*ffector *d*omain (DED) [5]. FADD binds to the intracellular portion of the oligomerized death receptor via DD interactions in FADD and the death receptor (e.g., Fas). FADD then recruits caspase-8 molecules via DEDs in both proteins.

Binding of caspase-8 molecules in the DISC brings them into close physical proximity in a process known as induced proximity [6, 7]. Induced proximity facilitates dimerization of caspase-8 molecules, at which point the caspase is active. Once dimerized, caspase-8 is auto-processed, that is, the catalytic domain of one caspase-8 molecule mediates cleavage of the other caspase-8 molecule between the large and small subunit. Further processing removes the prodomain generating a heterodimer comprised of the large and small catalytic subunits [8]. This cleavage is required to stabilize the active caspase-8 dimer [9].

Caspase-9 is similarly activated by induced proximity. Release of cytochrome c from mitochondria following *m*itochondrial *o*uter *m*embrane *p*ermeabilization (MOMP) induces the assembly of a complex known as the apoptosome (Fig. 1), which comprises cytochrome c, dATP, and APAF-1. The apoptosome functions as an activation platform for caspase-9, recruiting caspase-9 molecules via binding between protein–protein interaction motifs called CARDs (*ca*spase-*r*ecruitment *d*omains) that are present in the prodomain of caspase-9 and in the APAF-1 sequence [10].

Caspase-2 appears to be activated in an activation platform termed the PIDDosome (Fig. 1), which facilitates induced proximity and dimerization of caspase-2 monomers [11]. The PIDDosome consists of the PIDD and the adaptor protein RAIDD that interact via the DD in both their sequences. RAIDD, in turn, recruits caspase-2, and this binding is mediated by the CARD present in both proteins. Certain stimuli, such as heat shock and cytoskeletal disruption, have been shown to activate caspase-2 [12, 13], but it is unclear if such stimuli are dependent on PIDDosome assembly or if alternative caspase-2 activation platforms exist [14].

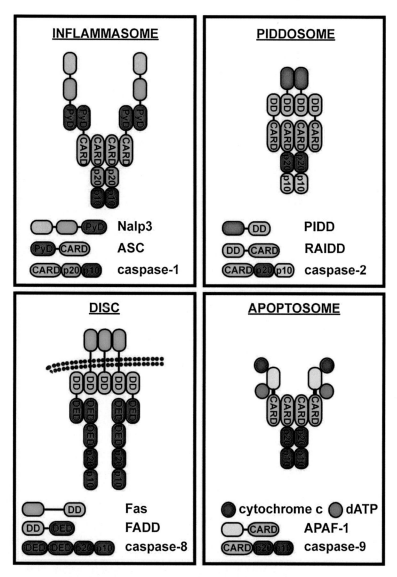

Fig. 1 Initiator caspases are activated by induced proximity upon recruitment to specific activation platforms. Examples of activation platforms that specifically recruit caspase-1, caspase-2, caspase-8, or caspase-9 are depicted. Caspase-1 is activated by inflammasomes, such as the Nalp3 inflammasome that is shown here. Proinflammatory stimuli, such as peptidoglycan, induce aggregation of Nalp3, which enables recruitment and clustering of the adaptor protein ASC. ASC, in turn, recruits caspase-1 monomers to the complex. Caspase-2 is similarly recruited to the PIDDosome comprising of PIDD and the adaptor protein RAIDD. Engagement of death receptors at the cell membrane, such as Fas, by their cognate ligands, such as FasL, results in assembly of the DISC that recruits caspase-8. The APAF-1 apoptosome is assembled in response to cytochrome c release from the mitochondria. Cytochrome c and dATP bind to APAF-1 leading to the assembly of the apoptosome and recruitment of caspase-9 monomers. The assembly of each of these activation platforms is governed by protein–protein interactions mediated by conserved protein binding motifs including the *c*aspase *r*ecruitment *d*omain (CARD), the *d*eath *d*omain (DD), the *d*eath *e*ffector *d*omain (DED), and the *py*rin *d*omain (PyD). Recruitment of initiator caspase monomers to their specific activation platform results in induced proximity, dimerization, and activation of each caspase

The inflammatory caspases are not primarily involved in apoptosis but they are similarly activated upon recruitment to activation platforms. Caspase-1 is activated by dimerization following recruitment to the inflammasome (Ref. [15] and Fig. 1). At least four distinct inflammasomes exist and each one appears to be assembled in response to specific pathogen products such as bacterial cell wall components or viral DNA [16]. Upon activation, caspase-1 cleaves the proinflammatory cytokines pro-interleukin-1β and pro-interleukin-18 to their mature forms and has fundamental roles on a variety of aspects of innate immune function [17].

Induced proximity is the most apical step in the activation of each initiator caspase. Caspase *Bi*molecular *F*luorescence *C*omplementation (BiFC) was developed to allow direct visualization this step following recruitment to activation platforms in real time, in single cells. BiFC is a method that uses split fluorescent proteins to measure protein–protein interactions in cells, and was first described for monitoring the interaction between the proteins Fos and Jun in live cells [18]. BiFC employs a fluorescent protein such as Venus (a brighter and more photostable version of *y*ellow *f*luorescent *p*rotein (YFP)) that is split into two fragments that are not fluorescent by themselves. When these protein fragments are fused to proteins that interact, the Venus fragments associate to reform the fluorescent molecule. Experimentally, this is detected by visualizing Venus fluorescence in live cells.

Caspase BiFC adapts this technique to measure the recruitment of initiator caspases to activation platforms (Fig. 2). Caspase BiFC was originally developed to interrogate the caspase-2 pathway [19] and has more recently been adapted to measure induced proximity of the inflammatory caspases -1, -4, -5, and -12 [20]. To achieve this, either the entire caspase or the portion of the caspase that binds to the activation platform (i.e., the prodomain that contains a CARD or DEDs) is fused to each fragment of Venus (Fig. 2a). In untreated cells, the caspase monomers maintain the Venus fragments separated from each other. The caspase-BiFC fusion proteins are recruited to the activation platform in response to an activating stimulus, such as vincristine for caspase-2 [19]. The resulting induced proximity of caspase BiFC fusion proteins promotes association of the Venus fragments, thereby restoring fluorescence (Fig. 2b, c). Thus, Venus fluorescence can be used as a measure of initiator caspase activation (Fig. 2d).

Caspase BiFC measures induced proximity as opposed to measuring cleavage of the caspase, which is a more downstream event. Executioner caspases are activated by cleavage and therefore monitoring disappearance of the full-length protein and the concurrent appearance of its cleavage products by western blot is a bona fide measure of activation [21]. However, initiator caspases are activated by dimerization and subsequent auto-processing stabilizes the active enzyme [21]. Importantly, cleavage of initiator caspases

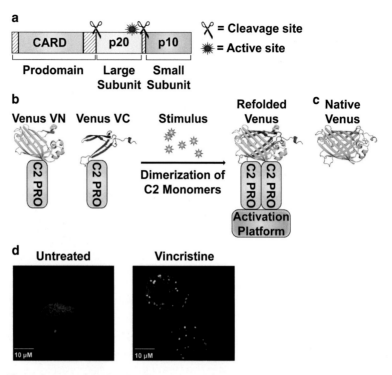

Fig. 2 Caspase Bimolecular Fluorescence Complementation measures induced proximity of initiator caspases. (**a**) Each initiator caspase is comprised of a large catalytic subunit (p20), a small catalytic subunit (p10), and a prodomain that contains one or two protein binding motifs; either a CARD, as illustrated, or two DEDs,. (**b**) Caspase BiFC fuses the prodomain of an initiator caspase to the N-terminus of Venus (Venus VN) or to the C-terminus of Venus (Venus C). Caspase-2 is shown here as an example. The initiator caspase-BiFC fusion proteins remain as monomers in unstimulated cells and the two halves of Venus do not refold. Following a caspase activating stimulus, the caspase fusion protein monomers are recruited to the activation platform. This allows for the two halves of Venus to be in sufficient proximity so that they refold and reconstitute the full length Venus protein, leading to fluorescence. (**c**) The protein structure of native Venus is shown for comparison. The molecular models of the protein structures were created using PyMOL v1.7.4.4 Edu with the amino acid sequences for Venus N (amino acids 1–173), Venus C (amino acids 155–238) and full length Venus. (**d**) The images show an example of caspase-2 BiFC in live cells induced by the cytoskeletal disruptor, vincristine. The caspase-2 Pro-BiFC components were expressed in Hela cells and the images were acquired using a spinning disk confocal microscope. The green fluorescence in the cells treated with vincristine (*right panel*) represents caspase-2 BiFC and is not detected in the untreated cells (*left panel*). Mitochondria are shown in red and represent expression of the reporter gene, DsRed-Mito. Nuclei were stained with Draq5 and are shown in *blue*

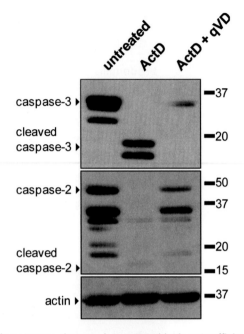

Fig. 3 Measuring caspase cleavage by western blot is not sufficient to determine activation of initiator caspases. In the example shown, *Act*inomycin *D* (ActD) induces cleavage of the executioner caspase, caspase-3, as measured by disappearance of the full length protein accompanied by appearance of the cleaved subunits. The initiator caspase, caspase-2, was also cleaved in the cells treated with ActD. However, ActD does not activate caspase-2 and this cleavage is likely mediated by caspase-3 rather than representing caspase-2 activation

in the absence of dimerization is not sufficient for activation [8, 22, 23]. In the example shown in Fig. 3, caspase-3 cleavage in lysates of cells treated with *Act*inomycin *D* (Act D) is easily detected. Caspase-2 cleavage is also detected in response to Act D. However, it has been shown that Act D does not activate caspase-2 [19, 24]. Therefore, this cleavage is most likely mediated by caspase-3 rather than representing the auto-processing step that is associated with activation of the initiator caspase. Single cell imaging of caspase BiFC, in contrast, allows for specific analysis of initiator caspase interactions at the level of the activation platform. This can provide a precise determination of the number of caspase BiFC events in a cell population, the subcellular location of caspase activation, and the kinetics of initiator caspase activation.

The method described here outlines how to introduce the caspase BiFC components into cells and how to visualize caspase BiFC in response to pro-apoptotic stimuli. Different approaches to acquiring and analyzing the data are detailed. These include determination of the percentage of caspase induced proximity at a single time point and visualization of activation platform assembly in single cells in real time using high resolution time-lapse confocal microscopy.

2 Materials

2.1 Plasmids

1. Plasmid constructs encoding the full length catalytically inactive caspase or the prodomain of the caspase fused to either the N-terminus of Venus (VN) or the C-terminus of Venus (VC):

   (a) BiFC plasmids for caspase-1, caspase-2, caspase-4, caspase-5, and caspase-12 were described in [19, 20]. These plasmids are available by request from the authors.

   (b) The BiFC plasmids, pBiFC.VC155 and pBiFC.VN173, are available from Addgene (#22011 and #22010). Additional caspases can be cloned into these plasmids by standard molecular cloning techniques.

2. Control plasmids: Plasmids containing point mutations that disrupt the binding interface between the caspase and its activation platform. Binding mutants for caspase-2 and caspase-1 were described in [19, 20] and are available by request from the authors.

3. Fluorescent reporter plasmid: A red or cyan fluorescent plasmid to be used as a reporter for transfection. This protocol uses DsRed-Mito (Clontech).

2.2 Cell Culture and Transfection

1. Cells can be obtained from the appropriate culture collections such as ATCC. HeLa cells were used in this protocol (ATCC).

2. Cell growth medium: Media and supplements should be chosen that are appropriate for each cell line. Hela cells are cultured in standard Dulbecco's Modified Eagle Media (DMEM), supplemented with fetal bovine serum (FBS, 10 % (v/v)), L-glutamine (2 mM), penicillin (100 U/mL), and streptomycin (100 μg/mL).

3. Imaging medium: Cell growth medium supplemented with Hepes (20 mM, pH 7.2–7.5) and 2-mercaptoethanol (55 μM).

4. Phosphate buffered saline (PBS).

5. Trypsin–EDTA.

6. Serum-free growth medium: This protocol uses Opti-MEM (Life Technologies).

7. Transfection reagent: This protocol uses Lipofectamine 2000 reagent (Life Technologies).

8. Glass bottom 3.5 cm dishes or 6-well plates, number 1.5 thickness (0.16–0.19 mm) (Mattek Corporation).

9. Fibronectin solution (1 mg/mL): Dilute in PBS to a final concentration of 0.1 mg/mL and store at 4 °C.

2.3 Induction of Apoptosis

1. qVD-OPh: resuspend in DMSO at 20 mM, store in aliquots at -20 °C, avoid repeated freeze/thaw.

2. Stimuli to induce apoptosis or caspase activation: Common inducers of apoptosis are listed below. All reagents listed should be stored in aliquots at –20 °C and repeated freeze–thaw should be avoided:

   (a) Actinomycin D: resuspend in DMSO at 1 mM.

   (b) TNFα: resuspend in sterile distilled water at 0.1 mg/mL.

   (c) Cycloheximide: resuspend in ethanol at 10 mg/mL.

   (d) Vincristine: resuspend in sterile distilled water at 20 mM.

2.4 Microscope Instrumentation

1. Microscope: In the protocol below we use a spinning disk confocal microscope for time-lapse or high resolution imaging, and an epi-fluorescence microscope for single time point analysis. Some useful microscope options include:

   (a) Motorized stage.

   (b) A heated stage or enclosed incubator to control the temperature and CO_2 concentration.

   (c) A camera to detect and image the cells: A CCD (*c*harge *c*oupled *d*evice) or EMCCD (*e*lectron *m*ultiplying *CCD*) camera offers high resolution and fast imaging.

   (d) Focal drift correction system such as Definite Focus (Zeiss).

2. Laser source: Venus is optimally excited by a 514 nm laser line (*see* **Note 1**). A second laser line is required to excite the fluorescent reporter (e.g., 458 nm for CFP or 561 nm for RFP).

3. Epi-fluoresence illumination:

   (a) Fluorescence filter sets consisting of an excitation filter, a dichroic beamsplitter, and an emission filter for the YFP or GFP wavelengths.

   (b) Fluorescence filter set to excite the fluorescent reporter such as RFP.

2.5 Data Processing

1. Microscope acquisition software.

2. Computer to run the microscope acquisition software: The computer should have a modern processor. A 64 bit computer is recommended for many types of acquisition software.

3. Image analysis software: This protocol uses ImageJ (NIH).

4. Graphing software such as Microsoft Excel.

3   Methods

3.1   Transfection of Caspase BiFC Plasmids

1. One day prior to transfection, coat 3.5 cm glass bottom dishes or a 6-well plate with embedded coverslips with fibronectin (0.1 mg/mL) (*see* **Note 2**). Add 1 mL per dish and incubate for at least 5 min at room temperature. Remove the fibronectin, wash wells once in PBS and remove PBS (*see* **Note 3**).

2. Plate 1×10^5 Hela cells per dish or well of a 6-well plate. Scale down as appropriate for smaller plate surface areas (*see* **Notes 4** and **5**).

3. The following day, add 12 μL Lipofectamine 2000 to 750 μL Opti-MEM in a sterile tube for 6×3.5 cm dishes or a 6-well plate and scale up or down as needed (*see* **Note 6**).

4. Incubate at room temperature for 5 min.

5. Add the appropriate amounts of BiFC plasmids and fluorescent reporter plasmid to a separate sterile microcentrifuge tube in a total of 100 μL Opti-MEM for each dish or well of a 6-well plate (*see* **Notes 7** and **8**).

6. Add 100 μL of the Lipofectamine solution from **step 3** to each plasmid solution.

7. Incubate at room temperature for 20 min to allow the DNA-lipid complexes to form.

8. Aspirate the growth medium from the cells and add 800 μL of serum free DMEM, pre-warmed to 37 °C, to each dish or well.

9. Add the DNA-lipid complexes (200 μL total) to each dish or well in a dropwise manner, followed by gently swirling of the dish to assure equal distribution throughout.

10. After 3 h, exchange the media with complete growth media, pre-warmed to 37 °C (*see* **Note 9**).

11. Incubate cells at 37 °C for at least 24 h to allow expression of the caspase BiFC components.

3.2   Induction of Apoptosis

1. Treat cells with the chosen death stimulus (*see* **Note 10**).

   (a) For drug treatments: Add the drug to imaging medium at the appropriate concentration and gently mix (*see* **Note 11**). Include a caspase inhibitor such as qVD-OPh (20 μM) if needed (*see* **Note 12**). Add the drug-containing medium to cells.

   (b) For heat shock: Exchange media for imaging media heated to the heat shock temperature. Add qVD-OPh (20 μM) to the heated media if using. Place the plate in an incubator set at the heat shock temperature for the appropriate time (*see* **Note 13**), and then return to 37 °C incubation.

2. Include an untreated control well or plate.

3. Incubate cells for the appropriate amount of time and proceed to Subheading 3.3 or immediately start the time-lapse and proceed to Subheading 3.4.

3.3 Caspase BiFC for Single Time Point Data Acquisition

1. Count the number of cells that are Venus (i.e., BiFC)-positive at the chosen time point using an epi-fluorescence microscope.

2. Visualize the cells under the filter that excites RFP and count all of the DsRed-positive cells in a field.

3. Visualize the same field under the filter that excites GFP and count the number of red cells that are Venus-positive (*see* **Note 14**).

4. Count 3 different areas of the plate, with each area containing at least 100 DsRed-positive cells.

5. Repeat each experiment at least three times to ensure reproducibility.

3.4 Caspase BiFC for Time-Lapse Microscopy

1. Turn on the microscope and set the temperature to 37 °C at least one hour prior to imaging the cells. This will give the microscope enough time to reach thermal equilibrium (*see* **Note 15**).

2. At least 30 min prior to imaging the cells, place the dish on the microscope stage using the correct adaptor.

3. Turn on the CO_2 source (*see* **Note 16**).

4. Locate transfected cells by looking for cells that express the transfection reporter. For example, if using DsRed-Mito, search for cells using the 561 nm laser.

5. Focus on the cells using a 60× or a 63× oil objective (*see* **Note 17**).

6. Determine the lowest amount of laser light and the shortest exposure time required to detect the Venus signal. Note the laser power and exposure times and keep consistent across experiments (*see* **Note 18**).

7. Determine the lowest amount of laser light and shortest exposure times required to detect the signal of the reporter plasmid. Note the laser power and exposure times and keep consistent across experiments.

8. Turn on focus drift correction system if available (*see* **Note 19**)

9. Choose a number of different positions within the plate or across multiple wells (*see* **Note 20**).

10. Set the time interval between each image capture and the duration of the experiment (*see* **Note 21**).

11. Visit each position and fine tune the focus.

12. Begin the time-lapse.

13. At the end of the experiment, save the data.

3.5 Data Analysis

3.5.1 Calculate the Percentage of Venus-Positive Cells at a Single Time Point

1. Using Excel or similar graphing software, enter the number of Venus-positive cells and the total number of transfected cells (e.g., DsRed-Mito-positive cells) for each area of the plate counted and for each experimental condition.

2. Calculate the percentage of Venus-positive cells for each area of the plate and for each experimental condition.

3. Average the percentage of Venus-positive cells of the three (or more) areas per condition, and calculate the standard deviation (*see* **Note 22**).

3.5.2 Calculate the Intensity of Venus-Positive Cells Over Time

1. Use ImageJ or available software to open a series of time-lapse images. Each series is comprised of the images of a single position taken at each time interval (this is called a stack in ImageJ *see* **Note 23**).

2. Using the polygon tool, draw a *region of interest* (ROI) around each DsRed-Mito-positive cell (*see* **Note 24**). In ImageJ, use the ROI manager under "Tools" in the "Analyze" menu. Add the ROIs to the ROI manager by pressing "+" or clicking "Add".

3. Select a small region in an area of the image where there are no cells: this will be the background measurement.

4. Once all ROIs have been generated, save them (*see* **Note 25**).

5. Measure the average intensity of Venus in each region in each frame of the time-lapse. In ImageJ: Go to "Measure" in the ROI manager. A Results window will open up. Open the "Set Measurements" window in the Results window and select "Mean gray value." To record the average intensities of multiple ROIs for the entire stack, select the "Multi Measure" option under "More" in the ROI manager.

6. Export the results to an Excel spreadsheet or similar graphing software.

7. For each cell, subtract the background fluorescence value at each time point.

8. Repeat this process for all the individual positions acquired in the experiment.

9. Average the values for all cells in each treatment group at each time point and graph the results (*see* **Note 26**).

10. Add error bars to each data point by calculating the standard error of the mean (SEM) of the cells for each time point.

4 Notes

1. A 488 nm laser can alternatively be used to image Venus, but in this case CFP-based plasmids cannot be used as a reporter for transfection because CFP is also excited at the 488 nm wavelength.

2. Standard plastic tissue culture plates can be used if the epifluorescence microscope is equipped with a long distance objective. Plastic dishes are not suitable for confocal microscopy as they are too thick for the majority of high numerical aperture objectives. In addition, plastic dishes tend to be autofluorescent because they are not sufficiently transparent. Therefore, No. 1.5 glass coverslips that have an average thickness of 0.17 mm are optimal for high resolution confocal imaging.

3. Cells do not adhere to glass surfaces as well as they do to plastic. Coating glass surfaces with fibronectin helps the cell adhere. The fibronectin solution can be reused a number of times. Alternatives to fibronectin include poly-L-lysine or collagen. This step can be omitted if cells are plated on plastic dishes for analysis by epi-fluorescence microscopy.

4. This protocol can be adapted to any adherent cell line. MCF-7 cells and mouse embryonic fibroblasts (MEF) have been used successfully [19, 20]. The transfection reagents and conditions should be optimized for each cell line (*see* **Note 6**).

5. Prepare a sufficient number of plastic or fibronectin-coated dishes or wells to include controls for the experiment such as an untreated control or cells expressing binding mutants of caspases.

6. The protocol outlined here works well for Hela cells, MCF7 cells and MEF, and has been optimized to limit toxicity induced by the transfection reagent. For other cell lines, test different transfection reagents and conditions to find the optimal parameters that give the desired transfection efficiencies with minimal cellular toxicity.

7. 20–40 ng of each caspase BiFC component in a 3.5 cm dish is usually sufficient to transfect Hela or MCF-7 cells. However, if the relative expression is lower, more plasmid should be transfected (100-200 ng). Cells that have lower transfection efficiencies, such as MEF, require higher amounts of plasmid (250–500 ng).

8. A fluorescent reporter plasmid is necessary to label the transfected cells because the cells will not all be Venus-positive. Choose a reporter that is a different and appropriate "color" based on the excitation/emission filters of the microscope that will be used for analysis. For example, if visualizing Venus using a GFP excitation/emission filter instead of a YFP excitation/emission filter on a standard epi-fluorescence microscope, CFP-based plasmids (e.g., CFP, Cerulean, mCyan) should not be used as transfection reporters. This is because CFP is also excited by the GFP filter. In this case, DsRed or a similar RFP-based plasmid (e.g., mCherry, mOrange) is a more suitable

transfection control. DsRed-Mito is used in this protocol and should be transfected at a lower amount than the BiFC plasmids, such as 10 ng per 3.5 cm dish.

9. If plasmids that are known to induce apoptosis are being co-transfected with the BiFC plasmids include a caspase inhibitor such as qVD-OPh (20 μM) in this media. This will prevent cell death induced by the pro-apoptotic protein but will not impact the BiFC (*see* **Note 12**).

10. Some common inducers include: heat shock (1 h at 45 °C for Hela cells) or vincristine (1 μM) to activate caspase-2; TNF plus cycloheximide (10 ng/ml/10 μg/ml) to activate caspase-8; or Actinomycin D (500 nM–1 μM) to activate caspase-9.

11. Imaging media includes Hepes and 2-mercaptoethanol. Hepes is required to buffer the pH. 2-mercaptoethanol prevents accumulation of reactive oxygen species (ROS) that can be generated by the confocal laser and are toxic to the cells. Phenol red-free media can be used to reduce artifacts due to auto-fluorescence but it is not essential.

12. Adding a pan-caspase inhibitor, such as qVD-OPh, is essential for single time point analysis to prevent the cells detaching from the surface of the dish or plate as they undergo apoptosis. It will not inhibit the BiFC as the probes have no catalytic activity, but will inhibit the execution phase of apoptosis. By inhibiting death, the addition of caspase inhibitors will also limit excessive movement and shape changes of the cells during time-lapse acquisition which greatly facilitates analysis. However, if caspase-dependent downstream effects are being monitored simultaneously, such as MOMP after caspase-2 activation or Annexin V binding, then the caspase inhibitor should be omitted.

13. Cells can alternatively be heat shocked in a water bath. Seal the plates with Parafilm and submerge in water that has been equilibrated to the required temperature.

14. The red cells are cells that have been transfected and are most likely expressing the BiFC plasmids. Cells that are Venus-positive are cells that are positive for BiFC and caspase induced proximity. Occasionally, cells that are not red will still be Venus-positive, but to maintain an objective count, do not include them in your analysis.

15. Cells should be maintained at constant temperature of 37 °C for the duration of the time-lapse experiment to prevent focus drift. Options to help maintain temperature throughout the time-lapse experiment include a temperature controller on the microscope stage, a microscope with an incubator enclosure, or an objective heater.

16. If this option is not available, overlay the media with mineral oil to prevent evaporation of the media from the dish.

17. A 60× or 63× objective is optimal for spinning disk confocal microscopy. However, a 40× objective can be used to capture more cells per field, although there will be a slight reduction in resolution.

18. There will be very little Venus fluorescence at the start of the time-lapse. Therefore, in a separate experiment or by using a positive control within the current experiment, determine the lowest amount of the laser light and exposure time required to image the cells that are BiFC positive. Choose settings where the Venus signal is visible but not saturated. Imaging cells with low levels of laser light and short exposure times will minimize phototoxicity. Control experiments should be carried out either simultaneously or using the same conditions on untreated cells to ensure that the imaging conditions are not phototoxic and that cell division proceeds normally.

19. Focus drift correction will prevent fluctuations in the z-plane by maintaining the distance between the objective and the coverslip constant. Thus, changes in focus due to temperature fluctuations can be avoided. Make sure that the correction system can update the z-plane for each position chosen. Otherwise the same z-plane will be used for all positions.

20. Multiple positions are only possible if the microscope has a motorized XY stage and multi-field capabilities. This allows multiple fields of view to be imaged in the same experiment. Multiple conditions can be assessed side by side if a multi-well chamber slide or if a slide holder that can hold more than one dish is used.

21. These parameters will depend on the timing of caspase activation. There is an increased potential for phototoxicity as the number of images increases. Therefore, if caspase activation or other cellular events following caspase activation is expected in a 16 h window, longer time intervals, such as 5–10 min, are recommended. If caspase activation occurs more rapidly, the time between frames can be reduced. Control experiments in untreated cells should be carried out to determine the maximum number of images that can be taken over a given time period without inducing toxicity or photobleaching.

22. A background level of Venus fluorescence of 10 % or lower in untreated cells is usually acceptable. If the results show a higher background it is likely because the plasmid concentrations are too high, leading to a greater probability of random complementation of the Venus fragments. If this is the case, titrate the amount of the caspase BiFC plasmid pair down to a level where the background is low but the specific signal is still detected.

23. There are different types of software available including Zen (Zeiss), Slidebook (3i), ImageJ (NIH), and Metamorph (Molecular Devices) that can be used for this type of analysis. ImageJ is a freely available and widely used software. Therefore, this protocol outlines specific steps using ImageJ, but other software programs will give identical results.

24. When drawing the ROI, try to minimize black space. It may be difficult to draw a suitable ROI that captures the cell throughout the time-lapse if the cell moves significantly or changes shape. An alternative approach is to draw a large ROI that encompasses all the cell's movements and use segmentation to select the cell within the ROI. Segmentation is a process that separates objects from the background based on differences in intensities and the results are usually represented as a mask. In ImageJ: segment the image by adjusting the threshold. Select "Threshold" under "Adjust" in the "Image" menu. Move the slider until the whole cell is high-lighted and select "Apply" to create the mask. Next "Create Selection" under "Selection" in the "Edit" menu. This will create an ROI that outlines the mask. Add this Selection ROI to the ROI manager. Select the Selection ROI and the ROI that encompasses the cell movements and use the AND function (in the ROI manager, click "More" and then "AND") to create a new ROI that outlines the cell within the chosen region. Measure the intensity of the masked region to follow changes in fluorescence intensity of the single cell throughout the time lapse.

25. In ImageJ, the ROIs are saved as a file. To save, go to the ROI manager, click "More" and then "Save".

26. This analysis generates a read-out for a population of cells. Single cell traces can also be generated, which can be useful for analyzing asynchronous caspase activation events.

References

1. Riedl SJ, Shi Y (2004) Molecular mechanisms of caspase regulation during apoptosis. Nat Rev Mol Cell Biol 5(11):897–907

2. Luthi AU, Martin SJ (2007) The CASBAH: a searchable database of caspase substrates. Cell Death Differ 14(4):641–650

3. Kischkel FC, Hellbardt S, Behrmann I et al (1995) Cytotoxicity-dependent APO-1 (Fas/CD95)-associated proteins form a death-inducing signaling complex (DISC) with the receptor. EMBO J 14(22):5579–5588

4. Boldin MP, Goncharov TM, Goltsev YV et al (1996) Involvement of MACH, a novel MORT1/FADD-interacting protease, in Fas/APO-1- and TNF receptor-induced cell death. Cell 85(6):803–815

5. Aravind L, Dixit VM, Koonin EV (1999) The domains of death: evolution of the apoptosis machinery. Trends Biochem Sci 24(2):47–53

6. Boatright KM, Renatus M, Scott FL et al (2003) A unified model for apical caspase activation. Mol Cell 11(2):529–541

7. Salvesen GS, Dixit VM (1999) Caspase activation: the induced-proximity model. Proc Natl Acad Sci U S A 96(20):10964–10967

8. Chang DW, Xing Z, Capacio VL et al (2003) Interdimer processing mechanism of procaspase-8 activation. EMBO J 22(16):4132–4142

9. Oberst A, Pop C, Tremblay AG et al (2010) Inducible dimerization and inducible cleavage reveal a requirement for both processes in caspase-8 activation. J Biol Chem 285(22):16632–16642

10. Riedl SJ, Salvesen GS (2007) The apoptosome: signalling platform of cell death. Nat Rev Mol Cell Biol 8(5):405–413

11. Tinel A, Tschopp J (2004) The PIDDosome, a protein complex implicated in activation of caspase-2 in response to genotoxic stress. Science 304(5672):843–846

12. Ho LH, Read SH, Dorstyn L et al (2008) Caspase-2 is required for cell death induced by cytoskeletal disruption. Oncogene 27(24):3393–3404

13. Tu S, McStay GP, Boucher LM et al (2006) In situ trapping of activated initiator caspases reveals a role for caspase-2 in heat shock-induced apoptosis. Nat Cell Biol 8(1):72–77

14. Bouchier-Hayes L, Green DR (2012) Caspase-2: the orphan caspase. Cell Death Differ 19(1):51–57

15. Martinon F, Burns K, Tschopp J (2002) The inflammasome: a molecular platform triggering activation of inflammatory caspases and processing of proIL-beta. Mol Cell 10(2):417–426

16. Stutz A, Golenbock DT, Latz E (2009) Inflammasomes: too big to miss. J Clin Invest 119(12):3502–3511

17. Creagh EM, Conroy H, Martin SJ (2003) Caspase-activation pathways in apoptosis and immunity. Immunol Rev 193:10–21

18. Shyu YJ, Liu H, Deng X et al (2006) Identification of new fluorescent protein fragments for bimolecular fluorescence complementation analysis under physiological conditions. BioTechniques 40(1):61–66

19. Bouchier-Hayes L, Oberst A, McStay GP et al (2009) Characterization of cytoplasmic caspase-2 activation by induced proximity. Mol Cell 35(6):830–840

20. Sanders MG, Parsons MJ, Howard AG et al (2015) Single-cell imaging of inflammatory caspase dimerization reveals differential recruitment to inflammasomes. Cell Death Dis 6:e1813

21. Boatright KM, Salvesen GS (2003) Mechanisms of caspase activation. Curr Opin Cell Biol 15(6):725–731

22. Baliga BC, Read SH, Kumar S (2004) The biochemical mechanism of caspase-2 activation. Cell Death Differ 11(11):1234–1241

23. Stennicke HR, Deveraux QL, Humke EW et al (1999) Caspase-9 can be activated without proteolytic processing. J Biol Chem 274(13):8359–8362

24. O'Reilly LA, Ekert P, Harvey N et al (2002) Caspase-2 is not required for thymocyte or neuronal apoptosis even though cleavage of caspase-2 is dependent on both Apaf-1 and caspase-9. Cell Death Differ 9(8):832–841

Chapter 5

In Vitro Use of Peptide Based Substrates and Inhibitors of Apoptotic Caspases

Gavin P. McStay

Abstract

Caspases are proteases that are essential components of apoptotic cell death pathways. There are approximately one dozen apoptotic caspases found in organisms where cells die via apoptosis. These caspases are responsible for initiation or execution of apoptosis through the proteolytic cleavage of specific substrates. These substrates contain specific motifs that are recognized and cleaved by caspases that result in alterations of substrate function that promotes the apoptotic phenotype. Analysis of caspase involvement, much like any other protease, can be followed using peptides corresponding to cleavage motifs of these substrates, which can be used as substrates, inhibitors, or affinity-based probes.

Different caspases have different substrates and therefore different motifs are recognized by each different caspase. However, these different caspases have a common amino acid recognition pattern containing an aspartic acid residue at the amino-side of the cleavage site. Therefore, caspase substrates have a certain overlap in the cleavage motif as this aspartic acid is found in almost every one. This means that certain peptide motifs are not exclusively cleaved by one single caspase. This lack of exclusive cleavage has brought the use of these motif-based probes into question and spurred the development of truly caspase-specific motifs. This chapter describes the use of peptide-based probes to measure caspase activity while highlighting the limitations of these reagents.

Key words Apoptosis, Caspase, Inhibitor, Substrate, Motif, Affinity-based probe, Protease

1 Introduction

The initiation and execution of the apoptotic pathway of cell death involves a critical family of proteases, known as caspases. These proteases are *c*ysteine-dependent *asp*artic acid-specific prote*ases* that are activated by specific apoptotic signaling cascades. In every organism that undergoes apoptotic cell death, caspases are critical components of the pathway. This can be in the role of an initiator or an executioner of the apoptotic pathway. Initiator caspases are activated by extrinsic death promoting ligands, such as Fas ligand or Tumor Necrosis Factor Related Apoptosis Inducing Ligand (TRAIL) in the case of caspase-8 and -10, or intrinsic signals, such as release of cytochrome *c* from the mitochondrial inter-membrane

Hamsa Puthalakath and Christine J. Hawkins (eds.), *Programmed Cell Death: Methods and Protocols*, Methods in Molecular Biology, vol. 1419, DOI 10.1007/978-1-4939-3581-9_5, © Springer Science+Business Media New York 2016

space to the cytosol, in the case of caspase-9. Initiator caspases undergo dimerization to form the active site resulting in the eventual cleavage of downstream substrates. The cleavage of these substrates ultimately results in the activation of executioner caspases via proteolytic cleavage allowing for the formation of the active site [1, 2]. Executioner caspases are then able to cleave specific substrates, which number in the range of a thousand different substrates [3–5]. The cleavage of these substrates ultimately causes the apoptotic phenotype allowing for rapid, ordered, and efficient dismantling of a cell followed by phagocytosis.

As apoptosis is implicated in many developmental processes and pathologies much research has gone into identifying which specific caspases are activated in specific apoptotic pathways. There are a variety of approaches that can be used to determine caspase involvement. Genetic approaches include disruption of the gene encoding the caspase, targeted degradation of the caspase messenger RNA or expression of genetic reporters, while small molecules can be used as exogenous inhibitors or substrates. The use of these small molecules allows assessment of caspase activity rapidly, without any need for genetic manipulation and can be utilized in many models of apoptosis. Their ease of use allowed for a rapid understanding of how and when apoptosis was occurring in cell culture, in vitro assays and in some cases animal models as they could be added before or after apoptosis induction.

The use of peptide-based probes for caspases arose from experiments performed to determine the favored motif of each caspase using peptide-based libraries [6–8]. In general, a tetrapeptide was used as a substrate or inhibitor and tested on isolated caspase to investigate how the enzyme and substrate acted on each other. Inhibitors are tetrapeptides conjugated to a moiety that cause them to act as competitive or irreversible inhibitors. Substrates are the same tetrapeptide motif conjugated to a moiety that when released from the peptide generates a signal that can be detected spectrophotometrically or by fluorescence based assays (Table 1).

All caspase cleavage motifs contain an aspartate residue at the amino-terminal side of the cleavage site. Therefore, all peptide-based reagents contain an aspartate residue immediately before the inhibiting or detectable moiety. The variation in these peptide-based motifs occurs before the aspartate and this will depend on the caspase intended to be monitored. The specific motif depends on the active site of each caspase which accommodates certain amino acids at certain places in the cleavage motif. As caspases generally have more than one substrate, a caspase active site must have conformational flexibility to accommodate multiple cleavage motifs that would be in the context of a number of substrate proteins. Therefore, this means a caspase active site has evolved to accommodate different motifs and is not exclusive to one motif in one substrate. This is the fundamental property of caspases that makes

Table 1
Peptide moieties used to probe caspase activity

Category	Moiety	Modification	Purpose
Covalent inhibitor	FMK	Fluoromethylketone	Irreversible inhibition of caspase by covalent attachment to active site cysteine
	OPH	2, 6-difluorophenyl ketone	
Competitive inhibitor	CHO	N-terminal aldehyde	Reversible inhibitor, no covalent attachment to active site cysteine
Chromogenic substrate	pNA	*p*-nitroaniline	Yellow product formed upon caspase cleavage monitored at 405 nm absorbance
Fluorogenic substrate	AFC	7-amino-4-trifluoromethylcoumarin	Fluorescent product (excitation 400 nm; emission 505 nm) generated upon caspase cleavage
	AMC	7-amino-4-methylcoumarin	Fluorescent product (excitation 350 nm; emission 450 nm) generated upon caspase cleavage

it difficult to use peptide-based reagents and interpret data based on these. Commercially available peptide-based caspase probes are generally marketed as being specific for a particular caspase and when used in assays for apoptosis the initial interpretation would be that this particular caspase is active in this pathway of apoptosis. However, when put into a context where multiple caspases are expressed at different concentrations it can be appreciated that another caspase could be responsible for some or all of the activity as determined using the caspase-specific peptide-based probe. Indeed, this has been observed in vitro and in cell culture [9–14].

The factors that influence whether a caspase could impact on an unintended peptide motif are related to protein abundance, mechanism of activation, enzymatic properties and substrate promiscuity. The enzymatic properties of each caspase have been determined to provide important parameters for each enzyme, such as the Michaelis-Menten constant (K_m) and turnover number (K_{cat}), which represent the concentration and speed at which an enzyme runs maximally. These are outlined in Table 2 for each apoptotic caspase and favored substrate. Table 3 outlines the inhibition constants for apoptotic caspases with a variety of reversible tetrapeptide based aldehyde inhibitors. These two tables demonstrate that a single caspase can act on multiple peptide motifs as a substrate and that some substrates can be acted on by multiple caspases. Despite having a preferred substrate, caspases generally cleave more than just one substrate. Initiator caspases, such caspase-8

Table 2
Enzymatic parameters of isolated caspases

Caspase	Preferred substrate	K_m (μM)	k_{cat} (s⁻¹)	k_{cat}/K_m (μM⁻¹ s⁻¹)	References
2 (ΔCARD)	Ac-VDVAD-AFC	25	0.6	0.024	[26]
	Ac-VDVAD-pNA	53	4.5	0.084	[7]
	Ac-DEHD-AMC			0.0003	[27]
3	Ac-DEVD-pNA	67.1	0.885	0.013	[28]
	Ac-DEVD-pNA	11	2.4	0.218	[7]
	Abz-GDEVD↓GVY(NO₂)D			0.2	[29]
	Ac-DEVD-AMC	10	14	1.4	[27]
6	Ac-VEID-pNA	30	5.0	0.168	[7]
6 (Δprodomain)	Ac-VEHD-AMC	170			[27]
7	Ac-DEVD-pNA	12	0.43	0.037	[7]
	Abz-GDEVD↓GVY(NO₂)D			0.033	[29]
	Ac-DEVD-AMC	100	6.3	0.063	[27]
8	IETD	Not described			
	Ac-DEVD-AMC	7	0.37	0.053	[27]
9	Ac-VEHD-AMC	780	0.1	0.00013	[27]
10	IETD	Not described			
10 (ΔCARD)	Ac-VEHD-AMC	42			[27]

Table 3
Inhibition of isolated caspases using peptide-based inhibitors

Caspase	Motif aldehyde inhibitor and K_i (nM)					Reference
	WEHD	YVAD	DEVD	IETD	AEVD	
2	>10,000	>10,000	1710	9400	>10,000	[8]
3	1960	>10,000	0.23	195	42	[8]
6	3090	>10,000	31	5.6	52	[8]
7	>10,000	>10,000	1.6	3280	425	[8]
8	21.1	352	0.92	1.05	1.6	[8]
9	508	970	60	108	48	[8]
10	330	408	12	27	320	[8]

or -9, generally cleave only a few substrates, while executioner caspases are capable of cleaving hundreds of substrates. This means that executioner caspases can cleave many different substrates in succession in one apoptotic cell, whereas initiator caspases cleave a small number of substrates that do not necessarily have to undergo cleavage rapidly.

The mechanism of activation of caspases also influences efficiency of substrate cleavage. An executioner caspase is only active once it has been cleaved allowing for conformational rearrangements leading to formation of the active site. In contrast, an initiator caspase forms an active site when monomers are brought together through activation platforms such as the apoptosome or the death inducing signaling complex (DISC). All caspases undergo autocleavage in the process of activating other caspases which in the case of initiator caspases removes the domains responsible for recruitment to activation platforms. In this instance, initiator caspases are located proximally in a large complex with other initiator caspases. When in close proximity, the active sites are able to cleave other close by caspases. However, executioner caspases are diffusible in the cytoplasm and this is when enzyme–substrate affinity and concentration dictate the extent of substrate cleavage. A caspase that cleaves multiple substrates throughout the cell must be able to recognize these substrates in a large cellular space, which is most likely due to enzyme concentration and substrate turnover.

The enzymatic capabilities of each enzyme cannot be considered in isolation, as every cell that can undergo apoptosis must express multiple caspase family members to undergo this process. Therefore, the endogenous expression level of each caspase must be taken into consideration and whether these expression levels are regulated in any way. As an example, the breast carcinoma cell line MCF-7 does not express caspase-3 protein due to a deletion in the gene [15]. Therefore, any proteolytic activity resulting in cleavage of the DEVD peptide must be due to another caspase. In this instance it would be due to the close homolog caspase-7 [16, 17]. In more subtle cases, differential caspase protein expression could result in unselectivity of motif-based probes. Expression of caspase-3 has been reported to be approximately 40,000 molecules per K562 cell, similar to the number of caspase-2, and -8 molecules and more than the number of caspase-9 molecules [18]. Therefore, when considering the use of peptide-based substrates details regarding these four parameters should also be considered. In most cases, a cell that is undergoing apoptosis and being monitored by a peptide-based substrate or inhibitor will be preferentially monitoring caspase-3. Therefore, use of any peptide-based substrate or inhibitor should be compared with the caspase-3/-7 motif DEVD. If the activity or effect of the DEVD-based probe is high than any activity of other peptide-based probes, this response is very likely to be at least partially or completely due to caspase-3.

This would then require other approaches to determine specificity of the caspase and substrate. One such approach is immunoprecipitation of the caspase and testing the activity of the isolated caspase without the confounding effects of other caspases present in the cell. Currently, more specific peptide-based caspase probes are being developed. These probes are based on the cleavage motifs of the caspase but now include unnatural amino acids to cause a stronger interaction between the probe and only one caspase. Still, the selectivity of these probes does not allow for the definite implication of one caspase alone in an apoptotic pathway, so complementary approaches must be performed in conjunction [12, 19–24].

2   Materials

1. Source of active caspase (recombinant caspase prepared according to Roschitzki-Voser et al. [25], in vitro activated cytosolic extract according to Subheading 3.1, apoptotic lysate prepared according to [9]).

2. Homogenization buffer: 10 mM HEPES pH 7.0, 5 mM MgCl$_2$, 0.67 mM DTT (added fresh from 1 M stock solution), complete protease inhibitors (Roche). Store at 4 °C.

3. Caspase assay buffer: 100 mM NaCl, 20 mM PIPES, pH 7.4, 1 mM EDTA, 0.1 % (w/v) CHAPS, 10 % (w/v) sucrose, 10 mM DTT (added fresh from 1 M stock). Store at 4 °C.

4. Peptide-based inhibitor: 10 mM stock prepared in DMSO. Store at –20 °C.

5. Peptide-based substrate: 20 mM stock prepared in DMSO. Store at –20 °C. Use at 100 μM final concentration by diluting 1 in 200 in caspase assay buffer.

6. Fluorescence plate reader capable of 400 nm excitation and 505 nm emission.

7. 96-well flat bottom fluorescence plates.

3   Methods

3.1   Preparation of Cytosolic Extract for Apoptosome Activation

1. Cells grown to confluence (see **Note 1**).

2. Harvest cells depending on whether adherent or suspension cells in a single conical tube (see **Note 2**).

3. Centrifuge at 200×*g*, 5 min, 4 °C.

4. Wash with ice-cold PBS.

5. Centrifuge at 200×*g*, 5 min, 4 °C.

6. Resuspend cell pellet in equal volume of ice-cold homogenization buffer (*see* **Note 3**).

7. Incubate on ice for 10 min.

8. Pass cell suspension through a 22 G needle 20 times (*see* **Note 4**).

9. Centrifuge at $21,000 \times g$, 30 min, 4 °C in a benchtop microcentrifuge tube.

10. Take supernatant.

11. Centrifuge at $21,000 \times g$, 30 min, 4 °C, in a benchtop microcentrifuge tube.

12. Take supernatant.

13. Centrifuge at $100,000 \times g$, 1 h, 4 °C, in a benchtop ultracentrifuge.

14. Take supernatant.

15. Pass supernatant through a 0.2 μm filter.

16. Determine protein concentration (*see* **Note 5**).

17. Adjust concentration to 10 mg/ml.

18. Aliquot 100 μg (10 μl) into individual microcentrifuge tubes.

19. Freeze at –80 °C.

3.2 Activation of Apoptosome in Cytosolic Extracts

1. Thaw required number of aliquots for the number of conditions to be tested.

2. Add 100 μM of cytochrome *c* and 1 mM dATP.

3. Incubate at 37 °C for 30 min.

3.3 Measuring Caspase Activity Using Fluorogenic Peptide Substrates

1. Add 100 μl of caspase assay buffer containing 100 μM fluorogenic substrate to 10 μl of your caspase-containing samples in a well of a 96-well flat bottom plate (*see* **Note 6**).

2. Place in fluorescent plate reader with excitation at 400 nm and emission at 505 nm, set to read each well every minute with shaking prior to each reading.

3. Readings can now be analyzed by determining the initial velocity of the enzyme catalyzed reaction, which is the linear part of the graph. This value from the chart can be compared amongst samples to demonstrate differences.

3.4 Immunoprecipitation of Caspases from Activated Cytosolic Extracts to Demonstrate Activity of Each Caspase with Each Substrate

The sample here is the activated cytosolic extract with cytochrome *c* and dATP.

1. Wash 20 μl protein-A/G beads with caspase assay buffer by adding 1 ml of caspase assay buffer and centrifuging at $1000 \times g$ for 30 s, repeat two more times.

2. Bind antibody to caspase of interest with protein-A/G beads for 1 h at 4 °C.

3. Wash as in **step 1**, three times.

4. Add activated cytosolic extract to the protein-A/G beads bound to the caspase antibody of interest overnight at 4 °C with end-over mixing.

5. Centrifuge at $1000 \times g$, 5 min, 4 °C.

6. Take supernatant and keep on ice until later use (immunodepleted supernatant).

7. Pass the immunodepleted supernatant through a 0.2 μm filter to remove any remaining protein-A/G beads (*see* **Note 7**).

8. Wash beads with 100 μl caspase assay buffer three times by centrifugation at $1000 \times g$, 1 min, 4 °C, and remove supernatant.

9. Add 100 μl of caspase assay buffer to protein-A/G beads and store on ice for later use (immunoprecipitated caspase).

10. Use the immunodepleted supernatant and immunoprecipitated caspase as samples to test each peptide-based substrate.

11. Follow the set up procedure in Subheading 3.3 above.

3.5 Use of Peptide-Based Inhibitors in Apoptosome Activation Assays: Inhibition Prior to Apoptosome Activation

This protocol is used to determine which inhibitors prevent caspase activation using cytosolic extracts prepared in Subheading 3.1.

1. Add desired concentration of inhibitor to 100 μg (10 μl) of cytosolic extract (*see* **Note 8**).

2. Add 1 mM dATP and 100 μM cytochrome *c* to the cytosolic extract.

3. Incubate at 37 °C, 30 min.

4. Determine caspase activity as described in Subheading 3.3.

3.6 Use of Peptide-Based Inhibitors in Apoptosome Activation Assays: Inhibition After Apoptosome Activation

This protocol is used to determine which caspases are active after apoptosome activation using cytosolic extracts prepared in Subheading 3.1.

1. Add 1 mM dATP and 100 μM cytochrome *c* to 100 μg (10 μl) cytosolic extract.

2. Incubate 37 °C, 30 min.

3. Add desired concentration of inhibitor to activated cytosolic extract (*see* **Note 8**).

4. Determine fluorescence using the method described in Subheading 3.3.

4 Notes

1. Cells should not be overgrown as this will cause cells to die and increase the amount of basal caspase activity in the samples. Approximately ten 10 cm dishes of adherent cells or five

175 cm² flasks results in a total number of cells appropriate for a good yield of cytosolic extract.

2. Adherent cells should be removed from the surface of the dish using standard trypsin conditions. Suspension cells are removed by simply transferring to an appropriate centrifuge tube.

3. The volume of cells can be estimated by adding a known amount of PBS to the cell pellet resuspend by tituration, noting the total volume and then subtract the volume of PBS added. A cell volume of approximately 2 ml gives a good yield of cytosolic extract.

4. Check the extent of cell disruption using trypan blue staining of the homogenate. Broken cells will allow trypan blue staining of the nucleus and approximately 90 % nuclear staining with trypan blue is desired. If this is not the case, pass the homogenate through the needle ten more times and check again with trypan blue. Continue until 90 % nuclear staining is observed.

5. Protein concentration can be determined by using absorbance at 280 nm or using colorimetric techniques such as the Biuret protein test.

6. Colorimetric substrates can be used instead of fluorogenic substrates. These are prepared in the same way. To determine the amount of colored product formation, follow absorbance at 405 nm using a spectrophotometer.

7. Use centrifugal filters appropriate for microcentrifuge tubes.

8. Concentrations appropriate for use range from 10–100 µM.

References

1. Boatright KM, Salvesen GS (2003) Mechanisms of caspase activation. Curr Opin Cell Biol 15:725–731. doi:10.1016/j.ceb.2003.10.009

2. Pop C, Salvesen GS (2009) Human caspases: activation, specificity, and regulation. J Biol Chem 284:21777–21781. doi:10.1074/jbc.R800084200

3. Mahrus S, Trinidad JC, Barkan DT, Sali A, Burlingame AL, Wells JA (2008) Global sequencing of proteolytic cleavage sites in apoptosis by specific labeling of protein N termini. Cell 134:866–876. doi:10.1016/j.cell.2008.08.012

4. Crawford ED, Seaman JE, Agard N, Hsu GW, Julien O, Mahrus S, Nguyen H, Shimbo K, Yoshihara HA, Zhuang M, Chalkley RJ, Wells JA (2013) The DegraBase: a database of proteolysis in healthy and apoptotic human cells. Mol Cell Proteomics 12:813–824. doi:10.1074/mcp.O112.024372

5. Shimbo K, Hsu GW, Nguyen H, Mahrus S, Trinidad JC, Burlingame AL, Wells JA (2012) Quantitative profiling of caspase-cleaved substrates reveals different drug-induced and cell-type patterns in apoptosis. Proc Natl Acad Sci U S A 109:12432–12437. doi:10.1073/pnas.1208616109

6. Thornberry NA, Rano TA, Peterson EP, Rasper DM, Timkey T, Garcia-Calvo M, Houtzager VM, Nordstrom PA, Roy S, Vaillancourt JP, Chapman KT, Nicholson DW (1997) A combinatorial approach defines specificities of members of the caspase family and granzyme B. Functional relationships established for key mediators of apoptosis. J Biol Chem 272:17907–17911. doi:10.1074/jbc.272.29.17907

7. Talanian RV, Quinlan C, Trautz S, Hackett MC, Mankovich JA, Banach D, Ghayur T, Brady KD, Wong WW (1997) Substrate speci-

ficies of caspase family proteases. J Biol Chem 272:9677–9682

8. Garcia-Calvo M, Peterson EP, Leiting B, Ruel R, Nicholson DW, Thornberry NA (1998) Inhibition of human caspases by peptide-based and macromolecular inhibitors. J Biol Chem 273:32608–32613

9. McStay GP, Salvesen GS, Green DR (2008) Overlapping cleavage motif selectivity of caspases: implications for analysis of apoptotic pathways. Cell Death Differ 15:322–331. doi:10.1038/sj.cdd.4402260

10. Pereira NA, Song Z (2008) Some commonly used caspase substrates and inhibitors lack the specificity required to monitor individual caspase activity. Biochem Biophys Res Commun 377:873–877. doi:10.1016/j.bbrc.2008. 10.101

11. Benkova B, Lozanov V, Ivanov I, Mitev V (2009) Evaluation of recombinant caspase specificity by competitive substrates. Anal Biochem 394(1):68–74

12. Maillard MC, Brookfield FA, Courtney SM, Eustache FM, Gemkow MJ, Handel RK, Johnson LC, Johnson PD, Kerry MA, Krieger F, Meniconi M, Muñoz-Sanjuán I, Palfrey JJ, Park H, Schaertl S, Taylor MG, Weddell D, Dominguez C (2011) Exploiting differences in caspase-2 and -3 S2 subsites for selectivity: structure-based design, solid-phase synthesis and in vitro activity of novel substrate-based caspase-2 inhibitors. Bioorg Med Chem 19:5833–5851. doi:10.1016/j.bmc.2011.08.020

13. Delgado ME, Olsson M, Lincoln FA, Zhivotovsky B, Rehm M (2013) Determining the contributions of caspase-2, caspase-8 and effector caspases to intracellular VDVADase activities during apoptosis initiation and execution. Biochim Biophys Acta. doi:10.1016/j. bbamcr.2013.05.025

14. Berger A, Sexton K, Bogyo M (2006) Commonly used caspase inhibitors designed based on substrate specificity profiles lack selectivity. Cell Res 16:961–963. doi:10.1038/ sj.cr.7310112

15. Jänicke RU, Sprengart ML, Wati MR, Porter AG (1998) Caspase-3 is required for DNA fragmentation and morphological changes associated with apoptosis. J Biol Chem 273:9357–9360

16. Liang Y, Yan C, Schor NF (2001) Apoptosis in the absence of caspase 3. Oncogene 20:6570–6578. doi:10.1038/sj.onc.1204815

17. Walsh JG, Cullen SP, Sheridan C, Lüthi AU, Gerner C, Martin SJ (2008) Executioner caspase-3 and caspase-7 are functionally distinct proteases. Proc Natl Acad Sci U S A 105:12815–12819. doi:10.1073/pnas.0707715105

18. Svingen PA, Loegering D, Rodriquez J, Meng XW, Mesner PW, Holbeck S, Monks A, Krajewski S, Scudiero DA, Sausville EA, Reed JC, Lazebnik YA, Kaufmann SH (2004) Components of the cell death machine and drug sensitivity of the National Cancer Institute Cell Line Panel. Clin Cancer Res 10:6807–6820. doi:10.1158/1078-0432. CCR-0778-02

19. Vickers CJ, González-Páez GE, Wolan DW (2014) Discovery of a highly selective caspase-3 substrate for imaging live cells. ACS Chem Biol 9:2199–2203. doi:10.1021/cb500586p

20. Vickers CJ, González-Páez GE, Litwin KM, Umotoy JC, Coutsias EA, Wolan DW (2014) Selective inhibition of initiator versus executioner caspases using small peptides containing unnatural amino acids. ACS Chem Biol 9:2194–2198. doi:10.1021/cb5004256

21. Xiao J, Broz P, Puri AW, Deu E, Morell M, Monack DM, Bogyo M (2013) A coupled protein and probe engineering approach for selective inhibition and activity-based probe labeling of the caspases. J Am Chem Soc 135:9130–9138. doi:10.1021/ja403521u

22. Berger AB, Witte MD, Denault JB, Sadaghiani AM, Sexton KM, Salvesen GS, Bogyo M (2006) Identification of early intermediates of caspase activation using selective inhibitors and activity-based probes. Mol Cell 23:509–521. doi:10.1016/j.molcel.2006.06.021

23. Edgington LE, van Raam BJ, Verdoes M, Wierschem C, Salvesen GS, Bogyo M (2012) An optimized activity-based probe for the study of caspase-6 activation. Chem Biol 19:340–352. doi:10.1016/j.chembiol.2011. 12.021

24. Edgington LE, Berger AB, Blum G, Albrow VE, Paulick MG, Lineberry N, Bogyo M (2009) Noninvasive optical imaging of apoptosis by caspase-targeted activity-based probes. Nat Med 15:967–973. doi:10.1038/nm.1938

25. Roschitzki-Voser H, Schroeder T, Lenherr ED, Frölich F, Schweizer A, Donepudi M, Ganesan R, Mittl PRE, Baici A, Grütter MG (2012) Human caspases in vitro: expression, purification and kinetic characterization. Protein Expr Purif 84:236–246. doi:10.1016/j.pep.2012.05.009

26. Tang Y, Wells JA, Arkin MR (2011) Structural and enzymatic insights into caspase-2 protein substrate recognition and catalysis. J Biol Chem 286:34147–34154. doi:10.1074/jbc. M111.247627

27. Garcia-Calvo M, Peterson EP, Rasper DM, Vaillancourt JP, Zamboni R, Nicholson DW, Thornberry NA (1999) Purification and catalytic properties of human caspase family members. Cell Death Differ 6:362–369. doi:10.1038/sj.cdd.4400497

28. Fang B, Boross PI, Tozser J, Weber IT (2006) Structural and kinetic analysis of caspase-3 reveals role for s5 binding site in substrate recognition. J Mol Biol 360:654–666. doi:10.1016/j.jmb.2006.05.041

29. Stennicke HR, Renatus M, Meldal M, Salvesen GS (2000) Internally quenched fluorescent peptide substrates disclose the subsite preferences of human caspases 1, 3, 6, 7 and 8. Biochem J 350(Pt 2):563–568

Chapter 6

Experimental In Vivo Sepsis Models to Monitor Immune Cell Apoptosis and Survival in Laboratory Mice

Marcel Doerflinger, Jason Glab, and Hamsa Puthalakath

Abstract

Sepsis is amongst the world's biggest public health problems with more than 20 million cases worldwide and a high morbidity rate of up to 50 %. Despite advances in modern medicine in the past few decades, incidence is expected to further increase due to an aging population and accompanying comorbidities such as cancer and diabetes. Due to the complexity of the disease, available treatment options are limited. Growing evidence links apoptotic cell death of lymphocytes and concomitant immune suppression to overall patient survival. In order to establish novel therapeutic approaches targeting this life threatening immune paralysis, researchers rely heavily on animal models to decipher the molecular mechanisms underlying this high impact disease. Here we describe variations of in vivo mouse models that can be used to study inflammation, cellular apoptosis, and survival in mice subjected to experimental polymicrobial sepsis and to a secondary infection during the immune suppressive secondary stage.

Key words Sepsis, Apoptosis, Immune suppression, "Two- hit" model, CLP, Cecal slurry, Intranasal, Intratracheal, *Pseudomonas*

1 Introduction

Sepsis is a complex inflammatory disease caused by systemic infection. With 20 million cases each year it is ranked second in the cause of death statistic in non-coronary Intensive Care Units and thus presents a great burden for the public health system [1]. Despite advances in medical treatment, sepsis incidence post-surgery trebled in the past 10 years and is estimated to further increase due to an aging population, associated comorbidities and expected rise in hospitalization [2]. In the early phase of a systemic infection, the hyper-inflammatory response, initiated by the host's immune system in order to fight the infection, can lead to a cytokine storm and septic shock. This results in hypotension, coagulation and metabolic changes potentially leading to tissue damage and organ failure [3]. As the condition continues, patients develop a state of immunosuppression in which, they are either unable to

Hamsa Puthalakath and Christine J. Hawkins (eds.), *Programmed Cell Death: Methods and Protocols*, Methods in Molecular Biology, vol. 1419, DOI 10.1007/978-1-4939-3581-9_6, © Springer Science+Business Media New York 2016

eliminate the primary infection and become vulnerable to nosocomial infections and latent viral reactivation [4]. Infections during this secondary phase account for the majority of deaths during sepsis [5]. Failure of more than 25 clinical trials targeting mainly the initial hyper-inflammatory phase of sepsis in the last two decades warrants novel therapeutic approaches to treat this high burden disease [2].

Apoptosis or programmed cell death of lymphocytes is an important mechanism contributing to the immunosuppression during sepsis, causing depletion of immune cells [6]. It was recently reported that lymphocyte apoptosis can be correlated with overall patient survival and therefore could be used as a diagnostic marker and, most importantly, as a target for novel therapeutics [7]. Despite great controversy about the translatability of animal studies into human context, researchers rely on rodent models for investigating the immunological response during sepsis. Nonsurgical methods including injections of endotoxin LPS (lipopolysaccharide), live bacteria, or cecal slurry as well as surgical methods such as Cecal Ligation and Puncture (CLP) have been described and are the most commonly used animal models to study experimental sepsis [8]. Since the emergence of immune suppression as a reliable parameter for patient survival, so-called "two-hit models" have been established to specifically target the animals' susceptibility to secondary infections during late stage sepsis-mediated immune paralysis [9].

In this chapter, we describe surgical (CLP) as well as nonsurgical induction of polymicrobial experimental sepsis in mice. Furthermore, we give recommendations and examples for the assessment of apoptotic cell death in the lymphocyte compartment.

Also, a "two-hit" model is described in which mice undergo the initial sepsis hit followed by infection by intubation or inhalation of *Pseudomonas aeruginosa* mimicking secondary infections in the immune suppressive phase responsible for a majority of sepsis related fatalities.

2 Materials

2.1 Induction of Polymicrobial Sepsis by Cecal Ligation and Puncture (CLP) or Cecal Slurry Injection

1. Rodent anesthesia machine (vaporizer) with nose-piece for mice.

2. Small animal surgery scissors, forceps, needle holder.

3. Heating pad.

4. Electric shaver/clippers.

5. Desk light.

6. Basin for saline solution.

7. Laboratory scale.

8. Normal saline solution (0.9 % w/v NaCl in H_2O).

9. 5 % dextrose solution (5 g glucose in 1 L H_2O).

10. Artificial tear eye ointment.

11. Betadine iodine solution.

12. Surgical drape.

13. Gauze.

14. Q tips.

15. 4–0 silk for ligature.

16. 4–0 silk suture with C-1 taper needle.

17. 3 M Vetbond.

18. 1 and 3 mL syringes.

19. 1 mL Eppendorf tubes and 10 mL Falcon tubes.

20. Needles (20–30 G).

21. Anesthetic: Isoflurane (2-chloro-2-(difluoromethoxy)-1,1,1 -trifluoro-ethane).

22. Analgesic: Buprenex (buprenorphine hydrochloride).

23. Antibiotics: Primaxin (imipenem/cilastin).

2.2 "Second Hit" Intranasal Infection and Intra-Tracheal Intubation

1. 37 °C microbiological incubator.

2. Spectrophotometer (for bacterial preparation).

3. Benchtop centrifuge.

4. Rodent anesthesia machine (vaporizer) with nose piece for mice.

5. Perspex stand with line/string for positioning mice.

6. Heating pad.

7. Desk light.

8. 12 V cold light source (for fiber optics).

9. Optic fibers (400 μM).

10. 200 μL pipettor.

11. LB broth (10 g Bacto-tryptone, 5 g yeast extract, 10 g NaCl in 1 L H_2O, adjust to pH 7.5, sterilize by autoclaving).

12. Phosphate buffered saline (PBS) solution (8 g NaCl, 0.2 g KCl, 1.44 g Na_2HPO_4, 0.24 g KH_2PO_4 in 1 L H_2O, adjust to pH 7.4, sterilize by autoclaving).

13. 1 mL Eppendorf tubes.

14. 1 mL spectrophotometer cuvettes.

15. 200 μL pipettor tips.

16. Flexible Teflon catheter/Cannula 20 G.

17. *Pseudomonas aeruginosa* ATCC 27853.

3 Methods

All methods outlined below are approved by the La Trobe University Animal Ethics committee and conducted according to SOP protocols. Wild type C57BL/6 mice are housed in the animal facility for at least 1 week prior to experiments in an environment under controlled temperature (22 °C), humidity (50 %), and lighting (12 h day–night cycle) with free access to drinking water and chow. During experiments general clinical observations of mice are performed daily; post-surgery, mice are checked every 6–8 h for the time of experiment. Due to the sepsis induction the mice are expected to become immune suppressed and thus sick. The animals' condition should be monitored throughout the course of the experiment—we recommend using monitoring sheets as outlined below (adapted from [10]). If the body condition of the mice reaches the critical score, the experiment is to be terminated and the animals sacrificed.

3.1 Cecal Ligation and Puncture (CLP)

Wichterman et al. were the first to describe Cecal Ligation and Puncture (CLP) in rats and proposed it as a reproducible animal model for sepsis 35 years ago [11]. In the following years, the CLP technique was subsequently modified for its use in mice by Baker et al. and is now considered as the "gold standard" model of experimental sepsis [12, 13]. The length of ligated cecum as well as the needle size and amount of punctures can be varied in order to titrate the severity of sepsis. For a more severe sepsis in C57BL/6 mice to study cellular apoptosis after 24 h, ligation of the cecum below the ileocecal valve and two punctures with a 25 G needle is recommended. For a mild sepsis in C57BL/6 mice resulting in low mortality (10–20 %) and induction of the prolonged immune suppressive secondary phase in surviving animals, ligation of the distal third of the cecum and one puncture with a 28 G needle is recommended.

1. Anesthetize C57BL/6 mice with 2–3 % isoflurane in vaporizer.

2. When anesthetized, move the animal onto the heating pad (*see* Fig. 1a).

3. Shave belly with electric clippers (*see* **Note 1**).

4. Lubricate eyes with artificial eye ointment using Q tips.

5. Insert head into nose cone to maintain 2–3 % isoflurane anesthesia.

6. Apply Betadine to belly and cut 0.5 cm mid-line incision to mouse's left side (*see* Fig. 1b).

7. Pull out cecum and lay flat onto gauze (keep wet with saline) (*see* **Note 2** and Fig. 1c).

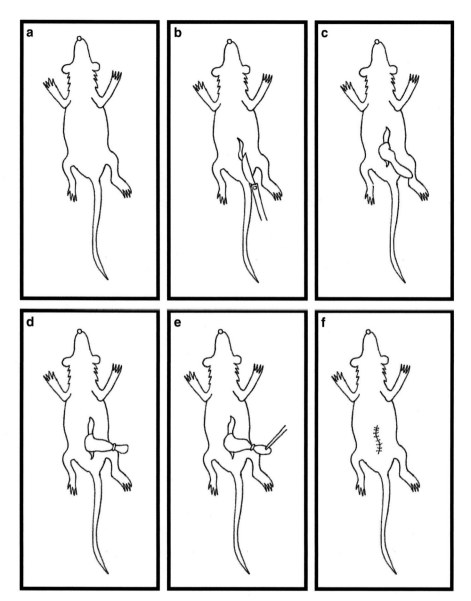

Fig. 1 CLP surgery procedure: (**a**) Anesthetize mouse, shave belly, and lubricate eyes (**b**) Cut 0.5 cm mid-line incision to mouse's left side (**c**) Pull out cecum and lay flat onto gauze (**d**) Ligate beyond the ileal–cecal junction (**e**) Puncture cecum with appropriate gauge needle and gently squeeze to extrude stool (**f**) Replace cecum, close muscle with continuous stitch and skin with wound glue

8. Tie off cecum beyond the ileal–cecal junction with 4–0 silk and cut off excess suture (*see* Fig. 1d).

9. Using fingers, force stool down to the distal end of cecum keeping pressure on to make puncture(s).

10. Puncture cecum with appropriate needle (25 G, two punctures for severe sepsis/28 G, one puncture for mild sepsis) and gently squeeze until stool is extruded (*see* **Note 3** and Fig. 1e).

11. Wipe clean and replace cecum using forceps into body cavity.

12. Close muscle with continuous stitch using 4–0 silk suture and C-1 taper needle (*see* Fig. 1f).

13. Dry area and close skin with wound glue (*see* **Note 4**).

14. Inject 1 mL buprenorphine (0.0015 mg/mL in saline solution) subcutaneous into the back of the mouse.

15. Note time of injury and place moist food pellets in cage.

16. Survey animal condition every 6–8 h using monitoring sheet (*see* Table 1).

17. Animals with scores exceeding ten are culled by CO_2 asphyxiation.
    Control mice undergo the same surgical procedure without ligation and puncture of the cecum.

3.2 Cecal Slurry (CS) Injection

The cecal slurry model adapted from Noble's Life science presents a nonsurgical alternative to CLP and does not require small animal surgical instruments. By variation of the amount of injected cecal slurry the severity of the septic response can be adjusted. For a more severe sepsis in C57BL/6 mice, injections of >1.5 g/kg body weight are recommended. For a mild sepsis resulting in low mortality (10–20 %) with induction of the prolonged immune suppressive secondary phase, injection with <1 g/kg cecal slurry is recommended.

3.2.1 Preparation of Cecal Slurry

1. Sacrifice donor mice by cervical dislocation or CO_2 asphyxiation.

2. Open the mouse abdomen with surgical scissors and dissect the colon.

3. Collect entire cecal content using forceps into 10 mL Falcon tube.

4. Determine wet weight and dilute the cecal slurry with 5 % dextrose solution to a concentration of 250 mg/mL (*see* **Note 5**).

5. Pass slurry through a 22 G needle and through a 100 μm cell strainer to break up solids to avoid blockage of syringe/needle during injection.

6. Directly use in experiment and dilute to desired concentration with 5 % dextrose or 1× PBS to a total volume between 100 and 200 μL. For long term storage, aliquot into 1 mL Eppendorf tubes or 10 mL Falcon tubes and freeze in –80 °C (*see* **Note 6**).

3.2.2 Injection of Cecal Slurry

1. Hold mouse to expose abdomen and inject cecal slurry (diluted into max. 200 μL total volume with 5 % dextrose solution) using a 22 G needle into the peritoneal cavity of the mouse (*see* **Notes 7** and **8**).

Table 1
Sepsis monitoring system

Parameter	Criteria	Scores
Appearance	Normal, smooth fur	0
	Roughened fur	1
	Wet fur	2
	Mucous eyes	3
Breathing pattern	Normal	0
	Fast	1
	Slow	2
	Weak and intermittent	3
Weight change	–5 %	0
	–15 %	1
	–20 %	2
	>–20 %	3
Behavior	Normal, agile, prying	0
	Slow movements, sitting position	1
	Dull, slouched, tottering movements	2
	Lateral position	3
Provoked reaction	Escape reaction when provoked	0
	Flight when approached by hand	1
	Flight when touched	2
	No flight reaction at all	3

2. Note the time of injection and place moist food pellets in the cage.

3. Survey animal condition after 8 h and upon signs of sickness every 2–3 h using the monitoring sheet as per Subheading 3.4.4 and Table 1.

4. Animals with scores exceeding ten are culled by CO_2 asphyxiation.
   Control mice undergo the same procedure but are injected with corresponding volume of 5 % dextrose solution, i.e., 200 μL.

3.3 Assessment of Immune Cell Apoptosis in Septic Mice

Apoptotic cell death of immune cells is a main contributor to immune suppression during sepsis [5]. In experimental mouse models, apoptosis was detected in various lymphoid tissues such as thymus and spleen [14]. A variety of approaches and assays to study apoptosis in the lymphocyte compartment of septic mice have been described previously, e.g., TUNEL stain, FACS-based analysis of apoptosis by staining for activated Caspases or with Annexin V-PI as well as Western blot and qRT-PCR analysis of genes and proteins involved in cell death [9, 14, 15]. Here, we give an example of a FACS-based analysis of lymphocyte apoptosis in mice 24 h after injection with cecal slurry by staining thymocytes with Annexin V-FITC and propidium iodide (*see* Fig. 2).

3.4 "Second Hit" Infection with *Pseudomonas aeruginosa*

Upon experimental sepsis in mice triggered by CLP or CS injection as "first hit," apoptosis of lymphocytes and the induction of an immune suppressive state of sepsis can be readily observed after 24 h [15]. Previous studies report that the immune suppressive state in mice lasts at for least 4 days followed by a partial reconstitution of the immune system [9]. Therefore, the optimum time

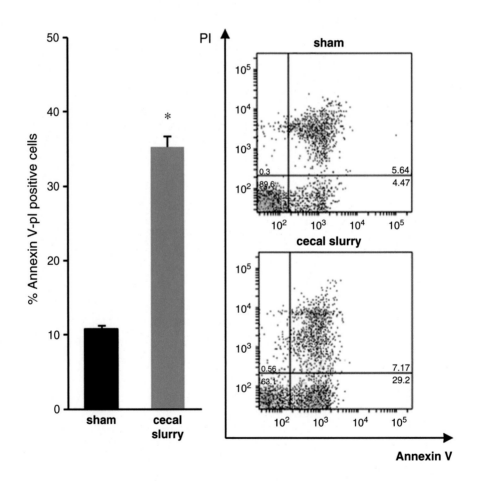

Error bars ±SD, * p<0.05

Fig. 2 Assessment of apoptotic cell death after cecal slurry injection: Thymic apoptotic cell death 24 h post-injection with cecal slurry was measured by FACS analysis of isolated thymocytes stained with Annexin V and PI. Error bars ± SD

point for the "second hit" is 72–96 h, as the mice are particularly vulnerable for secondary infections. *Pseudomonas aeruginosa* is a gram-negative bacterium that is one of the most common causes of pneumonia, a complication that frequently leads to fatalities in hospitals during ventilation of sepsis patients [16]. Thus, experimental induction of a secondary infection with *Pseudomonas aeruginosa* ATCC 27853 represents a widely accepted model for studying the immune suppressive phase of sepsis in mice.

All "second hit" infections with *Pseudomonas aeruginosa* ATCC 27853 as described below are conducted at 3 days past the initial septic insult. All procedures are carried out at room temperature in approved animal housing facilities.

3.4.1 Microbiologic Preparation of Pseudomonas aeruginosa

1. Inoculate *Pseudomonas* from glycerol stock and grow overnight at 37 °C with constant shaking to stationary phase in 3 mL LB broth.

2. Harvest bacteria by centrifugation ($10,000 \times g$ for 10 min), wash the pellet once and then resuspend in 1 mL 1× PBS.

3. Determine OD_{600} of undiluted, 1:2, 1:5, and 1:10 dilutions in 1× PBS using a spectrophotometer.

4. For intranasal and intratracheal infections, dilute bacteria to OD_{600} 0.7 in PBS, which corresponds to 3×10^9 CFU/mL.

3.4.2 Intratracheal Intubation and Infection

Intra-tracheal infection is a nonsurgical procedure, which allows precise delivery of pathogens into the lungs of immune suppressed mice in order to study the effect of secondary infections in sepsis. Endotracheal intubation of mice has been previously described [17] and the experimental protocol outlined below is modified from Das et al. [18]. For trained persons, the whole procedure should take no longer than 2 min per mouse and is highly reproducible. During the procedure the anesthetic level is to be monitored constantly by assessing skin and mucous membrane color and pedal reflex (response to stimuli).

1. Secure the 400 µm optic fiber within the 20 G cannula with its tip extending 4 mm beyond the tip of the cannula (*see* **Note 9**).

2. Anesthetize C57BL/6 mice with 2–3 % isoflurane in vaporizer and wait until it is breathing slowly.

3. Hang anesthetized mouse on the line attached to the perspex frame by its upper incisors (*see* **Note 10** and Fig. 3a).

4. Gently grasp and apply tension to the mouse's tongue while straightening the neck in order to visualize the vocal cords. Insert the 20 G cannula with the optic fiber through the vocal chords and into the trachea a distance of 5 mm (*see* **Note 11** and Fig. 3b, c).

5. Carefully remove the optic fiber leaving the cannula in the trachea.

6. Apply 20 µL of the *Pseudomonas* culture through the Teflon cannula into the trachea, causing the mouse to inhale it directly into the lungs (*see* **Note 12** and Fig. 3d).

7. Slowly remove the cannula from the trachea and remove the mouse from the suspension equipment.

8. Allow mice to recover on the heat pad and place back in their cages.

9. Note the time of infection and place moist food pellets in cage.

10. Survey animal condition every 6–8 h using monitoring sheet as per Subheading 3.4.4 and Table 1.

11. Animals with scores exceeding ten are culled by CO_2 asphyxiation.

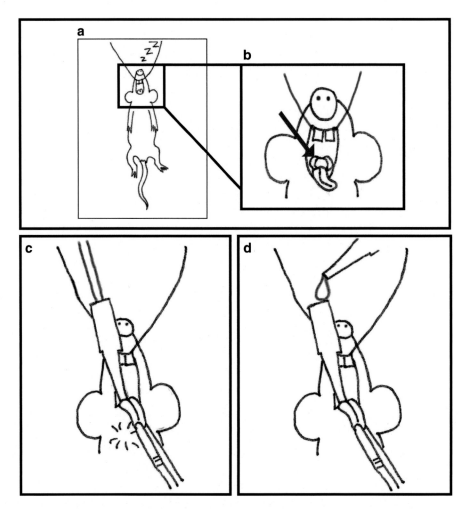

Fig. 3 Intratracheal infection: (**a**) Anesthetize mouse and hang it vertical using perplex stand. (**b**) Grasp tongue and locate the trachea with help of optical fiber. (**c**) Insert cannula into trachea. (**d**) Pipette Pseudomonas into cannula for mouse to automatically inhale

Control mice undergo the same procedure but are injected with 20 μL PBS solution.

3.4.3 Intra-Nasal Infection

A very simple noninvasive alternative to intra-tracheal infection is intranasal infection of anesthetized mice as described by *Muenzer* et al. [9]. However, a possible caveat of this method is the difficulty to ascertain successful delivery of the inhaled pathogen into the lungs of the animal.

1. Anesthetize C57BL/6 mice with 2–3 % isoflurane in vaporizer and wait until it is breathing slowly.

2. Take up 40 μl of *P. aeruginosa* (*see* Subheading 3.4.1) with a 200 μL pipettor and suitable tip.

3. Holding mouse in an upright position, dispense *Pseudomonas* solution slowly moving from one nostril to the other until the entire 40 μL is inhaled.

4. Hold in place for a full minute with mouse in an upright position (*see* **Note 13**).

5. Mice are allowed to recover on heat pad and placed back in their cages.

6. Note time of infection and place moist food pellets in cage.

7. Survey animal condition every 6–8 h using monitoring sheet as per Subheading 3.4.4 and Table 1.

8. Animals with scores exceeding ten are culled by CO_2 asphyxiation.
   Control mice undergo the same procedure but receive 40 μL PBS solution instead of bacterial suspension.

3.4.4 Sepsis Monitoring Sheet

Due to the sepsis induction the mice are expected to get immune suppressed and thus sick—we recommend the use of a monitoring sheet (adapted from Ref. [10]) to monitor the animals' condition throughout the course of the experiment. If the body condition reaches values >10, the mice are to be sacrificed and the experiment terminated (*see* Table 1).

4   Notes

1. Make sure the shaved area is big enough to allow incision, we recommend 2×2 cm.

2. Cecum is kept on wet gauze to ensure that it is kept moist throughout the entire procedure.

3. Extruding of cecal content indicates successful puncture.

4. As an alternative, clips can be used to close the wound.

5. Concentration of cecal slurry can be varied, however 250 mg/mL leads to reasonable total volumes of <200 µL for injection of 0.5–2 g/kg in mice of 20–30 g body weight.

6. Recent studies suggest preparation of the slurry in 15 % glycerol-PBS for increased bacterial viability and long-term storage [19].

7. Concentration of cecal slurry for optimal induction of sepsis and immune suppression has to be experimentally determined for mouse strain and genetic background by the researcher. However, for C57BL/6 aged 8–12 weeks we recommend <1 g/kg for a mild sepsis and >1.5 g/kg for a more severe sepsis.

8. *Critical Step*: Make sure needle reaches into the cavity and cecal slurry does not flow out while withdrawing needle.

9. Sharp edges of the fiber-optic cable should be blunted with emery paper as they can easily cut the trachea.

10. Once the mouse is suspended for the tracheal intubation procedure, it will not be inhaling any further anesthetic agent and will quickly wake; therefore all equipment has to be ready and work has to be done quickly to avoid the mouse waking prior to completion of the procedure.

11. *Critical Step*: make sure that the cannula is inserted in the trachea not the esophagus—movement of the vocal cords helps locating the trachea.

12. Make sure that the suspension was completely inhaled.

13. *Critical Step*: If any bacteria are expelled, try to make the mouse inhale them again.

Acknowledgement

This work was supported by the Research Grant #1085281 from the National Health and Medical Research Council, Australia and a La Trobe University Postgraduate Research Scholarship to M.D.

References

1. Angus DC, Linde-Zwirble WT, Lidicker J, Clermont G, Carcillo J, Pinsky MR (2001) Epidemiology of severe sepsis in the United States: analysis of incidence, outcome, and associated costs of care. Crit Care Med 29(7):1303–1310

2. Hotchkiss RS, Coopersmith CM, McDunn JE, Ferguson TA (2009) The sepsis seesaw: tilting toward immunosuppression. Nat Med 15(5):496–497

3. Cohen J (2002) The immunopathogenesis of sepsis. Nature 420(6917):885–891

4. Schefold JC, Hasper D, Reinke P, Monneret G, Volk HD (2008) Consider delayed immunosuppression into the concept of sepsis. Crit Care Med 36(11):3118

5. Hotchkiss RS, Coopersmith CM, Karl IE (2005) Prevention of lymphocyte apoptosis—a potential treatment of sepsis? Clin Infect Dis 41(Suppl 7):S465–S469

6. Hotchkiss RS, Nicholson DW (2006) Apoptosis and caspases regulate death and inflammation in sepsis. Nat Rev Immunol 6(11):813–822

7. Drewry AM, Samra N, Skrupky LP, Fuller BM, Compton SM, Hotchkiss RS (2014) Persistent lymphopenia after diagnosis of sepsis predicts mortality. Shock 42(5):383–391

8. Nemzek JA, Hugunin KM, Opp MR (2008) Modeling sepsis in the laboratory: merging sound science with animal well-being. Comp Med 58(2):120–128

9. Muenzer JT, Davis CG, Chang K, Schmidt RE, Dunne WM, Coopersmith CM, Hotchkiss RS (2010) Characterization and Modulation of the Immunosuppressive Phase of Sepsis. Infect Immun 78(4):1582–1592

10. Shrum B, Anantha RV, Xu SX, Donnelly M, Haeryfar SM, McCormick JK, Mele T (2014) A robust scoring system to evaluate sepsis severity in an animal model. BMC Res Notes 7:233

11. Wichterman KA, Baue AE, Chaudry IH (1980) Sepsis and septic shock--a review of laboratory models and a proposal. J Surg Res 29(2):189–201

12. Baker CC, Chaudry IH, Gaines HO, Baue AE (1983) Evaluation of factors affecting mortality rate after sepsis in a murine cecal ligation and puncture model. Surgery 94(2):331–335

13. Dejager L, Pinheiro I, Dejonckheere E, Libert C (2011) Cecal ligation and puncture: the gold standard model for polymicrobial sepsis? Trends Microbiol 19(4):198–208. doi:10.1016/j.tim.2011.01.001

14. Chang KC, Unsinger J, Davis CG, Schwulst SJ, Muenzer JT, Strasser A, Hotchkiss RS (2007) Multiple triggers of cell death in sepsis: death receptor and mitochondrial-mediated apoptosis. FASEB J 21(3):708–719

15. Hotchkiss RS, Tinsley KW, Swanson PE, Chang KC, Cobb JP, Buchman TG, Korsmeyer SJ, Karl IE (1999) Prevention of lymphocyte cell death in sepsis improves survival in mice. Proc Natl Acad Sci U S A 96(25): 14541–14546

16. Hotchkiss RS, Dunne WM, Swanson PE, Davis CG, Tinsley KW, Chang KC, Buchman TG, Karl IE (2001) Role of apoptosis in Pseudomonas aeruginosa pneumonia. Science 294(5548):1783

17. Brown RH, Walters DM, Greenberg RS, Mitzner W (1999) A method of endotracheal intubation and pulmonary functional assessment for repeated studies in mice. J Appl Physiol 87(6):2362–2365

18. Das S, MacDonald K, Chang HY, Mitzner W (2013) A simple method of mouse lung intubation. J Vis Exp 73:e50318

19. Starr ME, Steele AM, Saito M, Hacker BJ, Evers BM, Saito H (2014) A new cecal slurry preparation protocol with improved long-term reproducibility for animal models of sepsis. PLoS One 9(12):e115705

Chapter 7

Analysis of Cell Death Induction in Intestinal Organoids In Vitro

Thomas Grabinger, Eugenia Delgado, and Thomas Brunner

Abstract

The intestinal epithelium has an important function in the absorption of nutrients contained in the food. Furthermore, it also has an important barrier function, preventing luminal pathogens from entering the bloodstream. This single cell layer epithelium is quite sensitive to various cell death-promoting triggers, including drugs, irradiation, and TNF family members, leading to loss of barrier integrity, epithelial erosion, inflammation, malabsorption, and diarrhea. In order to assess the intestinal epithelium-damaging potential of treatments and substances specific test systems are required. As intestinal tumor cell lines are a poor substitute for primary intestinal epithelial cells, and in vivo experiments in mice are costly and often unethical, the use of intestinal organoids cultured from intestinal crypts provide an ideal tool to study cell death induction and mechanisms in primary intestinal epithelial cells. This protocol describes the isolation and culture of intestinal organoids from murine small intestinal crypts, and the quantitative assessment of cell death induction in these organoids.

Key words Apoptosis, Intestinal epithelial cells, Crypts, TNFα, Chemotherapy, MTT, Irradiation, Organoids, Enteroids

1 Introduction

The intestinal epithelium is a tissue with impressive dimensions and important physiological functions. The surface of the human small and large intestinal epithelium comprises several hundred square meters, representing the largest epithelial surface in the human body. The intestinal epithelium has an important function in the digestion and absorption of nutrients. While gastric and pancreatic enzymes contribute the major food digestive enzymatic machinery, the intestinal epithelium also secretes digestive enzymes, which further break down nutrients into absorbable monosaccharides, amino acids, and fatty acids. These and other food components, such as salts, vitamins, but also drugs, are then efficiently absorbed by epithelial cells and transported to the circulation for distribution in the body. Next to absorptive activities the intestinal

Hamsa Puthalakath and Christine J. Hawkins (eds.), *Programmed Cell Death: Methods and Protocols*, Methods in Molecular Biology, vol. 1419, DOI 10.1007/978-1-4939-3581-9_7, © Springer Science+Business Media New York 2016

epithelium has also important protective functions. The intestinal lumen is densely populated with an enormous number of harmless commensal bacteria, but also potentially pathogenic microbes and viruses. The intestinal epithelium, together with the mucus layer produced by the goblet cells, provides an important barrier for preventing pathogens from entering the bloodstream. Epithelial cells also transport secretory IgA to the luminal surface, which helps to protect the host from bacterial infections. Furthermore, Paneth cells at the bottom of the small intestinal crypts are a rich source of antibacterial proteins, such as lysozyme and defensins (reviewed in [1]). Last but not least, the intestinal epithelial layer not only contains diverse immature and mature epithelial cells, but is home of a large number of so-called intraepithelial lymphocytes, which contribute to immune regulation and host defense [2].

While this single cell layer epithelium is well adapted for efficient food uptake, it is also very vulnerable to damage, potentially leading to the breakdown of the epithelial barrier function, access of pathogens to circulation and other host tissues, and induction of harmful inflammatory processes and/or bacterial sepsis. Thus, gaps in the epithelial layer due to damage or loss of cells have to be immediately filled by new cells deriving from the fast diving intestinal stem and early progenitor cells [3, 4]. This fast diving cell population in the intestinal crypts is critical for the constant self-renewing of the intestinal epithelial layer. Due to their rapid proliferation and frequent DNA synthesis they are also prone to acquire DNA damage and mutations in critical genes, which may result in the development of cancer, such as colorectal tumors. This rapid proliferation rate also makes them highly susceptible to irradiation and chemotherapeutic drugs that target DNA [5].

The intestinal epithelium is one of the tissues with the highest proliferation rate, and also is a site where extensive physiological and pathophysiological cell death is ongoing. Since the epithelial layer is continuously exposed to damaging stimuli, there is a constant production of new epithelial cells, which are pushed from the bottom of the crypts to the top of the villus in the small intestine, and to the epithelial surface in the large intestine, where they are shed into the lumen. Thus, the life span of an intestinal epithelial cell is only a few days. Detachment of mature epithelial cells from the basal membrane leads to death by anoikis, a detachment-induced form of apoptosis (reviewed in [3]). Even though a very large number of intestinal epithelial cells die by anoikis every moment, only few dying cells are observed within the epithelial layer under physiological conditions [6]. Likely, the rapid shedding of dying cells into the lumen, and their replacement by new cells from the crypts ensures maintenance of the epithelial barrier function.

A completely different situation is seen under pathophysiological conditions. The high sensitivity of intestinal epithelial cells to a variety of drugs, treatments, and biological apoptosis induc-ers results in massive cell death induction in mature and immature epithelial cells in a variety of pathophysiological conditions [6, 7]. As mentioned above, intestinal crypt cells are amongst the fastest dividing cells in the human body, and the associated fast DNA replication makes them prime targets of chemotherapeutic drugs or irradiation [5, 8]. Thus, not surprisingly damage of the intesti-nal epithelium is one of the most common side effects of antican-cer treatments by chemotherapy or whole body irradiation. Similarly, the intestinal epithelium is also highly susceptible to a variety of immune cell effector molecules. For example, activation of T cells and macrophages leads to the expression of members of the TNF family, such as TNFα itself and Fas ligand (FasL), which have potent apoptosis-inducing activities in intestinal epithelial cells [7–9]. As a consequence, immune cells activation in mice in vivo upon injection of T cell-activating anti-CD3ε antibody or macrophage-activating lipopolysaccharide (LPS) results in FasL and TNFα expression, massive intestinal crypt cell death and mature epithelial cell shedding, and a breakdown of the epithelial barrier function. Immune cell-mediated intestinal epithelium damage is also frequently observed in the course of immunopath-ological disorders, such as intestinal graft-versus-host disease, inflammatory bowel disease, bacterial and viral infections, and sepsis [10, 11].

As the excessive permeabilization of the intestinal epithe-lium due to uncontrolled and extensive cell death induction may have devastating consequences, including death, a better under-standing of the signaling processes involved in cell death induc-tion in intestinal crypt and mature epithelial cells is a prerequisite for the development of novel strategies and treatments aiming at preventing such pathological processes. Unfortunately, thus far good model systems have been lacking. As mentioned above, upon isolation, primary intestinal epithelial cells rapidly die by apoptosis, making it difficult to study drug- or cytokine-induced cell death on top of this spontaneous cell death. As a conse-quence many functional studies have been either conducted in vivo in mouse models, which has its limits due to ethical and economical considerations, or in intestinal epithelium-derived tumor cells. However, tumor cells have been selected to survive and to grow, which makes them difficult to compare with mature, differentiated and apoptosis-sensitive intestinal epithe-lial cells. For example, most cell lines, including colorectal tumor cell lines, are intrinsically insensitive to TNFα, whereas primary intestinal epithelial cells are among the most sensitive cells in our body [8].

Thus, a model system more closely related to the in vivo situa-tion is required to study cell death in intestinal epithelial cells in vitro. Work pioneered by the research groups of Hans Clevers and Toshiro Sato, and subsequently adapted by others has lead to the development of 3D-cultures, which allow the growth of intestinal organoids from isolated intestinal crypts or intestinal stem cells [12–14]. These organoids show differentiation of stem cells into all mature epithelial cell types also observed in vivo, including mature absorptive epithelial cells, goblet cells and Paneth cells. Importantly, these organoids show comparable responses and sensitivities to apoptosis-inducing treatments as intestinal epithelial cells in vivo [8]. Thus, they provide an ideal tool to study mechanisms of and sensitivity to apoptosis induction in primary intestinal epithelial cells in an ex vivo model system with high relevance to the in vivo situation. Furthermore, by the use of intestinal crypts from various genetically modified mice the relevance of given gene products in the regulation of cell death in intestinal epithelial cells can be investigated [8]. This protocol describes the isolation of intestinal crypts form murine intestinal tissue, their outgrowth into intestinal organoids, and the quantitative assessment of cell death induction.

2   Materials

1. 1× Dulbecco's PBS (pH 7.4): 137 mM NaCl, 2.7 mM KCl, 8.1 mM Na_2HPO_4, 1.47 mM KH_2PO_4 in ddH_2O (filter sterilized). This is referred to as PBS in the Subheading 3.

2. N2 chemically defined serum free supplement (Invitrogen): Aliquot and store at –20 °C.

3. B27 chemically defined serum free supplement (Invitrogen): Aliquot and store at –20 °C.

4. Basal crypt medium: Advanced DMEM/F12 (Sigma), 0.1 % bovine serum albumin (BSA)
   2 mM L-glutamine 10 mM HEPES, 100 U/mL penicillin (Sigma), 100 µg/mL streptomycin (Sigma), 20 µg/mL nystatin, 1 mM N-acetyl cysteine (Sigma). Store at –20 °C in 48.5 mL aliquots. After thawing one aliquot, add 500 µL of 1× B27 supplement and 1 mL of 1× N2 supplement and store at 4 °C up to four weeks.

5. Complete crypt medium: Basal Crypt medium, freshly add 100 ng/mL murine EGF, 100 ng/mL murine Noggin, 500 ng/mL human R-spondin-1 (see **Note 1**).

6. Matrigel (growth factor reduced) (BD Biosciences): Thaw on ice overnight, store at –20 °C in 400 µL aliquots.

7. Human R-Spondin 1 (Preprotec): Dissolve in 1× Dulbecco's PBS, 0.1 % BSA at 100 µg/mL, store aliquots at –80 °C.

8. Murine Noggin (Preprotec): Dissolve in 1× Dulbecco's PBS, 0.1 % BSA at 100 µg/mL, store aliquots at –80 °C.

9. Murine EGF (Preprotec): Dissolve in 1× Dulbecco's PBS, 0.1 % BSA at 100 µg/mL, store aliquots at −80 °C.

10. MTT stock solution (Sigma): 5 mg/mL MTT (3-(4,5-dimeth ylthiazol-2-yl)-diphenyltetrazolium bromide) in ddH$_2$O (sterile filter), store at 4 °C

11. SDS-solution: 2 % sodium docecyl sulfate in ddH$_2$O.

12. Dimethylsulfoxide (DMSO).

13. Plain, clean surface to work with intestine (*see* **Note 2**).

14. 50 mL conical tubes, some with holes drilled in the lid (*see* **Note 3**).

15. Blunt end scissors.

16. Microscope (Minimum magnification: we use a 4× objective throughout. For detailed images we use up to 20× objectives).

17. 100 µm mesh cell strainer.

18. Coverslips (24 mm × 24 mm, 0.13–0.16 mm thickness).

19. 96-well flat-bottom plates.

20. 15 mL conical tubes.

21. p1000 pipette and tips.

3   Methods

3.1   Crypt Isolation

Keep crypts and solutions on ice at all stages of the procedure, unless stated otherwise.

1. Sacrifice a mouse (*see* **Note 4**), cut the small intestine (distal from stomach proximal from cecum) and store it in cold PBS until use. At this point, have Matrigel aliquot(s) thawed on ice.

2. Cut intestine open longitudinally and flatten it (*see* Fig. 1a, b) on a plain, clean surface with the cut (luminal) side facing up (*see* **Note 2**).

3. Use a coverslip to mechanically remove villi by gently scraping at angle along the intestine surface (*see* Fig. 1c and **Note 5**).

4. Cut the intestine in to pieces of 2–3 cm lengths and wash 3× with 20 mL PBS in a 50 mL conical tube by shaking vigorously for at least 10 s (*see* **Note 3**).

5. Incubate the intestinal tissue in 2 mM EDTA in 30 mL PBS in a 50 mL conical tube for 30 min at 4 °C, on an overhead rotator. Make sure that your PBS does not contain any Ca^{2+} or Mg^{2+} ions, which will saturate and inactivate EDTA.

6. Replace EDTA/PBS with normal PBS and gently shake (4–6×). Inspect the supernatant using a microscope (e.g., in a 96-well plate) (*see* Fig. 2a).

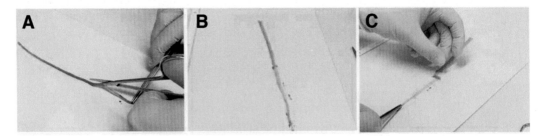

Fig. 1 Preparation of small intestine and villus removal during crypt isolation (*see* steps **2** and **3**). (**a**) Use scissors to open the small intestine longitudinally. The best technique is to put the lower part of the opened scissors into the intestinal tube and slide through the whole organ while fixing the beginning to the preparation surface with forceps. (**b**) Spread the intestine on a flat surface with the luminal side facing upwards. (**c**) Use a coverslip to physically remove villous structures by sliding it alongside the intestine

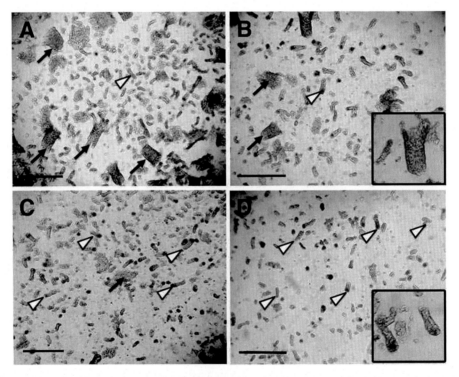

Fig. 2 Differential isolation of intestinal crypts by shaking and subsequent centrifugation step (*see* step **6–10**). (**a**) Villus-rich supernatant of EDTA-treated intestinal tissue pieces after initial shaking (*see* step **6**). (**b**) Supernatant after a second shaking step (*see* step **7**) with still some villus contained in the fraction (inlay: villus in detail). (**c**) Supernatant after a third shaking step (*see* step **7**). Only some villus fragments can be detected. (**d**) Final crypt-enriched fraction (*see* steps **9** and **10**) after straining and single cell removal by centrifugation (inlay: crypts in detail). *White arrowheads*: crypts (exemplary). *Black arrows*: villi (exemplary). Scale bar = 300 μm

7. If there are many villi and only few crypts (a threshold rate of 50 crypts to 1 villi can be considered as satisfactory for continuing), transfer supernatant to a new tube, add 20 mL of PBS to your tissue pieces, to shake them again (10–20×), this time more vigorously.

8. If the "contamination" with villi is still high, repeat **step 7** several times still the ratio crypt/villus is optimal, each time increasing the quantity and intensity of shaking. Use the microscope to decide if any additional shaking step is necessary (*see* Fig. 2b–d). Keep each fraction of your isolation on ice until you know which one you will proceed with.

9. Decide, which fraction has the best crypt/villus ratio and contains the highest number of intact crypts. Proceed with this fraction. Strain the fraction through a 100 μm mesh cell strainer (*see* **Note 6**). This will retain residual villi. Centrifuge the crypt enriched fraction at $100 \times g$ for 5 min.

10. Decant supernatant and add 10 mL PBS. Transfer into 15 mL conical tube and centrifuge at $60 \times g$ for 3 min. This will remove single cells.

11. Decant supernatant and resuspend pellet in 4–6 mL PBS. Use 25 μL for counting of crypt number in a flat bottom 96-well plate (*see* **Note 7**). Only count nice cup-shaped intact crypts, and calculate amount of crypts/mL (by multiplication with 40). If the quantity is too high for counting, use PBS to further dilute the resuspended crypts.

12. Take the appropriate volume with the total number of crypts to be seeded (100–300 crypts/well in a flat-bottom 96-well plate) and centrifuge at $100 \times g$, 5 min.

13. Completely remove supernatant and carefully resuspend crypts in Matrigel (we use 100–300 crypts in 8 μL of Matrigel) (*see* **Note 8**). Avoid creating air bubbles during resuspension. Seed crypts in a flat bottom 96-well plate (8 μL/well) by applying a droplet of crypt-containing Matrigel into the middle of each well. Be careful that the droplet does not get in contact with the wall of the well.

14. Let Matrigel polymerize at 37 °C in the incubator (15–20 min). Bring crypt medium to room temperature (*see* **Note 9**).

15. Add 80 μL of crypt medium to each well. Let crypts grow into organoids for 3 days at 37 °C and 5 % CO_2 in a humidified cell culture incubator (*see* Fig. 3).

3.2 Apoptosis Induction and Detection

1. Treat organoids as desired (*see* **Note 9**). Keep several wells untreated as control and treat at least one well with staurosporine (30 μM) as a positive control for cell death induction. This

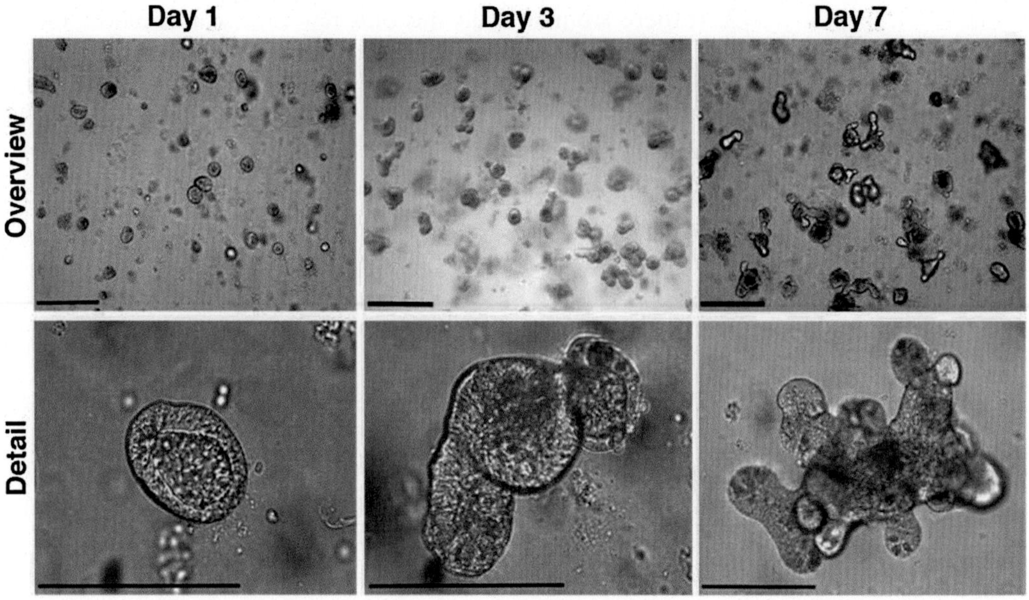

Fig. 3 Typical development of ex vivo cultured intestinal organoids. After one day of culture, crypts have closed to spheroids with a round, polar structure. At day 3 organoids form buds, crypt-like compartments, that contain Paneth cells, stem cells, and proliferating cells. Dead cells start to accumulate in the lumen of the shared central compartment, which consists of differentiated enterocytes. After 7 days of culture, a fully grown enteroid has formed with multiple crypt-like structures that feed into one shared central villus-like domain. Scale bar (overview) = 300 μm. Scale bar (detail) = 150 μm

will be needed later on for background subtraction (*see* Subheading 3.2, **step 6**).

2. Add 8 μL of MTT solution to each well (final concentration 500 μg/mL) and incubate for 1 h at 37 °C. Check for purple organoid staining with a microscope from time to time (*see* Fig. 4a–c).

3. Remove MTT containing medium completely and add 20 μL of a 2 % SDS solution. Incubate for 1 h at 37 °C to dissolve Matrigel (*see* **Note 10**). Occasionally agitate the plate gently.

4. Add 80 μL of DMSO into each well and incubate at 37 °C for 1 h to solubilize the metabolic product formazan.

5. Use a microtiter plate reader to determine absorbance at 562 nm. Subtract staurosporine-treated organoids (100 % death control) as background.

6. Use the following formula on background-subtracted OD values to calculate the percentage of dead cells:

$$\% \, \text{dead cells} = \left[1 - \frac{\text{OD}_{562}\left(\text{sample}\right)}{\text{OD}_{562}\left(\text{control}\right)} \right] \times 100.$$

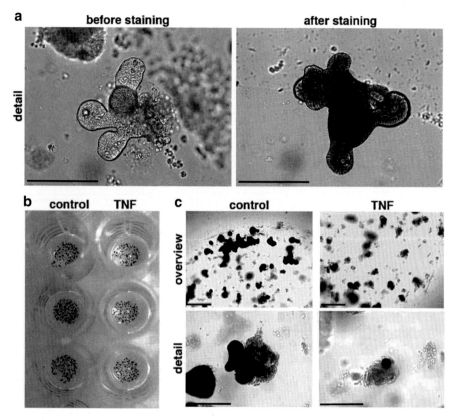

Fig. 4 Assessment of MTT reduction in intestinal organoids (*see* **step 17**). (**a**) Exemplary image of an organoid before the MTT staining process (*left*) and after 20 min of incubation at 37 °C (*right*). (**b**) Macroscopic image of stained organoids within a 96-well plate after control treatment or treatment with 30 ng/mL TNF. (**c**) Microscopic images of control or TNF-treated organoids after MTT staining. Scale bar (overview) = 300 μm. Scale bar (detail) = 150 μm

4   Notes

1. Some suppliers already offer ready-to-use complete crypt medium. These are in some cases even cheaper than buying the single components.

2. For a plain, clean surface, we use the lid of a styrofoam box covered with aluminum foil. For cutting the intestine, it is best to use scissors that have blunt tips to prevent piercing through the tissue. This enables you to cut the intestine open longitudinally in one sliding motion, without repeatedly trying to find the opening (*see* Fig. 1a).

3. To easily remove the PBS from the tissue pieces, we prepared a lid of a 50 mL conical tube with several holes drilled through it. These holes prevent the tissue pieces from being discarded while the buffer can go through. Just turn the conical tube

with the "special lid" around in a sink and apply some shaking movements to remove the PBS.

4. Commonly we use C57BL/6 mice, but other mouse strains also work fine. We usually do not use mice that are older than 16 weeks. The use of mice between 3 and 12 weeks of age gives a satisfactory yield of intestinal crypts.

5. Note 5 Using a coverslip will prevent you from applying too much pressure on the intestine, which will result in loss of crypts and a poor yield in the end. Just do not press so much that the coverslip breaks (*see* Fig. 1c).

6. If your fraction is already highly enriched in crypts and hardly contains any villi, you can skip the cell straining step at this point and proceed directly to the next step.

7. Crypts are too big to fit into a classical Neubauer cell counting chamber. We therefore count the crypts by pipetting 25 μL of the crypt-enriched supernatant into a flat bottom 96-well plate. The concentration of crypts/mL can easily be calculated by multiplication of (number of crypts in 25 μL) × 40.

8. Decanting is not sufficient to remove the supernatant completely. For obtaining a drier pellet, we usually first decant the supernatant and then remove the rest with a p1000 pipette. Be very careful to not disturb the pellet.

9. Make sure that the crypt medium is not cold, when it is added to the already solidified Matrigel droplets. If the medium is too cold, it will liquefy the Matrigel. As final volume, use the volume of Matrigel + medium. We usually treat the organoids 16 h or overnight, and assess cell death after that. To get a good idea about the kinetics and extend of cell death induction we recommend to start with a dose response of a standard cell death inducer (e.g., cisplatin, actinomycin D) and overnight incubation.

10. Do not cool the microtiter plate at this step. Otherwise this will cause the SDS to precipitate. If the Matrigel is already dissolved, you can see the purple stained organoids floating around under the microscope (and even by eye).

Acknowledgements

This work was supported by Research Grants from the German Science Foundation to TB. Thomas Grabinger received a fellowship from the RTG 1331 graduate school (supported by the German Science Foundation).

References

1. Kato T, Owen RL (2004) Structure and function of the intestinal mucosal epithelium. In: Mestecky J, Lamm ME, Strober W, Bienenstock J, McGhee JR, Mayer L (eds) Mucosal immunology, vol 1. Elsevier, San Diego, pp 131–151

2. Cheroutre H, Lambolez F, Mucida D (2011) The light and dark sides of intestinal intraepithelial lymphocytes. Nat Rev Immunol 11(7): 445–456

3. Stappenbeck TS, Wong MH, Saam JR, Mysorekar IU, Gordon JI (1998) Notes from some crypt watchers: regulation of renewal in the mouse intestinal epithelium. Curr Opin Cell Biol 10(6):702–709

4. Clevers H (2013) The intestinal crypt, a prototype stem cell compartment. Cell 154(2):274–284

5. Bowen JM, Gibson RJ, Cummins AG, Keefe DM (2006) Intestinal mucositis: the role of the Bcl-2 family, p53 and caspases in chemotherapy-induced damage. Support Care Cancer 14(7): 713–731

6. Piguet PF, Vesin C, Guo J, Donati Y, Barazzone C (1998) TNF-induced enterocyte apoptosis in mice is mediated by the TNF receptor 1 and does not require p53. Eur J Immunol 28(11): 3499–3505

7. Piguet PF, Vesin C, Donati Y, Barazzone C (1999) TNF-induced enterocyte apoptosis and detachment in mice: induction of caspases and prevention by a caspase inhibitor, ZVAD-fmk. Lab Invest 79(4):495–500

8. Grabinger T, Luks L, Kostadinova F, Zimberlin C, Medema JP, Leist M, Brunner T (2014) Ex vivo culture of intestinal crypt organoids as a model system for assessing cell death induction in intestinal epithelial cells and enteropathy. Cell Death Dis 5, e1228

9. Lin T, Brunner T, Tietz B, Madsen J, Bonfoco E, Reaves M, Huflejt M, Green DR (1998) Fas ligand- mediated killing by intestinal intraepithelial lymphocytes. Participation in intestinal graft-versus-host disease. J Clin Invest 101(3): 570–577

10. Dagenais M, Douglas T, Saleh M (2014) Role of programmed necrosis and cell death in intestinal inflammation. Curr Opin Gastroenterol 30(6):566–575

11. Brunner T, Mueller C (2003) Apoptosis in disease: about shortage and excess. Essays Biochem 39:119–130

12. Miyoshi H, Stappenbeck TS (2013) In vitro expansion and genetic modification of gastrointestinal stem cells in spheroid culture. Nat Protoc 8(12):2471–2482

13. Sato T, Clevers H (2013) Growing self-organizing mini-guts from a single intestinal stem cell: mechanism and applications. Science 340(6137):1190–1194

14. Sato T, Vries RG, Snippert HJ, van de Wetering M, Barker N, Stange DE, van Es JH, Abo A, Kujala P, Peters PJ, Clevers H (2009) Single Lgr5 stem cells build crypt-villus structures in vitro without a mesenchymal niche. Nature 459(7244):262–265

Chapter 8

In Vitro Differentiation of Mouse Granulocytes

Ramona Reinhart, Simone Wicki, and Thomas Kaufmann

Abstract

Granulocytes are central players of the immune system and, once activated, a tightly controlled balance between effector functions and cell removal by apoptosis guarantees maximal host benefit with least possible collateral damage to healthy tissue.

Granulocytes are terminally differentiated cells that cannot be maintained in culture for prolonged times. Isolating primary granulocytes is inefficient and challenging when working with mice, and especially so for the lowly abundant eosinophil and basophil subtypes. Here we describe an in vitro protocol to massively expand mouse derived myeloid progenitors and to differentiate them "on demand" and in large numbers into mature neutrophils or basophils.

Key words Granulocyte, Differentiation, Neutrophil, Basophil, Hoxb8, Mouse, In vitro

1 Introduction

Granulocytes, consisting of neutrophils, eosinophils, and basophils, are circulating immune cells belonging to the myeloid lineage. After maturation in the bone marrow the cells are released as terminally differentiated cells into the bloodstream where they are turned over rapidly due to their rather short life spans in the absence of activating stimuli [1]. Together with tissue resident mast cells, granulocytes are essential early players upon tissue damage and infection. Once activated, survival of granulocytes is typically prolonged and those activated cells release large amounts of immunostimulatory and/or potentially hazardous molecules [2–5]. It is therefore critical that granulocyte cell death is tightly controlled at sites of inflammation for proper immune function and the prevention of collateral tissue damage. Defects in granulocyte apoptosis have accordingly been implicated with numerous immunological disorders, including chronic inflammation, autoimmune diseases, allergies, and asthma [1].

Hamsa Puthalakath and Christine J. Hawkins (eds.), *Programmed Cell Death: Methods and Protocols*, Methods in Molecular Biology, vol. 1419, DOI 10.1007/978-1-4939-3581-9_8, © Springer Science+Business Media New York 2016

Investigation of molecular cell death mechanisms of primary granulocytes isolated from human or mouse is challenging, due to the cells' short life spans and nature to undergo spontaneous apoptosis in vitro, their terminal differentiation and/or their low frequencies found in vivo (basophils, eosinophils). Available cell models are largely restricted to malignantly transformed (leukemic) cell lines, such as the commonly used acute promyelocytic leukemia line HL60 [6]. Even though there are protocols published to differentiate hematopoietic stem cells (HSC) or progenitor cells (HPC) into the various granulocytic lineages, these are usually time consuming, often result in heterogenic populations and require isolation of primary HSC or HPC for each differentiation experiment [7–10].

Kamps and Häcker have pioneered a method to massively expand mouse-derived hematopoietic progenitors committed to the neutrophil/macrophage lineages, and, more recently, also to dendritic cells and lymphoid lineages [11, 12]. This is achieved by blocking cytokine driven leukocyte differentiation with conditionally overexpressed homeobox genes, such as Hoxb8 or HoxA9 [11]. Upon shutdown of the exogenous homeobox gene, differentiation resumes and, consequently, large amounts of mature and functional leukocytes of a specific type can be generated on demand. Conditional Hoxb8-immortalized progenitor cell lines seem to be genetically stable over prolonged times in culture [11] and can be handled, including further genetic manipulations, comparable to a standard cell line. As a consequence, this methodology saves experimental animals, allows the in vitro generation of cells from mice with severe or lethal phenotypes and allows generating cell amounts large enough for protein biochemistry or other methods requiring many cells. Overall, conditional Hoxb8 (HoxA9) immortalized cell lines are becoming an increasingly accepted model to generate mouse myeloid and lymphoid cells in vitro, facilitating their molecular analysis. To our knowledge, despite the high evolutionary conservation of Hox genes from mouse to human, no analogous protocol has been published yet to expand human myeloid or lymphoid progenitors.

Here we describe protocols to generate mouse neutrophils using conditional Hoxb8 (based on Wang et al. [11], with modifications respective to molecular tools and manipulations as described below) and our recently developed protocol to expand interleukin-3 (IL-3) driven mouse basophils using conditional Hoxb8 [13]. Several groups, including ours, have since used the 'Hoxb8 model system' to study cell death or autophagy mechanisms in neutrophils or basophils [13–21].

2   Materials

2.1   Production of Inducible Hoxb8 Lentiviral Particles

1. Complete DMEM medium: DMEM/GlutaMAX™ high glucose, supplemented with 10 % heat-inactivated fetal calf serum (FCS), 100 U/ml penicillin, 100 µg/ml streptomycin, and 50 µM 2-mercaptoethanol (2-ME).

2. X-tremeGENE HP DNA transfection reagent.

3. DNA plasmids for lentivirally delivered inducible Hoxb8: pMD2.G envelope vector (a kind gift from Didier Trono, Addgene plasmid #12259), pCMVδR8.2 packaging vector (a kind gift from Didier Trono, Addgene plasmid #12263), pF-5xUAS-SV40-Hoxb8(mm)-puro-Gev16 inducible Hoxb8 vector (*see* **Note 1**).

4. OptiMEM serum-free medium.

5. 1 M HEPES stock solution, sterile.

6. 10 cm cell culture dishes.

7. 20 ml syringes.

8. 0.2 µm syringe filters (polyethersulfone, PES).

2.2   Isolation of Lineage Marker Negative Cells

1. Staining buffer: sterile phosphate buffered saline (PBS), 3 % FCS, 0.1 % sodium azide (NaN₃).

2. IMag wash buffer: sterile PBS, 0.5 % bovine serum albumin BSA, 2 mM ethylenediaminetetraacetic acid (EDTA), and 0.1 % NaN₃.

3. Biotinylated anti-mouse lineage depletion cocktail, stored at 4 °C (*see* **Note 2**).

4. Anti-FcγRII/III (CD16/CD32) blocking antibody, clone 2.4G2, stored at 4 °C (*see* **Note 3**).

5. IMag™ Streptavidin Particles Plus-DM, stored at 4 °C.

6. Normal rat serum (NRS) stored at −20 °C.

7. IMag™ magnet.

8. 23G needles and syringes.

9. 15 ml and 50 ml sterile Falcon tubes.

10. Sterile, capped 5 ml round-bottom FACS tubes.

11. Cotton wool plugged and unplugged glass Pasteur pipettes, sterilized.

2.3   Generation and Cultivation of Inducible Hoxb8 Immortalized Cells

1. Complete RPMI medium: RPMI-1640 AQmedia™ supplemented with 10 % heat-inactivated FCS, 100 U/ml penicillin, 100 µg/ml streptomycin, and 50 µM 2-ME.

2. Mouse stem cell factor (SCF) from CHO/SCF (mm) conditioned medium [13], supplemented at 10 % (*see* **Note 4**).

3. Mouse interleukin-3 (IL-3) from WEHI-3B cell-conditioned medium [13], supplemented at 10 % (*see* **Note 5**).

4. 8 μg/ml working concentration: Polybrene (Hexadimethrinebromide).

5. 10 mM 4-hydroxytamoxifen (4-OHT): Prepare stock in ethanol. Prepare further dilution fresh on the day in complete RPMI. Protect from light. Store stock at –20 °C.

6. 1 μg/ml working concentration puromycin. Protect from light. Prepare working dilution in complete RPMI.

7. 6-well cell culture plates.

8. Culture flasks for suspension cells.

9. Sterile, capped 5 ml round bottom FACS Tubes.

10. 15 ml and 50 ml sterile Falcon Tubes.

11. Optional: recombinant mouse GM-CSF (working concentration 1 ng/ml) and recombinant mouse G-CSF (working concentration 10 ng/ml).

3 Methods

3.1 Production of Inducible Hoxb8 Lentiviral Particles

1. Seed approx. 6×10^6 HEK 293 T cells the day before transfection in a 10 cm plate in 10 ml complete DMEM medium and place cells in humidified incubator at 37 °C, 5 % CO_2 (*see* **Note 6**).

2. Mix a total of 5 μg of the following three DNA plasmids for co-transfection at a ratio of 2:5:3 = envelope vector—packaging vector—inducible Hoxb8 vector, with X-tremeGENE HP transfection reagent (*see* **Note 7**).

3. Dilute 1 μg pMD2.G, 2.5 μg pCMVδR8.2, and 1.5 μg pF-5xUAS-SV40-Hoxb8(mm)-puro-Gev16 in 500 μl OptiMEM.

4. Bring X-tremeGENE HP to room temperature (RT) prior to use and carefully add 10 μl to the diluted DNA. Pipette gently to mix.

5. Incubate the mixture at RT for 15 min.

6. During incubation time, carefully replace medium of the HEK 293 T cells with 10 ml of fresh complete DMEM.

7. Gently add the transfection complexes drop by drop to the HEK 293 T cells. Gently swirl the plate to ensure even distribution of transfection complexes within the medium.

8. The transfected, i.e., virus producing, HEK 293 T cells have to be kept in facilities approved for biosafety level 2 from this moment onwards.

9. After overnight incubation, replace the medium with 8 ml fresh complete DMEM containing 10 mM HEPES. Place cells in incubator at 37 °C, 5 % CO_2 for virus production.

10. Harvest first batch of virus-containing supernatant after 24 h and store at 4 °C. Gently add 8 ml fresh complete DMEM containing 10 mM HEPES.

11. Collect second batch of virus-containing supernatant after additional 12–24 h.

12. Pool first and second virus batch and pass through a 0.2 μm polyethersulfone (PES) syringe filter.

13. Proceed with freshly harvested virus to infect target cells (*see* Subheading 3.3), or snap-freeze in 1 ml aliquots and store at −80 °C for later use (*see* **Note 8**).

3.2 Isolation of Lineage Marker Negative Cells (See Note 2)

1. Prepare sterile buffers and place on ice (staining and IMag wash buffer).

2. Aseptically prepare a single-cell suspension from mouse bone marrow (*see* **Note 9**).

3. Sacrifice mouse, dissect femur and collect bone marrow by flushing ice cold staining buffer through both sides using a 23G needle until the bone becomes white.

4. Resuspend the bone marrow in 3 ml of staining buffer and filter through a Pasteur pipette plugged with cotton wool to obtain single cell suspension (*see* **Note 10**).

5. Centrifuge at $500 \times g$ for 5 min at 4 °C and resuspend cells in 1 ml staining buffer.

6. Count cells (*see* **Note 11**) and bring cell suspension to a concentration of 2×10^7 cells/ml by diluting with 2.4G2 hybridoma supernatant. Add 1 % normal rat serum.

7. Incubate on ice for 15 min (blocking step).

8. Add 5 μl of Mouse Lineage Depletion Cocktail per 1×10^6 bone marrow cells.

9. Incubate on ice for 15 min.

10. Add approximately tenfold excess volume of IMag wash buffer.

11. Centrifuge at $500 \times g$ for 5 min at 4 °C and carefully aspirate as much supernatant as possible.

12. Add 5 μl of Streptavidin Magnetic beads (IMag™ Streptavidin Particles Plus-DM, vortex well prior to use) per 1×10^6 bone marrow cells (*see* **Note 12**).

13. Mix thoroughly with a pipette and incubate for 30 min at 6–12 °C in the fridge.

14. Resuspend cells in 1 ml IMag wash buffer and transfer into sterile, round-bottom 5 ml capped FACS tubes; maximally 8×10^7 cells per tube (*see* **Note 13**).

15. Place the tube(s) into the BD IMagnet and incubate for 8 min.

16. By keeping the tube within the BD IMagnet, carefully remove the supernatant, corresponding to the lineage marker negative fraction, using a sterile glass Pasteur pipette and transfer to a new FACS tube.

17. Remove the tube with the lineage marker positive fraction from the BD IMagnet and gently resuspend cells by pipetting up and down in 1 ml IMag wash buffer.

18. Return tube to the BD IMagnet for another 8 min.

19. Repeat **step 16** and pool all lineage marker negative fractions.

20. Place the pooled negative fraction for a final round into the magnet for another 8 min and transfer the final lineage marker depleted fraction, containing the enriched hematopoietic progenitors, into a new sterile 15 ml Falcon tube.

21. Centrifuge at $500 \times g$ for 5 min at 4 °C and resuspend in 1 ml complete RPMI medium.

22. Count cells and subsequently use for infection with conditional Hoxb8 lentiviral particles. Alternatively, freeze cells in aliquots in 90 % FCS, 10 % DMSO or process for other downstream applications (*see* **Note 14**).

3.3 Generation and Cultivation of Inducible Hoxb8-Immortalized Cells

1. Seed 5×10^5 freshly isolated lineage marker negative cells in 4 ml complete RPMI medium supplemented with CHO/SCF (mm) conditioned medium (for generation of neutrophils), or with WEHI-3b conditioned medium as a source of mouse IL-3 (for generation of basophils), in a 6-well plate (*see* **Notes 4** and **5**). Place cells in incubator for 24 h (*see* **Note 15**).

2. Transfer cells in sterile capped FACS-tubes. Centrifuge at $500 \times g$ for 5 min, 4 °C.

3. Decant supernatant. Resuspend cells in 1 ml of conditional Hoxb8 lentivirus containing supernatant (*see* Subheading 3.1), supplemented with 8 μg/ml of polybrene.

4. Spin infect the cells at $300 \times g$ for 90 min at 30 °C.

5. Transfer the cells to a 6-well plate. Add 3 ml complete RPMI supplemented with SCF (for generation of neutrophils), or IL-3 (for generation of basophils), respectively, plus 100 nM of 4-OHT (*see* **Note 16**).

6. Start the selection of successfully transduced cells 2 days after infection by adding 1 μg/ml of puromycin.

7. Replace with fresh complete RPMI medium (containing either SCF or IL-3, 100 nM 4-OHT, and 1 μg/ml puromycin) every 3 days.

8. Puromycin resistant cells (termed SCF-^{cond}Hoxb8 or IL-3-^{cond}Hoxb8, respectively) typically grow out after 2–3 weeks, after which puromycin can be omitted from the medium (*see* **Note 17**).

9. Keep culturing SCF-^{cond}Hoxb8 and IL-3-^{cond}Hoxb8 cells in complete RPMI containing either SCF (10 % CHO/SCF-conditioned medium) or IL-3 (10 % WEHI3b-conditioned medium), and 100 nM 4-OHT (*see* **Note 18**).

10. Freeze cells for long-term storage in liquid nitrogen in 90 % FCS, 10 % DMSO.

11. OPTIONAL: subcloning or further selection of cell lines, as desired (e.g., selection of c-kit-negative fraction for basophils) (*see* **Note 19**).

12. Check expression of exogenous Hoxb8, as well as its disappearance upon removal of 4-OHT, by western blotting (*see* **Note 20**).

3.4 In Vitro Differentiation of SCF-^{cond}Hoxb8 Cells into Mature Neutrophils

1. Wash the cells twice with PBS to remove all traces of 4-OHT.

2. Seed cells at a density of 2.5×10^4 cells/ml in complete RPMI containing SCF (5 % of CHO/SCF conditioned medium) but no 4-OHT (*see* **Note 21**).

3. Nearly mature neutrophils, resembling bone marrow-derived primary neutrophils, are obtained after approx. 5 days (*see* **Note 22**). End-differentiated neutrophils will start to undergo spontaneous apoptosis, which is antagonized by activating cytokines, such as GM-CSF (*see* Fig. 1).

4. OPTIONAL: addition of recombinant mouse G-CSF (10 ng/ml) during the differentiation results in more mature neutrophils.

5. Check the morphology of differentiated neutrophils on cytospins and characterize surface marker expression by flow cytometry (*see* Fig. 1 and **Note 23**).

3.5 In Vitro Differentiation of IL-3-^{cond}Hoxb8 Cells into Mature Basophils

1. Wash the cells twice with PBS to remove all traces of 4-OHT.

2. Seed cells at a density of 7.5×10^4 cells/ml in complete RPMI containing IL-3 (10 % of WEHI-3b conditioned medium, or 500 pg/ml rec IL-3) but no 4-OHT.

3. It may be necessary to split the cells after 3–4 days (*see* **Note 22**).

4. End-differentiated basophils are obtained after 5–6 days. Basophils will start to undergo spontaneous apoptosis, which can be antagonized by addition of high concentrations (10 ng/ml) of rec. mouse IL-3 (*see* Fig. 2).

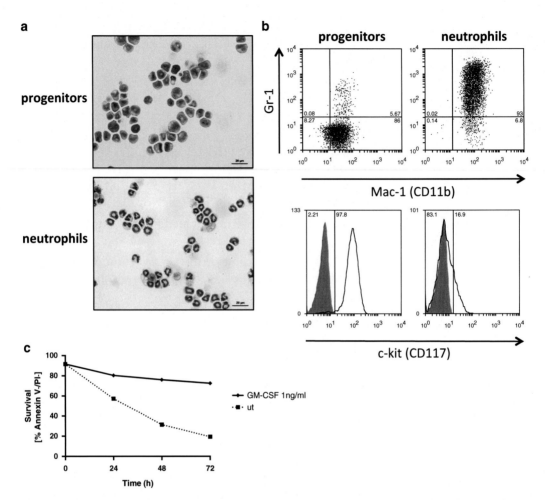

Fig. 1 Characterization of SCF-[cond]Hoxb8 neutrophils. (**a**) H & E stainings of cytospins showing SCF-[cond]Hoxb8 neutrophil progenitors and 5 day-differentiated neutrophils. (**b**) Surface marker analysis of progenitors and 5 day-differentiated neutrophils by flow cytometry. (**c**) Survival curve (FITC-Annexin V/propidium iodide exclusion) of in vitro (5 days) differentiated neutrophils, in the presence or absence of GM-CSF (1 ng/ml)

5. Check the morphology of differentiated basophils on cytospins and characterize surface marker expression (in particular FcεRI and c-kit) by flow cytometry (*see* Fig. 2 and **Note 24**).

4   Notes

1. Instead of a Hoxb8-estrogen receptor fusion protein and retroviral delivery described by Wang et al. [11], we conditionally express mouse wild-type Hoxb8 using a lentiviral system that lacks any exogenous Hoxb8 expression in the absence of 4-OHT [13, 22]. The pMD2.G envelope vector expresses the glycoprotein G of the vesicular stomatitis virus (VSV-G envelope protein), with very broad tropism.

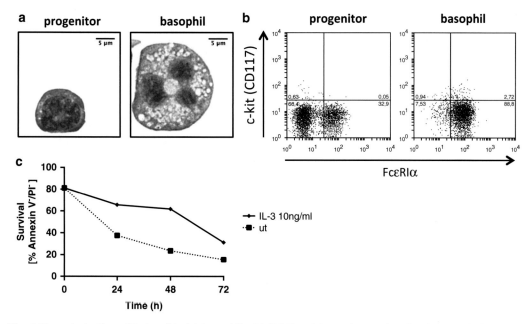

Fig. 2 Characterization of IL-3-^condHoxb8 basophils. (**a**) H & E stainings of cytospins showing a typical example of a IL-3-^condHoxb8 progenitor and a 6 day-differentiated basophil. (**b**) Surface marker analysis of progenitors and 6 day-differentiated basophils by flow cytometry. (**c**) Survival curve (FITC-Annexin V/propidium iodide exclusion) of in vitro (6 days) differentiated basophils, in the presence or absence of IL-3 (10 ng/ml)

2. We use the IMag™ Mouse Hematopoietic Progenitor Cell Enrichment Set-DM for negative selection of uncommitted hematopoietic progenitors. Any other magnetic bead based or FACS sorting based enrichment is suitable; however, negative selection is recommended. The majority of lineage surface marker positive cells, including T and B lymphocytes, monocytes/macrophages, granulocytes, and megakaryocyte-erythroid cells are eliminated from a bone marrow cell pool by using a supplied antibody cocktail (including biotinylated monoclonal antibodies directed against mouse CD3 ε chain (clone 145-2C11), CD11b (Mac-1) (clone M1/70), CD45R (B220) (clone RA3-6B2), Ly-6G/Ly-6C (Gr-1) (clone RB6-8C5), and TER-119 (Ly-76) (clone TER-199).

3. We use conditioned medium from the hybridoma cells (ATCC® HB-197™) for CD16/CD32 blocking purposes.

4. We routinely use conditioned medium from mouse SCF secreting cells (CHO/SCF(mm) cell line; kind gift from G. Häcker, Freiburg, Germany) as a source of mouse SCF. Alternatively, use recombinant mouse SCF (10–50 ng/ml).

5. We routinely use WEHI-3b cells as a source of mouse IL-3. IL-3 concentration of WEHI-3b conditioned medium is 2–5 ng/ml, as determined by ELISA.

6. Exact number of cells needs to be titrated. The aim is to have an optical cell confluence of 60–80 % at the time of transfection. Alternatively to seeding HEK 293 T the day before transfection, seed approx. 9×10^6 HEK 293 T cells in the morning for transfection in the late afternoon.

7. Allow X-tremeGENE HP DNA Transfection Reagent, DNA, and OptiMEM to equilibrate to RT prior to use and vortex gently. Due to interaction with plastic material, make sure to pipette X-tremeGENE HP reagent directly into liquid. Pipette mixture up and down just once as soon as X-tremeGENE HP is administered.

8. We get best infection results with fresh virus, as viral titers are diminished upon freezing/thawing. Virus may optionally be concentrated by ultracentrifugation. Viral titers can be determined by infection of suitable cell line (e.g., HeLa) using serial dilutions of virus-containing supernatant and puromycin selection.

9. Wang et al. have also reported the use of E13 derived fetal liver hematopoietic progenitors [11].

10. Alternatively use 70 μm nylon cell strainer to remove clumps of cells and/or debris.

11. We typically obtain ca. 60×10^6 bone marrow cells from 2 femora.

12. Streptavidin particles are magnetic nanoparticles that have streptavidin covalently conjugated to their surfaces.

13. Split into two FACS tubes if $>8 \times 10^7$ cells.

14. We typically obtain between 1 and 3×10^6 lineage marker negative cells (enriched in hematopoietic progenitors) from 2 femora.

15. In case cryopreserved lineage marker negative cells are used, pre-incubate cells for 36–48 h in cytokine containing medium (SCF or IL-3, respectively), prior to infection with conditional Hoxb8 lentivirus.

16. From this point on 4-OHT always needs to be present in the medium to guarantee expression of Hoxb8 and to maintain the progenitor state of the cells.

17. It is recommended to include a mock infection control and select with puromycin until all control cells have died.

18. As soon as SCF-^{cond}Hoxb8 cell lines are established and growing well, we reduce SCF to 5 % of CHO/SCF conditioned medium. IL-3 concentration is not changed (always 10 % of WEHI-3b conditioned medium, corresponding to 200–500 pg/ml IL-3).

19. Whereas SCF-^{cond}Hoxb8 cell lines usually do not need further selection or subcloning, IL-3-^{cond}Hoxb8 lines can show more heterogeneous populations based on FACS analysis of the surface markers FcεRI and c-kit (CD117). Sorting of the c-kit-negative population increases the percentage of basophil-committed progenitors.

20. Upon removal of 4-OHT from the medium, exogenous Hoxb8 protein should rapidly disappear after 2–3 days, as assessed by western blotting [13].

21. SCF-^{cond}Hoxb8 progenitors generated as described here retain their macrophage/neutrophil bi-phenotypic characteristics as described by Wang et al. [11]. Cells can accordingly also be differentiated into macrophages by adding 1 ng/ml of mouse GM-CSF during the differentiation. Macrophages will start to adhere after 3–4 days, at which time floating neutrophils can be washed away. Wang et al. [11] have further described that addition of IL-5 during differentiation results in a certain percentage of eosinophils.

22. Differentiating neutrophils will typically expand 3–4× (2–3× for basophils) during the 5 days (6 days for basophils). Depending on the starting cellular density, it may be necessary to split the cells at day 3 or 4 (keep density <1 × 10^6/ml at all times).

23. Additionally, neutrophils may be tested for the increased expression and activity of neutrophil myeloperoxidase (MPO) and matrix metalloproteinase 9 (MMP-9). Functional read-outs may include ROS activity assays, chemotaxis, phagocytosis, or bacterial killing assays [13, 21].

24. For further characterization of basophils: IgE-dependent or -independent degranulation (e.g., release of β-hexosaminidase, histamine), secretion of "Th2 cytokines" (IL-4, IL-13), and de novo production of lipid mediators (e.g., LTC4) [13]. A good marker to distinguish mouse basophils from mast cells is the expression of basophil-specific protease, mouse mast cell protease 8 (mMCP-8) by qPCR (*Mcpt8*), western blotting and/or immunofluorescent staining [13, 23].

Acknowledgements

This work was supported by the Swiss National Science Foundation, Nr. 31003A_149387 and Nr. 310030E150805 (part of the D-A-CH initiative from the SNF and the Deutsche Forschungsgemeinschaft (DFG), FOR-2036) and the 3R Research Foundation Switzerland (Nr. 127-11).

References

1. Geering B, Stoeckle C, Conus S, Simon HU (2013) Living and dying for inflammation: neutrophils, eosinophils, basophils. Trends Immunol 34(8):398–409

2. Mantovani A, Cassatella MA, Costantini C, Jaillon S (2011) Neutrophils in the activation and regulation of innate and adaptive immunity. Nat Rev Immunol 11(8):519–531

3. Karasuyama H, Mukai K, Obata K, Tsujimura Y, Wada T (2011) Nonredundant roles of basophils in immunity. Annu Rev Immunol 29:45–69

4. Hogan SP, Rosenberg HF, Moqbel R, Phipps S, Foster PS, Lacy P, Kay AB, Rothenberg ME (2008) Eosinophils: biological properties and role in health and disease. Clin Exp Allergy 38(5):709–750

5. Voehringer D (2013) Protective and pathological roles of mast cells and basophils. Nat Rev Immunol 13(5):362–375

6. Collins SJ, Gallo RC, Gallagher RE (1977) Continuous growth and differentiation of human myeloid leukaemic cells in suspension culture. Nature 270(5635):347–349

7. Dyer KD, Moser JM, Czapiga M, Siegel SJ, Percopo CM, Rosenberg HF (2008) Functionally competent eosinophils differentiated ex vivo in high purity from normal mouse bone marrow. J Immunol 181(6):4004–4009

8. Geering B, Stoeckle C, Rozman S, Oberson K, Benarafa C, Simon HU (2014) DAPK2 positively regulates motility of neutrophils and eosinophils in response to intermediary chemoattractants. J Leukoc Biol 95(2):293–303

9. Arinobu Y, Iwasaki H, Gurish MF, Mizuno S, Shigematsu H, Ozawa H, Tenen DG, Austen KF, Akashi K (2005) Developmental checkpoints of the basophil/mast cell lineages in adult murine hematopoiesis. Proc Natl Acad Sci U S A 102(50):18105–18110

10. Takao K, Tanimoto Y, Fujii M, Hamada N, Yoshida I, Ikeda K, Imajo K, Takahashi K, Harada M, Tanimoto M (2003) In vitro expansion of human basophils by interleukin-3 from granulocyte colony-stimulating factor-mobilized peripheral blood stem cells. Clin Exp Allergy 33(11):1561–1567

11. Wang GG, Calvo KR, Pasillas MP, Sykes DB, Hacker H, Kamps MP (2006) Quantitative production of macrophages or neutrophils ex vivo using conditional Hoxb8. Nat Methods 3(4):287–293

12. Redecke V, Wu R, Zhou J, Finkelstein D, Chaturvedi V, High AA, Hacker H (2013) Hematopoietic progenitor cell lines with myeloid and lymphoid potential. Nat Methods 10(8):795–803

13. Gurzeler U, Rabachini T, Dahinden CA, Salmanidis M, Brumatti G, Ekert PG, Echeverry N, Bachmann D, Simon HU, Kaufmann T (2013) In vitro differentiation of near-unlimited numbers of functional mouse basophils using conditional Hoxb8. Allergy 68(5):604–613

14. Koedel U, Frankenberg T, Kirschnek S, Obermaier B, Hacker H, Paul R, Hacker G (2009) Apoptosis is essential for neutrophil functional shutdown and determines tissue damage in experimental pneumococcal meningitis. PLoS Pathog 5(5), e1000461

15. Garrison SP, Thornton JA, Hacker H, Webby R, Rehg JE, Parganas E, Zambetti GP, Tuomanen EI (2010) The p53-target gene puma drives neutrophil-mediated protection against lethal bacterial sepsis. PLoS Pathog 6(12), e1001240

16. Kirschnek S, Vier J, Gautam S, Frankenberg T, Rangelova S, Eitz-Ferrer P, Grespi F, Ottina E, Villunger A, Hacker H, Hacker G (2011) Molecular analysis of neutrophil spontaneous apoptosis reveals a strong role for the pro-apoptotic BH3-only protein Noxa. Cell Death Differ 18(11):1805–1814

17. Morshed M, Hlushchuk R, Simon D, Walls AF, Obata-Ninomiya K, Karasuyama H, Djonov V, Eggel A, Kaufmann T, Simon HU, Yousefi S (2014) NADPH oxidase-independent formation of extracellular DNA traps by basophils. J Immunol 192(11):5314–5323

18. Echeverry N, Bachmann D, Ke F, Strasser A, Simon HU, Kaufmann T (2013) Intracellular localization of the BCL-2 family member BOK and functional implications. Cell Death Differ 20(6):785–799

19. Weber B, Schuster S, Zysset D, Rihs S, Dickgreber N, Schurch C, Riether C, Siegrist M, Schneider C, Pawelski H, Gurzeler U, Ziltener P, Genitsch V, Tacchini-Cottier F, Ochsenbein A, Hofstetter W, Kopf M, Kaufmann T, Oxenius A, Reith W, Saurer L, Mueller C (2014) TREM-1 Deficiency Can Attenuate Disease Severity without Affecting Pathogen Clearance. PLoS Pathog 10(1), e1003900

20. Klein M, Brouwer MC, Angele B, Geldhoff M, Marquez G, Varona R, Hacker G, Schmetzer H, Hacker H, Hammerschmidt S, van der Ende A, Pfister HW, van de Beek D, Koedel U (2014) Leukocyte attraction by CCL20 and its receptor CCR6 in humans and mice with

pneumococcal meningitis. PLoS One 9(4), e93057

21. Rozman S, Yousefi S, Oberson K, Kaufmann T, Benarafa C, Simon HU (2015) The generation of neutrophils in the bone marrow is controlled by autophagy. Cell Death Differ 22(3):445–456

22. Vince JE, Wong WW, Khan N, Feltham R, Chau D, Ahmed AU, Benetatos CA, Chunduru SK, Condon SM, McKinlay M, Brink R, Leverkus M, Tergaonkar V, Schneider P, Callus BA, Koentgen F, Vaux DL, Silke J (2007) IAP antagonists target cIAP1 to induce TNFalpha-dependent apoptosis. Cell 131(4):682–693

23. Ugajin T, Kojima T, Mukai K, Obata K, Kawano Y, Minegishi Y, Eishi Y, Yokozeki H, Karasuyama H (2009) Basophils preferentially express mouse Mast Cell Protease 11 among the mast cell tryptase family in contrast to mast cells. J Leukoc Biol 86(6):1417–1425

Chapter 9

Hydrodynamic Injection as a Method of Gene Delivery in Mice: A Model of Chronic Hepatitis B Virus Infection

Simon P. Preston, Marc Pellegrini, and Gregor Ebert

Abstract

Gene delivery methods are important for both therapeutic intervention and as tools in research to address specific questions. Hydrodynamic injection (HDI) is a method that facilitates the delivery and expression of genetic material in target cells, namely hepatocytes, through an intravenous injection. HDI has great utility for research involving cell death and signaling pathways essential in the processes of cancer, inflammation, and transplant therapy, as well as representing a valuable technique to establish hepatitis B virus (HBV) expression in hepatocytes. This chapter describes in detail how to generate a model of chronic HBV infection in immunocompetent mice using HDI as a delivery method.

Key words Hydrodynamic injection, HBV, In vivo transfection, Gene delivery

1 Introduction

The need for a quick and inexpensive way to deliver genetic material into mice was answered with the advent of the hydrodynamic injection (HDI). Viral delivery systems have been used with some efficacy for these purposes previously, but their production and purification are laborious [1].

HDI has been used successfully to modulate mRNA and protein expression in the livers of animals by delivery of a range of nucleic acid material including DNA [2], siRNA [3] and guide RNA for the purpose of CRISPR/Cas9 gene editing [4]. We, and others, have used this technique to induce liver specific hepatitis B virus (HBV) expression with great success in mice [5–7]. HDI has facilitated, for the first time, essential research into the immune mechanisms involved in chronic HBV infection, using a small animal model.

Although this method describes the injection of naked DNA, it is important to recognize that after HDI, hepatitis B virions are present in the blood of mice that may represent a potential health

Hamsa Puthalakath and Christine J. Hawkins (eds.), *Programmed Cell Death: Methods and Protocols*, Methods in Molecular Biology, vol. 1419, DOI 10.1007/978-1-4939-3581-9_9, © Springer Science+Business Media New York 2016

risk to those performing the research. Before commencing work, it is recommended that anyone involved in handling samples and/ or mice be vaccinated against HBV.

The method of HDI itself involves the delivery of a large bolus injection of phosphate buffered saline (PBS), containing your nucleic acid of choice, in a volume equivalent to 8 % body weight of the mouse. The injection is performed within 3–5 s, forcing the solution through the inferior vena cava. The large volume causes cardiac congestion, which results in retrograde movement of the solution at high pressure to the liver via the portal vein. The capillaries within the liver are lined with mainly endothelial cells; however, 6–8 % of this is made up of small pores called liver fenestrae that under homeostatic conditions have a diameter of 100 nm. It is these pores that are expanded due to the large hydrodynamic force induced by the injection that allows the genetic material to enter into the hepatocytes [8]. It is important to note that not all liver fenestrae are enlarged during this procedure, which may explain why only approx. 40 % of hepatocytes are transfected during HDI [9]. In our HBV model we can induce viral replication in approximately 10–20 % of hepatocytes [6]. Once the delivered genetic material has entered the hepatocytes and the hydrodynamic pressure subsides, the fenestrae close up, trapping the information inside the cells. Expression of genetic material peaks around 8 h post HDI [9], but this time differs in our model as we measure serum HBV-DNA derived from secreted virions, which peaks 3–7 days post injection [6]. From this point on, the duration of HBV expression is determined by both the genetic background of the mouse and the plasmid backbone [7, 9].

Using the method of HDI to deliver the HBV genome encoding plasmid, we generated two models of chronic HBV infection in immunocompetent mice. Induction of infection in C57BL/6 mice generates a model of partial viral control, in which serum HBV-DNA is first undetectable at 12 weeks post injection (Fig. 1). In this model we see stochastic flares of serum HBV-DNA after initial control, which resembles the oscillating pattern of viral control in humans with chronic HBV infection [10]. Importantly, this model allows the interrogation of immune and cell death signaling pathways involved in the host response to HBV due to the extensive suite of gene-targeted mice available on this genetic background [6]. Conversely, after induction of infection in C3H/HeJ mice (endotoxin sensitive strain), these animals are not able to control serum HBV-DNA and show persistently high viremia for at least 20 weeks post induction of infection [6] (Fig. 1). In both models we see the expression of serological markers of chronic HBV infection, including the production of anti-HBs antibodies concomitant with serum negativity of HBV-DNA and HBsAg.

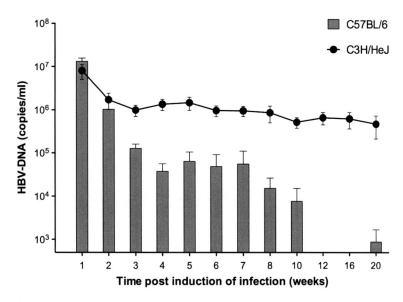

Fig. 1 Serial measurement of serum HBV-DNA levels in C57BL/6 or C3H/HeJ mice following induction of HBV Hepatitis B virus (HBV)infection ($n=6$–10 in each group). Graphs show means and SEMs

This chapter provides detailed instructions on how to prepare the HBV genome encoding plasmid and the technique of HDI itself, to generate a model of chronic HBV infection in immunocompetent mice.

2 Materials

2.1 Preparation of HBV Plasmid

The plasmid pAAV-HBV1.2 we use for the establishment of HBV expression by HDI contains a 1.2 over-length HBV genome spanning nucleotides 1400–3182/1–1987 flanked by inverted terminal repeats (ITR) of an adeno-associated virus (AAV) [6]. The plasmid encodes for HBV genotype A, serotype adw. Whilst some plasmids used for HDI require strong promoters such as CMV, endogenous HBV promoters regulate the expression of viral open reading frames in pAAV-HBV1.2. Most importantly for this model, the AAV-ITR sequences flanking the HBV genome augment persistent virus production. This is likely due to the coexistence of different forms of the HBV genome within hepatocytes: episome formation, integration of HBV-DNA into the hepatocyte genome as well as existence of extrachromosomal concatemers, consistent with the known characteristics of AAV-ITR sequences [11].

1. Plasmid-DNA (pAAV-HBV1.2).

2. Chemically competent *Escherichia coli*.

3. 1.5 ml eppendorf tube.

4. Wet ice.

5. Water bath at 42 °C.

6. Super optimal broth (SOB) medium: For 1 L measure approx. 900 ml deionized H_2O and add 20 g tryptone, 5 g yeast extract, 2 ml of 5 M NaCl, 2.5 ml of 1 M KCl, 10 ml of 1 M $MgCl_2$, 10 ml of 1 M $MgSO_4$. Adjust to 1 L with deionized H_2O and sterilize by autoclaving. Store at room temperature or 4 °C.

7. Lysogeny broth (LB) medium with added carbenicillin (50 µg/ml) for selection: For 1 L dissolve 10 g tryptone, 5 g yeast extract, and 10 g NaCl in approx. 900 ml deionized H_2O, adjust pH to 7.0 using 1 N NaOH and adjust volume to 1 L with deionized H_2O. Sterilize by autoclaving. Allow medium to cool down to 50 °C and add antibiotic for bacterial selection (50 µg/ml). Store at room temperature or 4 °C.

8. LB agar plates for bacterial growth with added carbenicillin (50 µg/ml) for selection. Prepare LB medium and add 15 g/L agar before autoclaving. Allow medium to cool down to 50 °C and add antibiotic for bacterial selection (50 µg/ml). Pour into petri dishes and allow to set at room temperature. Invert plates and store at 4 °C.

9. Incubator at 37 °C.

10. Endotoxin-free plasmid purification kit: Qiagen (Hilden, Germany) (*see* **Note 1**).

2.2 Hydrodynamic Injection

1. Mice.

2. Scales to weigh mice.

3. Hypodermic sterile single use 3 ml syringe Luer lock (Terumo).

4. Hypodermic sterile single use PrecisionGlide™ needle 25GX5/8 (BD).

5. Heat lamp (*see* **Note 2**).

6. Injection cone and plug.

3   Methods

3.1 Preparation of HBV Plasmid

1. Transformation: Thaw chemically competent bacteria quickly in your hand and put them back on ice. Transfer 100 ng of plasmid-DNA into an Eppendorf tube and put on ice. Add 50 µl of bacteria. Incubate for 5 min on ice. Induce heat-shock for 90 s in 42 °C water bath. Incubate for 2 min on ice. Add 500 µl of SOB-medium and incubate for 30 min at 37 °C. Streak out 100 µl on LB agar plate, invert plate and incubate at 37 °C overnight (*see* **Note 3**).

2. Bacterial culture: Pick a colony and transfer into 250 ml LB medium including antibiotic for selection. Shake culture at 180 rpm and 37 °C overnight.

3. Plasmid purification: Use endotoxin-free plasmid purification kit and follow manufacture's instructions. Solubilize plasmid-DNA in TE buffer at room temperature overnight and adjust concentration to 1 µg/µl. Prepare 50 µl and 100 µl aliquots of plasmid-DNA and store aliquots at –20 °C. Verify plasmid by restriction enzyme digestion and gel electrophoresis (*see* **Note 4**).

3.2 Hydrodynamic Injection

It is recommended that only researchers who routinely perform intravenous (iv) injections attempt this technique. The description of HDI below builds on practical knowledge that is only gained from iv experience. Additionally, it is recommended that injections be carried out under sterile conditions and according to animal ethics approval.

1. Weigh mice on scales to calculate the volume of solution that must be injected per mouse (8 % v/w).

2. Mice can now be grouped by the volume they are each to receive (*see* **Note 5**).

3. Prepare grouped solutions such that every mouse receives 10 µg plasmid-DNA in the appropriate volume of 1× PBS.

4. Warm up mice under heat lamp for approx. 15 min.

5. Prepare one syringe per mouse with the appropriate volume of DNA/PBS solution (*see* **Notes 6** and **7**).

6. Place mouse in injection cone with plug (*see* **Note 8**).

7. Inject appropriate DNA/PBS solution by intravenous injection (iv) through the tail vein within 3–5 s (*see* **Notes 9–11**).

8. Monitor mice for 10–15 min to ensure recovery (*see* **Note 12**).

4 Notes

1. Other endotoxin-free plasmid purification kits are suitable for plasmid preparation, but make sure purified DNA is endotoxin free (with < 0.1 EU/µg DNA) to minimize immune activation after hydrodynamic injection.

2. Note that the indicated time given in this protocol to heat mice up is dependent on the lamp type and height that the lamp is set at. Values given in this protocol are based on a Philips infrared industrial heat incandescent globe-BR125 IR 250 W E27 Red 1CT. The height of the globe from the bench surface is 48 cm and the distance from the globe to the mouse box lid is 35 cm.

3. To increase bacterial colony number, spin down transformed bacteria in 500 µl SOB medium at 150 g for 2 min, discard approx. 400 µl of SOB medium, resuspend bacterial pellet in remaining volume and streak out the full volume on agar plates.

4. To upscale plasmid quantity we inoculate multiple maxi-preparation cultures from a mini-preparation.

5. The volume calculated is rounded to one decimal place for grouping. For each tube that contains the DNA/PBS mix, it is helpful to calculate 3 ml of extra solution.

6. To fill the syringes, we attach the needle to the Luer lock syringe first and then draw up the DNA/PBS solution through the needle (use of the Luer lock syringes is essential as large pressure is generated during the injection procedure). Ensure there are no bubbles in the solution.

7. We generally allow approx. 1 min to fill each syringe. We take this into account when deciding when to put mice under the heat lamp.

8. To provide a stable setup, in which mouse movement is minimized, we use a plug to close the injection cone and restrict space for the mouse. This is made from a mini petri dish (diameter similar to that of cone opening) with tissues to form a cone that is kept in place with tape. Two pieces of tape are then attached to the plug, in which one piece is attached to the cone for the duration and the other is attached to the cone only when the mouse is within the cone (creating a kind of trap door).

9. To ensure comparable maximal plasmid-DNA transfer into hepatocytes, injections must be conducted without any interruption. If the needle migrates out of the vein lumen and continuous injection is hindered, the procedure should be abandoned.

10. After slightly inserting the needle, push the plunger to test whether solution can be injected. If that is the case, fully unload the volume with constant pressure. If the solution cannot be released, we suggest not forcing the solution through the vein, since this might result in limited transfer of material into hepatocytes. For ethical reasons we only attempt to inject a mouse two times per lateral vein. There is no preferential side for injection. We start with the first attempt for injection at the most distal part of the visible vein. If a second attempt is required, this should be done at a more proximal site. If two attempts for injection failed on one side, we switch to the other lateral vein. To avoid leakage of solution through punctures, we usually transfer the mouse back under the heat lamp for an additional 5 min to allow puncture clotting.

11. It is personal preference, but we find it very helpful to turn the lights in the safety cabinet off when performing the injections. The tail vein is then more visible.

12. Directly after the injection mice show reduced mobility for approx. 15 min whilst they adjust to the large amount of injected fluid that dilutes their circulating red cells.

Acknowledgments

We thank Pei-Jer Chen and Ding-Shin Chen for constructing the HBV encoding plasmid for hydrodynamic injection.

References

1. Bouard D, Alazard-Dany D, Cosset FL (2009) Viral vectors: from virology to transgene expression. Br J Pharmacol 157(2).153–165. doi:10.1038/bjp.2008.349

2. Hofman CR, Dileo JP, Li Z, Li S, Huang L (2001) Efficient in vivo gene transfer by PCR amplified fragment with reduced inflammatory activity. Gene Ther 8(1):71–74. doi:10.1038/sj.gt.3301373

3. Song E, Lee SK, Wang J, Ince N, Ouyang N, Min J, Chen J, Shankar P, Lieberman J (2003) RNA interference targeting Fas protects mice from fulminant hepatitis. Nat Med 9(3):347–351. doi:10.1038/nm828

4. Xue W, Chen S, Yin H, Tammela T, Papagiannakopoulos T, Joshi NS, Cai W, Yang G, Bronson R, Crowley DG, Zhang F, Anderson DG, Sharp PA, Jacks T (2014) CRISPR-mediated direct mutation of cancer genes in the mouse liver. Nature 514(7522):380–384. doi:10.1038/nature13589

5. Ebert G, Allison C, Preston S, Cooney J, Toe JG, Stutz MD, Ojaimi S, Baschuk N, Nachbur U, Torresi J, Silke J, Begley CG, Pellegrini M (2015) Eliminating hepatitis B by antagonizing cellular inhibitors of apoptosis. Proc Natl Acad Sci U S A 112(18):5803–5808. doi:10.1073/pnas.1502400112

6. Ebert G, Preston S, Allison C, Cooney J, Toe JG, Stutz MD, Ojaimi S, Scott HW, Baschuk N, Nachbur U, Torresi J, Chin R, Colledge D, Li X, Warner N, Revill P, Bowden S, Silke J, Begley CG, Pellegrini M (2015) Cellular inhibitor of apoptosis proteins prevent clearance of hepatitis B virus. Proc Natl Acad Sci U S A 112(18):5797–5802. doi:10.1073/pnas.1502390112

7. Huang LR, Wu HL, Chen PJ, Chen DS (2006) An immunocompetent mouse model for the tolerance of human chronic hepatitis B virus infection. Proc Natl Acad Sci U S A 103(47):17862–17867. doi:10.1073/pnas.0608578103

8. Zhang G, Gao X, Song YK, Vollmer R, Stolz DB, Gasiorowski JZ, Dean DA, Liu D (2004) Hydroporation as the mechanism of hydrodynamic delivery. Gene Ther 11(8):675–682. doi:10.1038/sj.gt.3302210

9. Liu F, Song Y, Liu D (1999) Hydrodynamics-based transfection in animals by systemic administration of plasmid DNA. Gene Ther 6(7):1258–1266. doi:10.1038/sj.gt.3300947

10. Ganem D, Prince AM (2004) Hepatitis B virus infection—natural history and clinical consequences. N Engl J Med 350(11):1118–1129. doi:10.1056/NEJMra031087

11. Nakai H, Yant SR, Storm TA, Fuess S, Meuse L, Kay MA (2001) Extrachromosomal recombinant adeno-associated virus vector genomes are primarily responsible for stable liver transduction in vivo. J Virol 75(15):6969–6976. doi:10.1128/JVI.75.15.6969-6976.2001

Chapter 10

Isolation of Cardiomyocytes and Cardiofibroblasts for Ex Vivo Analysis

George Williams Mbogo, Christina Nedeva, and Hamsa Puthalakath

Abstract

Heart failure (HF) is a common clinical endpoint to several underlying causes including aging, hypertension, stress, and cardiomyopathy. It is characterized by a significant decline in the cardiac output. Cardiomyocytes are terminally differentiated cells and therefore, apoptotic death due to beta adrenergic (β-AR) signaling contributes to high attrition rate of these cells. Past treatments of HF offer some survival benefit to patients (e.g., the beta blockers), but at the expense of blocking the compensatory beta-adrenergic signaling in surviving cells. One prerequisite for developing new therapeutics is to be able to grow cardiomyocytes ex vivo, and test their apoptotic response to drugs. Here we describe methods for isolation and culturing of neonatal and adult calcium tolerant cardiomyocytes. Similarly, cardiofibroblasts can also be isolated using the same protocol and subsequently, immortalized with SV40 T-Antigen for ex vivo studies.

Key words Cardiomyocytes, Heart failure, Ex vivo models, Cardiofibroblasts, Apoptosis

---

1 Introduction

Various risk factors such as aging, hypertension, valvular disease, myocardial infarction, cardiomyopathy, myocarditis, among other causes, predispose patients to heart failure (HF) [1]. The disease is characterized by a gradual decline in cardiac output, especially in response to oscillations in metabolic demands produced by other organs of the body. In an acute phase, these effects can be reversed by the existing compensatory mechanisms such as neuro-hormonal system and physiological hypertrophy [2]. However, chronic activity of these compensatory mechanisms can revert the heart homeostatic phenotype to that of a pathogenic state, which in turn contributes to worsened prognosis in heart failure. Overall perturbations in the functionality of the heart are associated with significantly high odds of morbidity and mortality, with damage to the myocardium (contractile portion of the heart) a major determinant of disease severity and progression.

Hamsa Puthalakath and Christine J. Hawkins (eds.), *Programmed Cell Death: Methods and Protocols*, Methods in Molecular Biology, vol. 1419, DOI 10.1007/978-1-4939-3581-9_10, © Springer Science+Business Media New York 2016

At the cellular and molecular level, apoptotic damage to the myocardium, being a tissue composed of a syncytium of contractile cardiomyocytes, is responsible for end stage HF. Apoptotic loss of terminally differentiated cardiomyocytes cannot be replenished and therefore leads to the loss of contractile function of the organ [3, 4]. Indeed, there are various lines of evidence corroborating the contribution of the cardio-damaging effects of the myocardium, as well as successive chronic cardiomyocyte apoptosis leading to progression of HF [5–7]. Current treatments are aimed at reversing these physiological perturbations to that of a normal homeostatic state [8]. Almost all the existing HF treatments block receptor activity for (Beta-blockers or β-blockers). This has been part of the standard care of HF for several decades, however, median 5-year survival has not improved beyond 50 %. Additionally, the undesirable side effects of this class of drugs warrant investigation into novel therapeutics to target and treat HF.

A thorough understanding of the molecular mechanisms underlying disease progression is paramount for the development of new therapeutics. Recently, Lee and colleagues [9] have deciphered the molecular mechanisms leading to the apoptosis of cardiomyocytes. This group provided unequivocal evidence implicating the role of pro-apoptotic BH3 only protein BIM-mediated apoptosis in several HF disease models. The phenotype of the model was associated with progression of HF, while *BIM* ablation offered protection from the disease. A clear understanding of the transcriptional regulation of *BIM* during beta-Adrenergic Receptor (β-AR) activation provides an opportunity for developing novel therapeutics through high throughput screening strategies. Validation of drug hits through these screens requires reliable and amenable tools especially ex vivo cell culture methods.

To study cardiac function in healthy and disease conditions, viable animal models such as the mouse and rat have been extensively used for the past 50 years [10]. The availability of such models has provided valuable information regarding the pathophysiology of heart failure, which has led to therapeutic development. Despite strength of evidence derived from whole organism studies, it is technically challenging to derive information on molecular events leading to the disease progression using in vivo models. Besides, the expense, large animal numbers and sophisticated technical requirements associated with animal experiments limit their use in translational science [11]. Therefore, ex vivo experiments in primary and/or cardiomyocyte cell lines can provide reliable and reproducible alternatives for in vivo experimental studies.

Primary cardiomyocyte isolation has been a fundamental procedure vital towards addressing questions in physiology and molecular biology concerning the heart. However, the process has endured technical difficulties, which in turn forced researchers toward the use of cell lines [12, 13]. The use of cell lines has its own drawbacks as there are a limited number of established cardiomyocyte cell lines, including atrial tumor derived HL-1 [14] and embryonic rat ventricular myocyte derived H9C2 [15]. Such lines have been used in studying various cardiac functions such as atrial fibrillation [16], heart metabolism [13] and hypertrophy [12]. Of note, the use of H9C2 cells as an ex vivo model of cardiac studies has some limitations as they are a proliferating cell line, compared to the non-proliferating primary cardiomyocytes [12]. Therefore, their representativeness to an in vivo condition may confound results. There is increasing demand to harness, adapt, and standardize methods for isolation and establishment of primary cardiomyocytes ex vivo for both short and long term experimentation. In doing so, this will provide reliable and reproducible data with a high degree of homogeneity which translates to an in vivo animal environment during HF modeling. Finally, isolating primary cardiomyocytes/fibroblasts from mouse strains with specific gene(s) ablation will greatly help to understand the role of those genes in cellular processes.

In this chapter, we describe the current methods for isolation and establishment of adherent, rod-shaped cardiomyocytes at two developmental stages. Particularly, the methods for isolation of neonatal and adult calcium tolerant cardiomyocytes from mice (a main model for most genetic manipulation studies in the lab) will be emphasized here. The rationale for choice of cell type, i.e., neonatal or adult cardiomyocytes depends mainly on the experiment to be performed. The advantages of harboring neonatal cardiomyocytes include: simpler isolation procedure and culturing methods and they are the system of choice for studying myofibrillogenesis and myofibrillar functions. On the other hand, adult cardiomyocytes are widely accepted as a good model for cardiac cellular physiology and pathophysiology, as well as for pharmaceutical intervention. Genetically modified mice preclude the need for complicated cardiomyocyte infection processes to generate the desired genotype, which are inefficient due to cardiomyocytes' terminal differentiation. Furthermore, these cells are prone to calcium transients that impair their survival in ex vivo settings; therefore, calcium tolerant cardiomyocytes isolation methods have to be well optimized. The protocol described here is adapted in part from Jacobson and Piper [17] and O'Connell et al. [18] with minor modifications for calcium tolerant adult cardiomyocytes. Specifically, isolation of neonatal cardiomyocytes, as well as cardiofibroblasts for a wide range of ex vivo studies, including cell signaling networks and apoptosis regulation.

2   Materials

2.1   Material for Isolation of Neonatal Mouse Cardiomyocytes

1. Sterile Hanks' Balanced Salt solution with calcium (Hanks' medium).

2. Pre-sterilized scalpel for dissection.

3. Sterile trypsin, aliquots stored at 4 °C.

4. Dulbecco's Modified Eagle's Medium (complete DMEM) 1 g/L D-glucose, L-glutamine, 110 mg/L sodium pyruvate; supplemented with 10 % fetal calf serum (FCS).

5. Hanks/Collagenase II: 2.5 g collagenase II in 30 ml Hanks' medium.

6. Incubators at 37 °C at either 2 or 10 % CO_2.

7. 100 μM 5-bromo-2′-deoxyuridine (BRDU) (10 mM stocks stored –20 °C).

8. 0.1 μM Vitamin C (100 μM stocks stored at –20 °C).

9. 96-well flat-bottom plates.

10. 50 ml conical tubes.

11. 10 ml conical tubes.

12. All pipette types ranging 2–1000 μl and sterile tips.

2.2   Materials for Immortalization of Neonatal Mouse Cardiofibroblasts

1. Lentiviral constructs expressing pSV40 large T antigen, pCMVδ8.2 and pCAG (immortalization, packaging and mouse lentiviral receptor constructs, respectively).

2. 5 mg/ml Polybrene (1000×).

3. 0.1 % gelatin coated 6-well tissue culture plates (coating of gelatin to be performed by experimenter prior to use of 6-well plates) (*see* **Note 1**).

4. Fugene 6 Transfection Reagent.

5. 3 μg/ml puromycin (3 mg/ml stocks stored at 4 °C).

6. Freeze mix: FCS with 10 % dimethyl sulfoxide (DMSO).

7. Serum free DMEM (SFM).

2.3   Material for Isolation and Culture of Primary Adult Mouse Cardiomyocytes

1. 10× perfusion buffer: 70.3 g of NaCl, 11 g of KCl, 0.82 g of KH_2PO_4, 0.85 g of Na_2HPO_4 3 g of $MgSO_4$-$7H_2O$, 100 ml of 1 M Na-HEPES per liter. To make fresh single use 1× perfusion buffer, dilute 10× buffer in water and add 0.39 g of $NaHCO_3$, 3.75 g of taurine, 1 g of 2,3-butanedione monoxime (BDM), 1 g of glucose per liter (*see* **Note 2**).

2. Digestion buffer: 1× perfusion buffer supplemented with 2.4 mg/ml Collagenase II.

3. Myocyte stopping buffer: 1× perfusion buffer, supplemented with 10 % FCS, 12.5 μM $CaCl_2$ (*see* **Note 2**).

4. Myocyte culture medium (1× final concentration): To make 50 ml, add 47 ml of Minimum Essential Medium Eagle with Hanks' Balanced Salt solution, 0.5 ml Bovine Serum Albumin (1 mg/ml), 0.5 ml Penicillin (100 IU/ml), 1 ml BDM (10 mM), 0.5 ml of ITS cocktail comprising of: 5 μg/ml Insulin, 5 μg/ml Transferrin and 5 ng/ml Selenium/ITS (1 g/L D-glucose, L-Glutamine, 110 mg/L Sodium pyruvate, supplemented with 10 % FCS) (*see* **Note 2**).

5. 5 % isoflurane anesthetic.

6. 100 IU/ml heparin in PBS.

7. 27 gauge needles.

8. 0.5 ml insulin needles.

9. Fine tip forceps.

10. Small clamp.

11. 5-0 suture nylon thread.

12. Bright field light dissection microscope (minimum magnification 4×).

13. 0.22 μm micrometer filters.

14. 100 μm strainer.

15. Sterile 1× phosphate buffered saline (10× PBS stocks).

16. 96-well tissue culture plates.

17. The peristaltic pump with associated tubing immersed in 37 °C water bath.

18. Hemocytometer.

19. 100 ml Schott bottle.

20. 2 mM ATP (200 mM stock).

21. Refrigerated centrifuge 4 °C.

3   Methods

3.1 Method for Isolation of Neonatal Mouse Cardiomyocytes and Cardiofibroblasts

1. Decapitate 0–2 days old neonatal mice under 10× magnification on a dissecting microscope.

2. Remove the heart and place it on a 10 cm dish containing 15–20 ml of Hanks' medium (*see* Fig. 1 and **Note 3**).

3. Remove surrounding tissues using a scalpel, this includes the atria (*see* Fig. 1).

4. Transfer the heart onto a 10 cm dish containing 20 ml fresh Hanks' medium.

5. Make minor cuts on the ventricles to increase the surface area for enzymatic digestion (*see* Fig. 1).

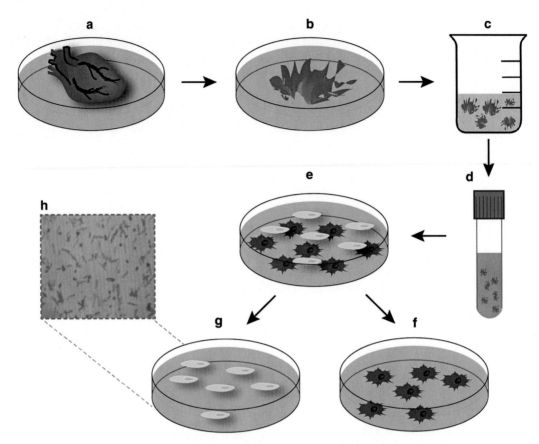

Fig. 1 Schematic for isolation of neonatal cardiomyocytes and cardiofibroblasts. (**a**, **b**) Remove the hearts and place them on a 10 cm dish containing 15–20 ml of Hanks' medium twice. (**c**) Slice the ventricles with a scalpel to increase the surface area for enzymatic digestion. (**d**) Transfer the heart tissues to a digestion bottle (*blue cup bottles*) containing 30 ml of Hanks' solution with 2.5 % trypsin and shake overnight at 4 °C. (**e**) Cellular homogenate treated with equal amounts of DMEM supplemented with 10 % FCS. (**f**) Cardiofibroblasts, and (**g**) cardiomyocytes fractions. (**h**) Cardiomyocytes appearance under ×4 magnification with light microscope

6. Transfer the heart tissue for digestion into 50 ml conical tubes containing 30 ml of Hanks' medium with 2.5 % trypsin (*see* **Note 3**).

7. Gently shake on an orbital shaker at 4 °C (usually performed in the cold room) for overnight tissue digestion.

8. Add equal amounts of DMEM supplemented with 10 % FCS to stop digestion.

9. Incubate in a water bath at 37 °C with agitation for 10 min.

10. Discard the supernatant after incubation, and add 7.5 ml of Hanks' medium/Collagenase II.

11. Incubate at 37 °C in a water bath with agitation for further 10 min.

12. Harvest the supernatant into a 50 ml conical tube containing 5 ml pre-warmed complete DMEM.

13. Incubate the tube at 37 °C and 10 % CO_2.

14. Repeat the collagenase II digestion in **step 10**, four times, and each time harvest the supernatants in individual tubes as described in **step 12** above.

15. Spin all four tubes at $151 \times g$ for 5 min at 4 °C.

16. Discard the supernatants and resuspend the cell pellet in 2 ml of complete DMEM per tube. At this stage the solution contains both cardiomyocytes and cardiofibroblasts.

17. Transfer the cell suspension from the four tubes into a 10 cm culture dish and incubate it at 37 °C and 2 % CO_2 for 60 min.

18. Transfer the cell suspension containing both cardiofibroblasts and cardiomyocytes into a fresh 10 cm dish. At this point, all the fibroblasts would have adhered to plastic.

19. Add 100 μM BRDU and 0.1 μM vitamin C to the culture and incubate again at 37 °C and 2 % CO_2.

20. To the 10 cm dish with adhered cardiofibroblasts from **step 17**, add fresh 10 ml complete DMEM and incubate at 37 °C and 10 % CO_2

21. Maintain these cells in culture for approximately 72 h before proceeding with further experimentation.

3.2 Immortalization of Neonatal Mouse Cardiofibroblasts With SV40 T-Antigen

1. To generate lentiviral particles; seed 1.5×10^6 HEK 293 T cells per 10 cm plate and incubate them for 24 h at 37 °C in 10 % CO_2.

2. Transiently co-transfect HEK 293 T cells from **step 1** above with pCMVδR.2 (the packaging plasmid), pCAG4 (the mouse lentiviral receptor) and pSV40 T antigen (the immortalization construct) at a ratio of 5: 2: 3 respectively using Fugene 6 (ratio of 1: 3 DNA to Fugene 6 respectively) (*see* **Note 4**).

3. Harvest the lentiviral supernatant from HEK 293 T cells 48 h post-transfection and store it at −80 °C or use immediately.

4. Seed isolated primary cardiofibroblasts at density of 1×10^5 cells/well in a 0.1 % gelatin coated 6-well plate and incubate at 37 °C and 10 % CO_2 for 24 h.

5. Infect adhered cardiofibroblasts from **step 4**, with 3 ml of filter-sterilized lentiviral supernatant supplemented with 1× polybrene.

6. Spin the cells to be infected cells at 37 °C at $808 \times g$ for 45 min.

7. Incubate the infected cells for 48 h in conditions as described above in 1.

8. To the infected cells from **step 7** above, add fresh DMEM containing 3 μg/ml of puromycin to select for stably transfected cells and grow cells to confluency for approximately 1–4 days.

9. To maintain immortalized cells, treat cells with 2 ml of trypsin and expand plate to a gelatin free 10 cm plate along with fresh DMEM and incubate at 37 °C and 10 % CO_2.

10. Spin the remaining cell suspension from **step 9** at $151 \times g$ at 4 °C and add freezing mix to the pellet and store them at –80 °C for approximately 3 months and long term in liquid nitrogen (keep cells on ice for ~30 min before storage at –80 °C).

3.3 Isolation and Culture of Primary Adult Mouse Cardiomyocytes

3.3.1 Harvesting of the Mouse Heart

1. Anesthetize the mouse with 5 % isoflurane inhalation and confirm anesthesia onset using pinch reflexes.

2. Transfer the anesthetized mouse to a surgery area and fix it on a dissecting board (*see* Fig. 2).

3. Inject it with 0.5 ml heparin (100 IU/ml PBS) intraperitoneally, and allow the heparin to circulate for 2–3 min.

4. Wipe the chest with 70 % ethanol while checking pinch reflexes (*see* **Note 3**).

5. Make a small incision at the level of the pubis for entrance of the scissors and proceed with a median longitudinal cut superiorly to the chin and separate the skin carefully from the underlying musculature (*see* Fig. 2b).

6. Open the thoracic cavity by cutting through the diaphragm and the costal cartilages at their point of union on both sides until reaching the articulation of the sternum with the ribs. Raise the inferior part of the sternum with the clamp to expose the heart (*see* Fig. 2c).

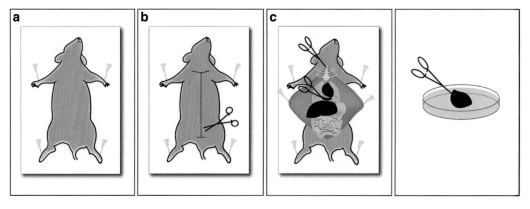

Fig. 2 Schematic of surgical procedures of mouse heart harvesting. (**a**) Pin mouse on board (**b**) Open the mouse starting at the pubis to the chin. (**c**) Open the thoracic cavity by cutting through the diaphragm and the costal cartilages at their point of union. (**d**) Harvested heart on 10 cm petri dish ready for perfusion

7. Once the heart is exposed, lift it gently using fine curved head forceps. Identify the pulmonary vessels and the aorta, cut the transverse aorta between the carotid arteries and immediately place the heart in a 10 cm dish containing 10 ml of perfusion buffer at room temperature (*see* Fig. 2d).

8. Remove the extraneous tissues (thymus and lungs) and transfer the heart into a second 10 cm dish with perfusion buffer at room temperature.

3.3.2 Heart Cannulation, Perfusion and Enzymatic Digestion

1. Place the 10 cm dish with the heart under a dissecting microscope (*see* **Note 3**).

2. Slide the aorta onto a 27 gauge needle cannula with the help of fine tip forceps, insert the tip of the cannula just above the aortic valve as fast as possible (less than 60 s) (*see* Fig. 3).

3. Attach a small clamp to the end of the aorta on the cannula to keep the heart in place and tighten the junction with 5-0 suture nylon thread tied to the cannula (*see* Fig. 3).

4. Set up the perfusion apparatus assembling the perfusion system with two pre-warmed buffer reservoirs at 37 °C in a water bath; one with the perfusion buffer and the other with the digestion buffer in 50 ml conical tubes (*see* Fig. 3 and **Note 2**).

5. Keep all the tubings at 37 °C in water bath before pumping perfusion fluids through the peristaltic pump (*see* Fig. 3 and **Note 5**).

6. Connect the outlet of the pump to a cannula open end directly above the aorta.

7. Turn on the peristaltic pump and perfuse the heart with 1× perfusion buffer for 4 min at the rate of 4 ml/min to flush blood away from the vasculature, and to remove extracellular calcium so as to stop contractions (*see* Fig. 4, and **Note 2**).

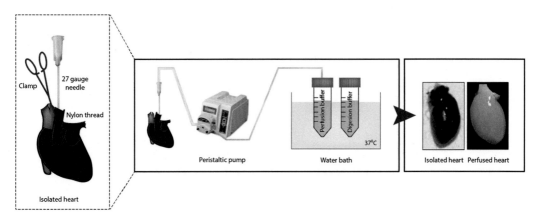

Fig. 3 Schematic for cannulation and perfusion setup. The heart is cannulated with a syringe a 27 gauge needle through the aorta and the well perfused becomes swollen and turns slightly pale, and flaccid

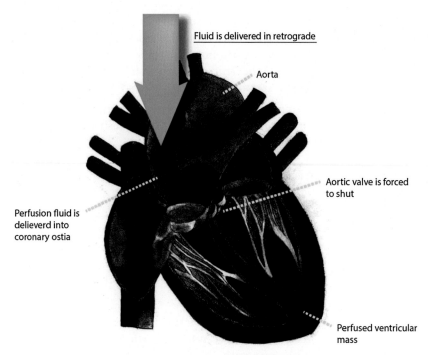

Fluid is delivered in retrograde

Aorta

Aortic valve is forced
to shut

Perfusion fluid is
delieverd into
coronary ostia

Perfused ventricular
mass

Fig. 4 Schematic for heart perfusion. Retrograde introduction of perfusion and digestion buffers to shut the aortic valve and to allow complete ventricular perfusion

8. Switch the perfusion buffer with myocyte digestion buffer and perfuse the heart for another for 3 min at the same flow rate as in **step 7** above (*see* **Notes 2, 5** and **6**).

9. Add 15 μl of 100 mM CaCl₂ to the myocyte digestion buffer reservoir and continue with the perfusion for another 8 min at 4 ml/min.

10. Perfuse the heart until it appeared swollen with slight color change from red to pale (*see* Fig. 3).

11. Stop digestion if the heart feels spongy upon pinching.

12. *Critical step*: Turn off the peristaltic pump during the fluid swap to avoid bubble

13. Cut the perfused heart just below the atria to release it from the cannula.

14. Place the ventricles in a 10 cm sterile dish containing 5 ml CaCl₂ and topped up with myocyte digestion buffer.

15. To isolate cardiomyocytes from the ventricles, place them in a 10 cm dish containing 5 ml CaCl₂ topped up with digestion buffer and cut them into small pieces.

16. Transfer the cut tissues together with the enzyme buffer to a sterile digestion bottle (100 ml Schott bottle containing a 1 cm magnetic stir bar) and stir using a magnetic stirrer gently

for 15 min, while keeping the bottle immersed in a 37 °C water bath to ensure further tissue digestion (*see* **Note 3**).

17. Pass the cell suspension through a 100 μm strainer to filter out tissue debris and retain single cell suspension of cardiomyocytes (*see* **Note 7**).

18. Transfer the suspension from **step 17** above to a 10 ml yellow cap tube, and aliquot 10 μl for cell counting in a hemocytometer.

19. Allow the remaining myocytes to sediment by gravity for 1–2 min at room temperature in the 10 ml tube.

20. Spin the cell suspension at $151 \times g$ for 5 min.

21. Discard the supernatant and resuspend the cell pellet in 10 ml myocyte stopping buffer and add 100 μl of 200 mM ATP to the tube, up to the final concentration of 2 mM (*see* **Note 3**).

22. Up-titrate the calcium concentration from 12.5 μM to 1.2 mM in a 3-step procedure at room temperature as described in **step 23**.

23. Label three 10 ml tubes and with increasing calcium concentrations as follows: Tube 1: 100 μM calcium, 10 μl of 100 mM $CaCl_2$ in 10 ml myocyte stopping buffer; Tube 2: 400 μM calcium, 40 μl of 100 mM $CaCl_2$ in 10 ml myocyte stopping buffer; Tube 3: 900 μM calcium, 90 μl of 100 mM $CaCl_2$ in 10 ml myocyte stopping buffer.

24. Spin the myocytes at $151 \times g$ for 5 min at 4 °C.

25. Collect the cardiofibroblasts contained in the supernatant and keep it on ice and resuspend the pellet containing cardiomyocytes in each tube gently using a 1.5 mm plastic pipette (*see* **Note 3**).

26. Allow the myocytes to stand for 2 min and spin as described in **step 24**.

27. Repeat **steps 24–26** for all the three tubes.

28. Resuspend the final cell pellet in 1 ml of myocyte culturing medium and seed the cells on a laminin coated (10 μg/ml in PBS) plate (*see* **Note 8**).

4   Notes

1. To make 0.1 % Gelatin, add 100 mg of gelatin powder to 100 mL of 1× PBS. Dissolve the gelatin with gentle heating, and filter the solution with a 0.2 μM filter. To coat plates, add sufficient volume to cover plate surface, and incubate at 37 °C for 1 h or overnight at 4 °C. Aspirate off the gelatin before use.

2. Adjust the pH of 1× perfusion buffer with sterile HCl. Equilibrate the perfusion buffer, the myocyte stopping buffer, the digestion buffer and the culture medium at 37 °C before starting heart perfusion. Filter-sterilize all the reagents with 0.22-µm filter before use.

3. Strict aseptic techniques must be adhered to throughout all processes and all centrifugation fractions must be kept on ice at all times.

4. For transient transfection, make a total DNA mix of 4 µg and 12 µl of Fugene 6 Transfection Reagent in a ratio of 1:3 respectively, in 100 µl of serum free media (SFM). Add Fugene to SFM but not vice versa for efficient transfection.

5. Wash the perfusion set up with 40 ml of 1 M HCl, followed by 50 ml of distilled water and 40 ml of 70 % ethanol, followed by 50 ml of distilled water at a flow rate of 4 ml/min before perfusion.

6. Introduce perfusion and digestion buffers in retrograde to force the aortic valve to shut and to allow the perfusion fluid to enter the coronary artery via the coronary ostia, thus resulting into complete ventricular perfusion (*see* Fig. 4). Turn off the peristaltic pump during the fluid swap to avoid bubble formation. The myocyte digestion buffer can be collected and reused the process is complete.

7. To enhance cardiomyocytes isolation from the ventricles, mix the digestion mixture by pipetting up and down with a plastic transfer pipette.

8. To make laminin at 10 µg/ml laminin, adjust concentration from 2 mg/ml stock frozen aliquot and make it up with 1× PBS for all the wells to be seeded. For coating plates (*see* **Note 1**).

Acknowledgements

This work was supported by the National Health and Medical Research Council Grant No. 1085281 to HP. GWM and CN are supported by La Trobe University Post graduate Research scholarships.

References

1. Kovacs A, Papp Z, Nagy L (2014) Causes and pathophysiology of heart failure with preserved ejection fraction. Heart Fail Clin 10(3):389–398

2. Butler J, Fonarow GC, Zile MR, Lam CS, Roessig L, Schelbert EB, Shah SJ, Ahmed A, Bonow RO, Cleland JG, Cody RJ, Chioncel O, Collins SP, Dunnmon P, Filippatos G, Lefkowitz MP, Marti CN, McMurray JJ, Misselwitz F, Nodari S, O'Connor C, Pfeffer MA, Pieske B, Pitt B, Rosano G, Sabbah HN, Senni M, Solomon SD, Stockbridge N, Teerlink JR, Georgiopoulou VV, Gheorghiade M (2014) Developing therapies for heart failure

with preserved ejection fraction: current state and future directions. JACC Heart Fail 2(2):97–112

3. Sliwa K, Mayosi BM (2013) Recent advances in the epidemiology, pathogenesis and prognosis of acute heart failure and cardiomyopathy in Africa. Heart 99(18):1317–1322

4. Shiojima I (2012) Chronic heart failure: progress in diagnosis and treatment. Topics: I. Progress in epidemiology and fundamental research; 2. Molecular mechanisms of chronic heart failure. Nihon Naika Gakkai Zasshi 101(2):314–321

5. Vatta M, Stetson SJ, Perez-Verdia A, Entman ML, Noon GP, Torre-Amione G, Bowles NE, Towbin JA (2002) Molecular remodelling of dystrophin in patients with end-stage cardiomyopathies and reversal in patients on assistance-device therapy. Lancet 359(9310):936–941

6. Abbate A, Sinagra G, Bussani R, Hoke NN, Merlo M, Varma A, Toldo S, Salloum FN, Biondi-Zoccai GG, Vetrovec GW, Crea F, Silvestri F, Baldi A (2009) Apoptosis in patients with acute myocarditis. Am J Cardiol 104(7):995–1000

7. Gurtl B, Kratky D, Guelly C, Zhang L, Gorkiewicz G, Das SK, Tamilarasan KP, Hoefler G (2009) Apoptosis and fibrosis are early features of heart failure in an animal model of metabolic cardiomyopathy. Int J Exp Pathol 90(3):338–346

8. Reed BN, Sueta CA (2015) A practical guide for the treatment of symptomatic heart failure with reduced ejection fraction (HFrEF). Curr Cardiol Rev 11(1):23–32

9. Lee YY, Moujalled D, Doerflinger M, Gangoda L, Weston R, Rahimi A, de Alboran I, Herold M, Bouillet P, Xu Q, Gao X, Du XJ, Puthalakath H (2013) CREB-binding protein (CBP) regulates beta-adrenoceptor (beta-AR)-mediated apoptosis. Cell Death Differ 20(7):941–952

10. Zaragoza C, Gomez-Guerrero C, Martin-Ventura JL, Blanco-Colio L, Lavin B, Mallavia B, Tarin C, Mas S, Ortiz A, Egido J (2011) Animal models of cardiovascular diseases. J Biomed Biotechnol 2011:497841

11. McGonigle P, Ruggeri B (2014) Animal models of human disease: challenges in enabling translation. Biochem Pharmacol 87(1):162–171

12. Watkins SJ, Borthwick GM, Arthur HM (2011) The H9C2 cell line and primary neonatal cardiomyocyte cells show similar hypertrophic responses in vitro. In Vitro Cell Dev Biol Anim 47(2):125–131

13. Zordoky BN, El-Kadi AO (2007) H9c2 cell line is a valuable in vitro model to study the drug metabolizing enzymes in the heart. J Pharmacol Toxicol Methods 56(3):317–322

14. Claycomb WC, Lanson NA Jr, Stallworth BS, Egeland DB, Delcarpio JB, Bahinski A, Izzo NJ Jr (1998) HL-1 cells: a cardiac muscle cell line that contracts and retains phenotypic characteristics of the adult cardiomyocyte. Proc Natl Acad Sci U S A 95(6):2979–2984

15. Kimes BW, Brandt BL (1976) Properties of a clonal muscle cell line from rat heart. Exp Cell Res 98(2):367–381

16. Rao F, Deng CY, Wu SL, Xiao DZ, Yu XY, Kuang SJ, Lin QX, Shan ZX (2009) Involvement of Src in L-type Ca2+ channel depression induced by macrophage migration inhibitory factor in atrial myocytes. J Mol Cell Cardiol 47(5):586–594

17. Jacobson SL, Piper HM (1986) Cell cultures of adult cardiomyocytes as models of the myocardium. J Mol Cell Cardiol 18(7):661–678

18. O'Connell TD, Rodrigo MC, Simpson PC (2007) Isolation and culture of adult mouse cardiac myocytes. Methods Mol Biol 357:271–296

Chapter 11

Detection of Cell Death in *Drosophila* Tissues

Deepika Vasudevan and Hyung Don Ryoo

Abstract

Drosophila has served as a particularly attractive model to study cell death due to the vast array of tools for genetic manipulation under defined spatial and temporal conditions in vivo as well as in cultured cells. These genetic methods have been well supplemented by enzymatic assays and a panel of antibodies recognizing cell death markers. This chapter discusses reporters, mutants, and assays used by various laboratories to study cell death in the context of development and in response to external insults.

Key words Drosophila cell death reporters, TUNEL, Acridine orange, Immunostaining, DEVD cleavage assay

1 Introduction

Drosophila is a highly popular experimental model for studying various aspects of biological problems as they can be easily genetically manipulated and have relatively short life cycles. The *Drosophila* community over the last hundred years has built a formidable library of mutants, transgenes, transgenic reporter lines, and genetic techniques that allow study of many processes and pathways. This includes programmed cell death (PCD), which makes possible the metamorphosis from larvae to adult flies, and also plays many other important roles in development.

Similar to other organisms, cell death pathways in *Drosophila* can be activated in response to DNA damage and excess stress imposed in various subcellular compartments by extrinsic factors. While the apoptotic cascade in *Drosophila* culminates in the activation of initiator and effector caspases, the upstream components vary from canonical apoptotic genes in mammals. There are seven known caspases: Dredd [1], Dronc [2], and Strica [3] are initiator caspases; Drice [4], DCP-1 [5], DECAY [6], and DAMM [7] are effector caspases. These caspases are synthesized as inactive zymogens, but gain activity after proteolytic processing. In *Drosophila*, a family of proteins called inhibitors of apoptosis proteins (IAPs)

Hamsa Puthalakath and Christine J. Hawkins (eds.), *Programmed Cell Death: Methods and Protocols*, Methods in Molecular Biology, vol. 1419, DOI 10.1007/978-1-4939-3581-9_11, © Springer Science+Business Media New York 2016

play particularly prominent roles in regulating caspase activity [8]. There are three known IAPs—DIAP1 [8], DIAP2 [9, 10], and BRUCE [11–13]. There is certain degree of tissue-specificity and mutual redundancy between the caspases and IAPs and this is extensively reviewed elsewhere [14–16]. When death-inducing signals are received by cells, IAPs are inactivated by the IAP-antagonists: *hid* [17], *reaper* [18], *grim* [19], and *sickle* [20–22] (Fig. 1). Inhibition of IAPs trigger apoptosis in most somatic cells of *Drosophila*, as caspases are constantly undergoing proteolytic activation in living cells, only to be inhibited by IAPs. Specifically, the *Drosophila* initiator caspase Dronc constitutively forms a complex with the adaptor protein Dark, even without cytochrome c released from the mitochondria [23]. In living cells, the small amount of activated caspases engage in negative feedback, with the help of IAPs. In cells doomed to die, inhibition of IAPs by IAP-antagonists leads to the stable activation Dronc and Dark [24]. This leads to the activation of effector caspases such as Drice, which subsequently orchestrate apoptosis by cleaving various nuclear and cytoplasmic proteins.

A dramatic case of programmed cell death (PCD) associated with normal development occurs during metamorphosis. In this case, cell death is largely under the control of ecdysone, a steroid hormone. Spikes in ecdysone levels are seen at the interface of the major development stages: embryo hatching, larval molting, pupariation, and pupation [25, 26]. The most morphological change and correspondingly the largest ecdysone spike occurs during pupation when entire larval organs, such as the gut and salivary gland, are eliminated by PCD to give way to adult tissues. It is thus not surprising that ectopic misexpression of ecdysone and other apoptotic pathway proteins has dire developmental consequences [27, 28].

This overview of cell death in *Drosophila*, while simplistic, provides the basis for methods detailed in this chapter. Subheading 2 of this chapter discusses genetic and genotoxic methods of

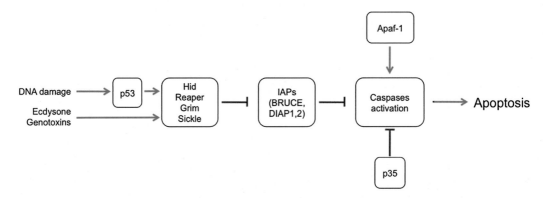

Fig. 1 A schematic showing various manipulatable elements of the cell death pathway in *Drosophila*: See text

inducing or blocking cell death. Subheading 3 discusses cell biological and biochemical assays to visualize and quantify cell death. Both sections discuss methods that can be used in larvae, adults and in cultured cells. Overall, this chapter attempts to illuminate what makes *Drosophila* a comprehensive model for studying cell death: the ability to finely regulate expression of genes with spatial and temporal control, and the variety of physiological contexts that can be simulated.

2   Materials

1. Fly stocks: Commonly used fly stocks and suggested sources are described in Table 1.

2. Media for culturing S2, S2R+, and SL2 cells: Schneider's Insect Cell Medium (Life Technologies), 10 % fetal bovine serum, penicillin (100 U/ml), streptomycin (100 U/ml).

3. Phosphate buffer saline (PBS): 137 mM NaCl, 2.7 mM KCl, 10 mM Na_2HPO_4, 1.8 mM KH_2PO_4, pH 7.4. 10× stock can be made and stored at room temperature.

Table 1
Commonly used fly lines for modulating and observing cell death

Fly Lines	Available sources
hid, reaper, grim deletion—Df(3L)H99	BDSC, Kyoto DGGR
dronc[RNAi]	VDRC, TRiP
drice[RNAi]	
dcp-1[RNAi]	
dredd[RNAi]	
diap1[RNAi]	TRiP
Gain of function	
uas-*hid*	BDSC
uas-*reaper*	BDSC, Kyoto DGGR
gmr-*hid*	
heat shock-*hid*	
Reporters	
rpr-*lacZ*	Laboratories of Hermann Steller, Andreas Bergmann and John Abrams
hid-lacZ	

4. Ringer's solution: 116 mM NaCl, 1.2 mM KCl, 1 mM CaCl$_2$, pH 7.4.

5. Fixative for tissue staining: 4 % paraformaldehyde, 1×PBS, made fresh (*see* **Note 1**).

6. Phosphate buffer Tween (PBT): 0.1 % Tween 20 or Triton X-100 (*see* **Note 2**), PBS.

7. Blocking buffer for immunostaining: 10 % donkey serum or 3 % BSA in PBT.

8. Acridine Orange (AO) stain: 1.25 µg/ml AO, 50 % heptane.

9. Lysis buffer for larval tissue: 50 mM Tris, 1 mM EDTA, 10 mM EGTA, 10 µM digitonin.

10. 2× reaction buffer for DEVD assay: 50 mM HEPES pH 7.4, 20 mM MgCl$_2$, 200 mM NaCl, 0.1 % NP40.

11. Fixative for cells: 10 % formaldehyde, PBS.

3 Methods

3.1 Tools for Manipulating Cell Death

This section aims to give an overview of methodologies used to either block or induce cell death. Genetic methods are useful when precise control is needed over tissue and cell type while genotoxic methods can be used to induce organism wide cell death. Chemical methods are mostly used in cell culture studies, often to corroborate results seen in vivo.

3.1.1 Genetic Tools for Blocking or Inducing Cell Death in Drosophila

The four pro-apoptotic genes, *hid*, *reaper*, *grim*, and *sickle* are clustered together in a genetic locus on the 3L chromosome [14, 29]. Various deletions of this locus have been employed to block cell death but the most commonly used strain is a third chromosome deficiency, Df(3L)H99, which deletes *hid*, *reaper*, and *grim*. Mild phenotypes have been observed when only one copy of the H99 locus remains [30–34] but homozygotic mutants abolish virtually all apoptosis [17–19, 35]. Conversely, induction of IAP-antagonists in response to DNA damage or stress (proteotoxicity, viral infection, etc.) is mediated by p53, overexpression of which can be used to induce cell death [36, 37].

Cell death can also be inhibited by deletion or RNAi knockdown of caspases such as Dronc, Drice, DCP-1, and Dredd [9], etc. On the other hand, cell death can be induced by deleting or RNAi-mediated knockdown of anti-apoptotic genes such as DIAP1 [9] using the GAL4-UAS system [38]. In this system, a promoter typically drives the expression of a transcription factor, GAL4, which in turn induces the expression of any gene downstream of a UAS element. This two-component system allows for tissue-specific control of transgene expression. A further level of temporal control can be added using a number of genetic tools. For example, one strategy employs the GAL80 transcription factor, which inhibits GAL4 activity. GAL80 temperature sensitive (GAL80^{TS}) mutants are widely used to grow animals at GAL80-permissive temperatures to the desired developmental stage before upshifting to a non-permissive temperature where GAL80 activity is inhibited and hence GAL4 activity is restored [39]. Another popular system for temporal control of gene expression is the Geneswitch system, where GAL4 is fused to a mifepristone-responsive promoter [40]. Thus GAL4 will be active only in the presence of mifepristone, which can be regulated via diet. Often GAL4-driven expression of a transgene in the entire tissue results in various side effects, for which experimental controls are difficult to design. To circumvent this issue, transgene expression can be induced in small sections of the tissue using clonal analysis. Mosaic analysis with a repressible cell marker (MARCM) utilizes site-specific recombinases under the control of inducible promoters in combination with the GAL4-GAL80 system to generate subpopulations of cells that express a given transgene [41]. Surrounding tissue that did not undergo recombination serves as a great internal control for the experiment.

In addition to the methods described above, there are several other genetic tricks that provide various degrees of control over GAL4 expression and these are discussed in detail elsewhere [42].

Transgenic lines that fuse pro-apoptotic genes such as *reaper* or *hid* directly to tissue specific promoters have also been used widely to induce apoptosis [17] and this has been the basis of many screens to identify new modulators of cell death. For example, overexpression of *hid* in the *Drosophila* eye using the GMR-promoter results in ablation of the eye (Fig. 2). This phenotype is readily visible and hence convenient to score. Our laboratory and others have utilized this system to identify many different components of the cell death machinery [9]. Table 1 lists transgenic fly lines which can be been to induce or block cell death.

In cultured S2, SL2, and S2R+ cells (Life Technologies, *see* Subheading 2, **item 2**), *hid*, *reaper*, or *grim* overexpression can induce apoptosis [43–45]. Overexpression constructs are typically under the control of an inducible promoter and are transiently transfected into cells using a lipid-based transfection reagent such as Effectene (Qiagen).

3.1.2 Genotoxic Methods to Induce Cell Death

The most common method of inducing DNA damage in flies is by ionizing irradiation. In adults, 40 Gy of gamma-ray radiation is an efficient way to induce massive apoptosis and for larvae, a lower dose (20 Gy) maybe sufficient. The animals are allowed to recover for 2–3 h prior to dissecting out desired tissues to observe for markers of apoptosis. UV irradiation at 254 nm and 40 mJ/cm^2 can also be used to induce apoptosis in embryos, adult and larval tissues [46]. Recovery time before dissection may vary from 4 to 10 h.

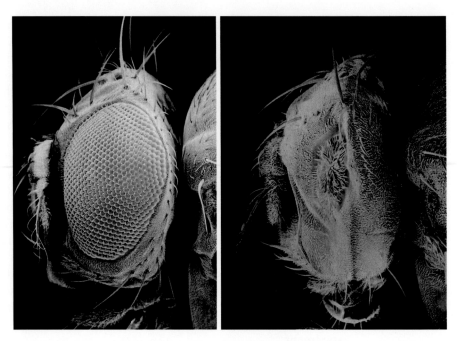

Fig. 2 High resolution image of *Drosophila* eye: Overexpression of the pro-apoptotic gene, *hid,* using the eye-specific promoter *gmr* results in eye ablation (*right*) when compared to wild type eyes (*left*)

In cultured cells cells, apoptosis can be induced by any of the following methods: 1 μM staurosporine or 10 μM ecdysone for 24 h, 200 nM actinomycin D for 3–8 h, UV-irradiation at 200 mJ/cm^2 followed by 4–12 h recovery [43, 45, 47]. When using irradiation, cells should be cultured in UV-transparent plates to allow for exposure.

3.2 Detecting Cell Death

Reflecting their roles as initiators of apoptosis in *Drosophila*, the expression pattern of *reaper*, *grim*, and *sickle* coincides with induction of apoptosis [19–21, 35]. Based on these observations, reporters have been made that can be used to monitor cell death in your favorite tissue. The first described *reaper-lacZ* reporter was generated by placing an 11-kb radiation *reaper* upstream element driving *lacZ* expression [48]. Subsequently other groups have whittled down this region to generate more fine-tuned reporters that respond to specific types of injury and stress [36, 49]. Subheading 3.2.3 describes how to examine *lacZ* expression using antibody staining. A similar reporter for *hid* has also been employed to monitor cell death in tissues [50]. Apoptotic pathways can be cell-type or tissue-type specific [51], so it maybe worth trying multiple methodologies when experimenting with a less explored tissue. Apoptotic cells can be visualized using the TUNEL labeling or by staining for specific apoptotic marker.

3.2.1 Collecting Samples of Defined Ages for Analysis

An important control to consider when designing experiments is to ensure that all animals being compared in the study, embryo, larvae, or adult, are of similar age and thus at similar developmental stages. Age is especially relevant when studying apoptosis in the developing embryo.

1. Eggs are collected in large cages on grape juice or apple juice plates smeared with yeast paste to encourage laying. Plates can be switched out in timed intervals so as to control the age of the embryo. For accurate staging, the first plate must be discarded since flies can carry embryos in their oviducts for 30 min before laying them.

2. Eggs have a protective proteinaceous covering called the chorion. Embryos must be dechorionated to allow for penetration of both staining agents and for effective lysis. Add 66 % bleach to the apple juice plates and incubate for 3 min.

3. Collect embryos in a mesh basket and thoroughly rinse with tap water for at least 3 min before transferring to a microfuge tube containing PBS (Subheading 2, **item 3**).

4. Dissect larvae and adults in PBS (Subheading 2, **item 3**) or Ringer's solution (Subheading 2, **item 4**). When dissecting several samples, store tissues in a dissection dish or microfuge tube containing PBS (Subheading 2, **item 3**).

5. For staining experiments, leave the carcass of the organism attached until just before mounting. This allows samples to sink to the bottom of tubes making washes and solution changes easier. For imaging apoptosis in S2 cells, it is best to culture them in poly-lysine coated multi-chamber slide (Nunc). The chamber allows for the entire staining process to be done on the slide and can be removed prior to mounting.

3.2.2 TUNEL Assay

TUNEL is an abbreviation for Terminal deoxynucleotidyl transferase-mediated dUTP nick-end labeling (TUNEL) and is based on the detection of fragmented DNA ends, which are characteristic of apoptotic cells. A deoxynucleotidyl transferase (TdT) is used to add modified nucleotides to double- and single-stranded DNA breaks. The modified nucleotides, such as digoxigenin-dUTP, can then be detected using specific antibodies. Several commercial kits provide complete panels of reagents with different dUTP. This protocol describes staining of tissues or cells using Apoptag (Millipore).

1. Incubate sample in fixative (Subheading 2, **item 5**) for 20 min (*see* **Note 1**).

2. Wash fixed samples in PBT (Subheading 2, **item 6**) for 5 min thrice. This step permeabilizes samples using the detergent, to allow for reagent penetration.

3. Equilibrate samples in the buffer provided at room temperature.

4. Incubate in appropriate volume (till tissue is submerged) of the TdT reaction mixture (2:1 mixture of reaction buffer and TdT enzyme) for 1 h at 37 °C.

5. Stop the reaction by the adding the stop buffer (diluted 34:1 with distilled water). Agitate for 15 s and incubate at room temperature for 10 min.

6. Wash three times with PBT, 5 min each.

7. Incubated with a fluor-conjugated anti-digoxigenin antibody according to manufacturer's instructions for 2 h at room temperature.

8. If desired, counterstain with DAPI (300 nM final concentration) for 5 min at room temperature toward the end of the previous incubation (*see* **Note 3**).

9. Wash three times with PBT, 10 min each.

10. Mount samples on glass slide in a suitable medium such as Vectashield (Life Technologies) or 50 % glycerol (sterile) and visualize using a fluorescent or confocal microscope (Fig. 3).

3.2.3 Immunostaining of Apoptotic Cells

While TUNEL assays for a specific aspect of apoptosis, i.e., DNA damage, there are other markers of apoptosis that can be stained for. Table 2 lists a subset of antibodies that are commercially available or generated by various laboratories to detect pro-apoptotic proteins. The choice of apoptotic marker is dictated by how apoptosis was induced and in which cell type. Unlike mammalian cells, cytochrome *c* release is not necessary for apoptosis in *Drosophila* so anti-cytochrome *c* antibodies are generally not used

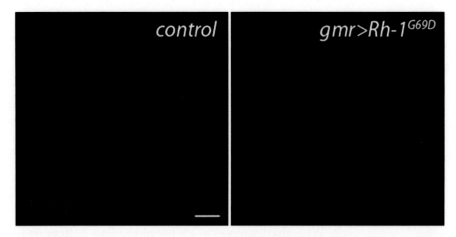

Fig. 3 TUNEL staining of third instar larval eye imaginal discs: Discs expressing mutant Rh-1^G69D that imposes stress in the endoplasmic reticulum show increased TUNEL staining (*right*) in comparison to the wild type control (*left*)

[52]. If using a *lacZ* reporter line, then an anti-βgal antibody can be used to visualize *rpr* or *hid* transcriptional activity. Staining for apoptotic markers can also be performed in combination with TUNEL (*see* **Note 4**).

1. Incubate sample in fixative (Subheading 2, **item 5**) for 20 min (*see* **Note 1**).

2. Wash fixed samples in PBT (Subheading 2, **item 6**) for 5 min thrice (*see* **Note 2**). This step permeabilizes samples using the detergent, to allow for reagent penetration.

3. Incubate samples in blocking buffer (Subheading 2, **item 7**) for an hour at room temperature. Remove blocking buffer and rinse with PBT three times. Remove as much PBT as possible before proceeding to next step.

4. Add primary antibody solution (Table 2) diluted appropriately in PBT and incubate either for 2 h at room temperature or overnight at 4 °C (*see* **Note 5**).

5. Wash samples three times with PBT, 5 min each.

6. Incubate with fluor-conjugated secondary antibody (Alexa Fluor, Life Technologies) diluted 1:1000 in PBT for 1 h at room temperature. If combining with TUNEL staining, include anti-digoxigenin antibody in the secondary antibody solution.

7. Counterstain and mount as described in TUNEL staining.

Table 2
Antibodies for cell death markers

Antibody	Source
cleaved-dcp-1	Cell Signaling Technology
cleaved caspase 3	Cell Signaling Technology
hid	Santa Cruz, Hermann Steller, Hyung Don Ryoo
DIAP1	AbCam, Bruce Hay, Hyung Don Ryoo
Dronc	Sharad Kumar
Drice	Sharad Kumar
ATP5a (mitochondrial marker)	AbCam
Cytochrome C	Zymed
anti-βgal	Sigma

3.2.4 Acridine Orange Staining

Acridine Orange (AO) is a cell-permeable organic compound that is retained selectively by dying cells because of their internal pH imbalance [53]. Since AO is used primarily in live tissues, options for counterstaining samples with other markers are limited. AO staining is popularly used for screening large numbers of embryos [35] but can also be used in other tissues such as imaginal discs [54]. While the staining protocol for AO is brief, so is the time available to observe the tissue after staining.

1. For embryos: Add sufficient amount of AO stain (Subheading 2, **item 8**) to dechorionated embryos and shake vigorously. Incubate for 5 min at room temperature. Mount the embryos from the interphase on to a glass slide and visualize in the green or red channel within 15 min.

2. For tissues: Incubate freshly dissected tissues (not fixed, *see* **Note 6**) in 0.5–10 µg/ml AO for 5 min at room temperature. The concentration of AO varies depending on tissue type and can be optimized. Mount tissues on glass slide and visualize as above.

3.3 Bioch mical Assays

3.3.1 Colorimetric Assay for Caspase Activity

A direct approach for quantifying cell death activity is by measuring *hid* and *reaper* levels by either western blotting (Table 2) or RT-PCR (Table 3). A broader method is measuring caspase activity. One method to assess caspase activity is by western blotting for cleaved caspase (anti-Dronc, -Drice) using antibodies described in Table 2. A more precise assessment of cell death can be obtained by measuring caspase activity using synthetic substrates. Several synthetic substrates have been developed and their catalysis can be measured colorimetrically or by fluorescence. DEVD-pNA is a peptide caspase substrate whose sequence is similar to the caspase cleavage site in PARP. Cells or tissues prepared as described in Subheading 3.2.1 can be used in the assay as follows.

1. Add appropriate amount of lysis buffer (Subheading 2, **item 9**) to samples and vortex intermittently for 10 min to enable complete lysis. Larger tissues or whole organisms can be homogenized in the lysis buffer using a pestle.

Table 3
Primers for amplification of pro-apoptotic genes [56]

Gene	Primer sequence
reaper	AGTCACAGTGGAGATTCCTGG
	TGCGATATTTGCCGGACTTTC
hid	ACGGCCATCCGAATCCGAAC
	TGCTGCTGCCGGAAGAAGAAGTT

2. Clear lysate by centrifuging at $10,000 \times g$ for 10 min at 4 °C. Lysates can be stored in –70 °C for several weeks with minimal loss in caspase activity.

3. Mix lysate with equal volume of 2× reaction buffer (Subheading 2, **item 10**). Add 10 μM (final) Z-DEVD-pNA or Ac-DEVD-pNA (SCBT) and incubate for 37 °C for 1 h.

4. Measure absorbance at 400 nM (or other wavelength depending on substrate). Samples maybe diluted to obtain readings in a linear range if necessary.

3.3.2 Viability Assays in S2 Cells

An alternative to staining for cell death in cultured cells is to assess viability by using an exclusion dye such as trypan blue or an inclusion dye such as crystal violet. It is important to start with equal number of cells per sample so as to be able to compare values directly. Crystal violet staining is more suitable for high throughput experiments since it can be read using a plate reader. Crystal violet stains all cells adherent to the plate so it relies on the assumption that apoptotic cells will be washed away. Cells cultured in 6-well dishes and transfected or treated with appropriate apoptotic inducers are assayed as follows.

1. Aspirate culture media, add 250 μl fixative (Subheading 2, **item 11**), and incubate for 15 min at room temperature.

2. Rinse three times with distilled water.

3. Incubate with 250 μl 0.5 % crystal violet solution for 30 min at room temperature on a shaker.

4. Rinsing with tap water ten times to remove excess dye. Leave plates inverted on a paper towel to drain water completely.

5. Elute with 500 μl of 10 % acetic acid for 10 min at room temperature on a shaker.

6. Diluted eluate in water 100-fold before measuring absorbance at 595 nm.

Trypan blue staining is more suitable for kinetic monitoring of samples. It is based on membrane impermeability of the blue dye thus negatively marking live cells. A suspension of S2 cells in PBS is mixed with trypan blue (Sigma) solution such that the final concentration of trypan blue is 0.04 %. After incubating for 5 min at room temperature, the number of alive versus dead cells can be counted using a hemocytometer or a cell counter (Bio-Rad). Other cell viability assays are reviewed in [55].

4   Notes

1. The concentration of paraformaldehyde (or formaldehyde) in the fixative can be adjusted depending on the samples being stained. Although 4 % paraformaldehyde is standard, there are antibody or TUNEL labeling protocols that use as low as 1 % of fixative.

2. The precise condition for antibody labeling may vary from tissues to tissues. For example, larval imaginal disc epithelium is an accessible tissue, and the PBT solution contains typically 0.2 % of either Tween-20 or Triton X-100. However, embryos and adult eyes and brains are less permeable to antibodies, and require a higher concentration of detergent (we suggest 0.3 %). If the quality of the antibody labeling is unsatisfactory, one might want to vary the concentration of detergent.

3. Samples can be counterstained with a cytoskeletal marker such as fluor-conjugated Phalloidin (Life Technologies) to observe cellular and nuclear morphology. Early apoptotic nuclei have a condensed appearance. Mitochondrial markers show shattered-glass-like patterns in apoptotic cells. Staining for pro-apoptotic proteins such as *hid* show their localization to the mitochondria and appear as a punctate pattern.

4. 4: This technique can also be performed on samples already incubated with TdT (but before anti-digoxigenin staining).

5. The primary antibody incubation step has scope for optimization in antibody dilution ratios, incubation times and temperatures. Typically, primary antibody solutions can be reused by storing at 4 °C in 0.01 % sodium azide for up to a month.

6. It is important to remember that acridine orange or trypan blue labeling should not be performed with fixed tissues. These assays work only when tissues contain live cells. By contrast, TUNEL or antibody labeling assays require prior fixation of tissues.

References

1. Chen P, Rodriguez A, Erskine R et al (1998) Dredd, a novel effector of the apoptosis activators reaper, grim, and hid in Drosophila. Dev Biol 201:202–216. doi:10.1006/dbio.1998.9000

2. Dorstyn L, Colussi PA, Quinn LM et al (1999) DRONC, an ecdysone-inducible Drosophila caspase. Proc Natl Acad Sci U S A 96:4307–4312. doi:10.1073/pnas.96.8.4307

3. Doumanis J, Quinn L, Richardson H, Kumar S (2001) STRICA, a novel Drosophila melanogaster caspase with an unusual serine/threonine-rich prodomain, interacts with DIAP1 and DIAP2. Cell Death Differ 8:387–394. doi:10.1038/sj.cdd.4400864

4. Fraser AG, Evan GI (1997) Identification of a Drosophila melanogaster ICE/CED-3-related protease, drICE. EMBO J 16:2805–2813. doi:10.1093/emboj/16.10.2805

5. Song Z, McCall K, Steller H (1997) DCP-1, a Drosophila cell death protease essential for development. Science 275:536–540. doi:10.1126/science.275.5299.536

6. Dorstyn L, Read SH, Quinn LM et al (1999) DECAY, a novel Drosophila caspase related to

mammalian caspase-3 and caspase-7. J Biol Chem 274:30778–30783. doi:10.1074/jbc.274.43.30778

7. Harvey NL, Daish T, Mills K et al (2001) Characterization of the Drosophila caspase, DAMM. J Biol Chem 276:25342–25350. doi:10.1074/jbc.M009444200

8. Hay BA, Wassarman DA, Rubin GM (1995) Drosophila homologs of baculovirus inhibitor of apoptosis proteins function to block cell death. Cell 83:1253–1262. doi:10.1016/0092-8674(95)90150-7

9. Leulier F, Ribeiro PS, Palmer E et al (2006) Systematic in vivo RNAi analysis of putative components of the Drosophila cell death machinery. Cell Death Differ 13:1663–1674. doi:10.1038/sj.cdd.4401868

10. Vucic D, Kaiser WJ, Harvey AJ, Miller LK (1997) Inhibition of reaper-induced apoptosis by interaction with inhibitor of apoptosis proteins (IAPs). Proc Natl Acad Sci U S A 94:10183–10188. doi:10.1073/pnas.94.19.10183

11. Vernooy SY, Copeland J, Ghaboosi N et al (2000) Cell death regulation in Drosophila: conservation of mechanism and unique insights. J Cell Biol 150:F69–F76. doi:10.1083/jcb.150.2.F69

12. Domingues C, Ryoo HD (2012) Drosophila BRUCE inhibits apoptosis through non-lysine ubiquitination of the IAP-antagonist REAPER. Cell Death Differ 19:470–477. doi:10.1038/cdd.2011.116

13. Arama E, Agapite J, Steller H (2003) Caspase activity and a specific cytochrome C are required for sperm differentiation in Drosophila. Dev Cell 4:687–697

14. Orme M, Meier P (2009) Inhibitor of apoptosis proteins in Drosophila: gatekeepers of death. Apoptosis 14:950–960. doi:10.1007/s10495-009-0358-2

15. Kumar S (2007) Caspase function in programmed cell death. Cell Death Differ 14:32–43. doi:10.1038/sj.cdd.4402060

16. Ryoo HD, Baehrecke EH (2010) Distinct death mechanisms in Drosophila development. Curr Opin Cell Biol 22:889–895. doi:10.1016/j.ceb.2010.08.022

17. Grether ME, Abrams JM, Agapite J et al (1995) The head involution defective gene of Drosophila melanogaster functions in programmed cell death. Genes Dev 9:1694–1708

18. White K, Tahaoglu E, Steller H (1996) Cell killing by the Drosophila gene reaper. Science 271:805–807

19. Chen P, Nordstrom W, Gish B, Abrams JM (1996) grim, a novel cell death gene in Drosophila. Genes Dev 10:1773–1782. doi:10.1101/gad.10.14.1773

20. Christich A, Kauppila S, Chen P et al (2002) The damage-responsive Drosophila gene sickle encodes a novel IAP binding protein similar to but distinct from reaper, grim, and hid. Curr Biol 12:137–140. doi:10.1016/S0960-9822(01)00658-3

21. Srinivasula SM, Datta P, Kobayashi M et al (2002) sickle, a novel Drosophila death gene in the reaper/hid/grim region, encodes an IAP-inhibitory protein. Curr Biol 12:125–130. doi:10.1016/S0960-9822(01)00657-1

22. Wing JP, Karres JS, Ogdahl JL et al (2002) Drosophila sickle is a novel grim-reaper cell death activator. Curr Biol 12:131–135

23. Yu X, Wang L, Acehan D et al (2006) Three-dimensional structure of a double apoptosome formed by the Drosophila Apaf-1 related killer. J Mol Biol 355:577–589. doi:10.1016/j.jmb.2005.10.040

24. Shapiro PJ, Hsu HH, Jung H et al (2008) Regulation of the Drosophila apoptosome through feedback inhibition. Nat Cell Biol 10:1440–1446. doi:10.1038/ncb1803

25. Handler A (1982) Ecdysteroid titers during pupal and adult development in Drosophila melanogaster. Dev Biol 93:73–82. doi:10.1016/0012-1606(82)90240-8

26. Warren JT, Yerushalmi Y, Shimell MJ et al (2006) Discrete pulses of molting hormone, 20-hydroxyecdysone, during late larval development of Drosophila melanogaster: correlations with changes in gene activity. Dev Dyn 235:315–326. doi:10.1002/dvdy.20626

27. Baehrecke EH (2000) Steroid regulation of programmed cell death during Drosophila development. Cell Death Differ 7:1057–1062. doi:10.1038/sj.cdd.4400753

28. Denton D, Aung-Htut MT, Kumar S (2013) Developmentally programmed cell death in Drosophila. Biochim Biophys Acta. doi:10.1016/j.bbamcr.2013.06.014

29. Abbott MK, Lengyel JA (1991) Embryonic head involution and rotation of male terminalia require the Drosophila locus head involution defective. Genetics 129:783–789

30. Miguel-Aliaga I, Thor S (2004) Segment-specific prevention of pioneer neuron apoptosis by cell-autonomous, postmitotic Hox gene activity. Development 131:6093–6105. doi:10.1242/dev.01521

31. Fichelson P, Gho M (2003) The glial cell undergoes apoptosis in the microchaete lineage of Drosophila. Development 130:123–133

32. Macías A, Romero NM, Martín F et al (2004) PVF1/PVR signaling and apoptosis promotes

the rotation and dorsal closure of the Drosophila male terminalia. Int J Dev Biol 48:1087–1094. doi:10.1387/ijdb.041859am

33. Wichmann A, Jaklevic B, Su TT (2006) Ionizing radiation induces caspase-dependent but Chk2- and p53-independent cell death in Drosophila melanogaster. Proc Natl Acad Sci U S A 103:9952–9957. doi:10.1073/pnas.0510528103

34. Choi YJ, Lee G, Park JH (2006) Programmed cell death mechanisms of identifiable peptidergic neurons in Drosophila melanogaster. Development 133:2223–2232. doi:10.1242/dev.02376

35. White K, Grether ME, Abrams JM et al (1994) Genetic control of programmed cell death in Drosophila. Science 264:677–683

36. Brodsky MH, Nordstrom W, Tsang G et al (2000) Drosophila p53 binds a damage response element at the reaper locus. Cell 101:103–113. doi:10.1016/S0092-8674(00)80627-3

37. Thomas SE, Malzer E, Ordóñez A et al (2013) p53 and translation attenuation regulate distinct cell cycle checkpoints during endoplasmic reticulum (ER) stress. J Biol Chem 288:7606–7617. doi:10.1074/jbc.M112.424655

38. Brand AH, Perrimon N (1993) Targeted gene expression as a means of altering cell fates and generating dominant phenotypes. Development 118:401–415

39. McGuire SE, Le PT, Osborn AJ et al (2003) Spatiotemporal rescue of memory dysfunction in Drosophila. Science 302:1765–1768. doi:10.1126/science.1089035

40. Osterwalder T, Yoon KS, White BH, Keshishian H (2001) A conditional tissue-specific transgene expression system using inducible GAL4. Proc Natl Acad Sci U S A 98:12596–12601. doi:10.1073/pnas.221303298

41. Lee T, Luo L (1999) Mosaic analysis with a repressible cell marker for studies of gene function in neuronal morphogenesis. Neuron 22:451–461

42. Del Valle RA, Didiano D, Desplan C (2012) Power tools for gene expression and clonal analysis in Drosophila. Nat Methods 9:47–55. doi:10.1038/nmeth.1800

43. Kanuka H, Hisahara S, Sawamoto K et al (1999) Proapoptotic activity of Caenorhabditis elegans CED-4 protein in Drosophila: implicated mechanisms for caspase activation. Proc Natl Acad Sci U S A 96:145–150

44. Juhász G, Sass M (2005) Hid can induce, but is not required for autophagy in polyploid larval Drosophila tissues. Eur J Cell Biol 84:491–502. doi:10.1016/j.ejcb.2004.11.010

45. Abdelwahid E, Yokokura T, Krieser RJ et al (2007) Mitochondrial disruption in Drosophila apoptosis. Dev Cell 12:793–806. doi:10.1016/j.devcel.2007.04.004

46. Zhou L, Steller H (2003) Distinct pathways mediate UV-induced apoptosis in Drosophila embryos. Dev Cell 4:599–605

47. Means JC, Muro I, Clem RJ (2006) Lack of involvement of mitochondrial factors in caspase activation in a Drosophila cell-free system. Cell Death Differ 13:1222–1234. doi:10.1038/sj.cdd.4401821

48. Nordstrom W, Chen P, Steller H, Abrams JM (1996) Activation of the reaper gene during ectopic cell killing in Drosophila. Dev Biol 180:213–226. doi:10.1006/dbio.1996.0296

49. Jiang C, Lamblin AF, Steller H, Thummel CS (2000) A steroid-triggered transcriptional hierarchy controls salivary gland cell death during Drosophila metamorphosis. Mol Cell 5:445–455

50. Fan Y, Lee TV, Xu D et al (2010) Dual roles of Drosophila p53 in cell death and cell differentiation. Cell Death Differ 17:912–921. doi:10.1038/cdd.2009.182

51. Steller H (2008) Regulation of apoptosis in Drosophila. Cell Death Differ 15:1132–1138. doi:10.1038/cdd.2008.50

52. Dorstyn L, Read S, Cakouros D et al (2002) The role of cytochrome c in caspase activation in Drosophila melanogaster cells. J Cell Biol 156:1089–1098. doi:10.1083/jcb.200111107

53. Robbins E, Marcus PI (1963) Dynamics of acridine orange-cell interaction. I. interrelationships of acridine orange particles and cytoplasmic reddening. J Cell Biol 18:237–250

54. Bonini NM, Leiserson WM, Benzer S (1993) The eyes absent gene: genetic control of cell survival and differentiation in the developing Drosophila eye. Cell 72:379–395

55. Riss TL, Moravec RA, Niles AL, et al. (2004) Cell viability assays. Assay Guidance Manual

56. Morishita J, Kang MJ, Fidelin K, Ryoo HD (2013) CDK7 regulates the mitochondrial localization of a Tail-anchored proapoptotic protein, hid. Cell Rep 5:1481–1488. doi:10.1016/j.celrep.2013.11.030

Chapter 12

Methods to Study Plant Programmed Cell Death

Joanna Kacprzyk, Adrian N. Dauphinee, Patrick Gallois, Arunika HLAN Gunawardena, and Paul F. McCabe

Abstract

Programmed cell death (PCD) is a critical component of plant development, defense against invading pathogens, and response to environmental stresses. In this chapter, we provide detailed technical methods for studying PCD associated with plant development or induced by abiotic stress. A root hair assay or electrolyte leakage assay are excellent techniques for the quantitative determination of PCD and/or cellular injury induced in response to abiotic stress, whereas the lace plant provides a unique model that facilitates the study of genetically regulated PCD during leaf development.

Key words Programmed cell death, Root hair assay, Electrolyte leakage, Developmental programmed cell death, Leaf morphogenesis, Abiotic stress

1 Introduction

Programmed cell death (PCD) is defined as the sequence of potentially interruptible events leading to controlled destruction of the cell [1, 2]. PCD is a fundamental process that can be activated throughout the entire plant life cycle. PCD occurs as an essential component of several highly specialized developmental programs, such as xylogenesis [3], leaf shape remodeling [4], suspensor elimination during embryogenesis [5], or during organ senescence [6, 7]. PCD is also a facet of plant responses to biotic and abiotic stresses and can be activated in response to fungal, bacterial, or viral infections. The rapid death of plant cells surrounding the site of avirulent pathogen infection, known as the hypersensitive response, is thought to restrict pathogen growth [8]. PCD also occurs in response to abiotic stresses such as cold, waterlogging, salinity, and hypoxia [1]. While high levels of cell death may result

Electronic supplementary material: The online version of this chapter (doi:10.1007/978-1-4939-3581-9_12) contains supplementary material, which is available to authorized users. Videos can also be accessed at http://link.springer.com/chapter/10.1007/978-1-4939-3581-9_12.

Hamsa Puthalakath and Christine J. Hawkins (eds.), *Programmed Cell Death: Methods and Protocols*, Methods in Molecular Biology, vol. 1419, DOI 10.1007/978-1-4939-3581-9_12, © Springer Science+Business Media New York 2016

in the death of whole organism it has been suggested that judicious activation of PCD during stress may be a part of an adaptation response that eliminates cells, tissues, or organs that could potentially render the plant more vulnerable to adverse conditions [9].

In this chapter, we present three methods to monitor and quantify PCD in plants. Firstly, we describe a system for studying a developmental PCD that sculpts the leaf morphology of the lace plant (*Aponogeton madagascariensis*). The lace plant is an aquatic monocot producing leaves with perforated lamina, formed by the genetically regulated PCD of redundant cells during leaf development [4]. Next we describe two techniques for quantitative determination of PCD induced by abiotic stress that can be used to determine death rates in practically all plant species. The root hair assay can be used for determination of PCD rates in root hairs, here used with perennial ryegrass (*Lolium perenne*) and PCD induced by salinity or cold, while electrolyte leakage is used to measure cellular damage in leaf discs, here used with Arabidopsis with PCD induced by oxidative stress.

2 Materials

2.1 Growth of Lace Plant and Live Cell Imaging of Developmental PCD

1. Murashige and Skoog (MS) half-strength media: dissolve 0.4 mg/l thiamine HCL, 100 mg/l myo-inositol, 30 g/l sucrose, and 2.15 g/l MS basal salt mixture (e.g., #M5524, Sigma-Aldrich, St Louis, Missouri, USA) in distilled water and adjust to pH 5.7 with 1 M NaOH prior to autoclaving. Half-strength solid MS is prepared by adding 1 % agar to the aforementioned mixture. Following autoclave-sterilization store excess materials at 4 °C and bring to room temperature before use.

2. Axenic cultures: establish according to Gunawardena et al. [12] and propagate in Magenta GA-7 Plant Culture boxes (bioWORLD, Dublin, OH, USA).

3. Filtered freshwater aquarium with gravel substrate and full spectrum LED light canopy (Fluval, Baie d'Urfé, QC, Canada).

4. Fertilizers: 1 mg/l monopotassium phosphate, 3 mg/l CSM + B plantex and 10 mg/l potassium nitrate (Aquarium Fertilizers, Napa, CA, USA).

5. Plastic and glass slides: Clear nonglare polystyrene (Plaskolite, Columbus, OH, USA) and glass slides of thicknesses of 1.25 and 1 mm, respectively, are amenable to imaging whole lace plant window stage leaves. Glass provides higher resolution than the polystyrene (Fig. 1), but the polystyrene is easily customizable with common tools and the greater depth allows for larger leaves to be observed in comparison.

6. Grooved microscope slides: Lace plant leaves feature a thin lamina (4–5 cell layers thick) with parallel longitudinal and

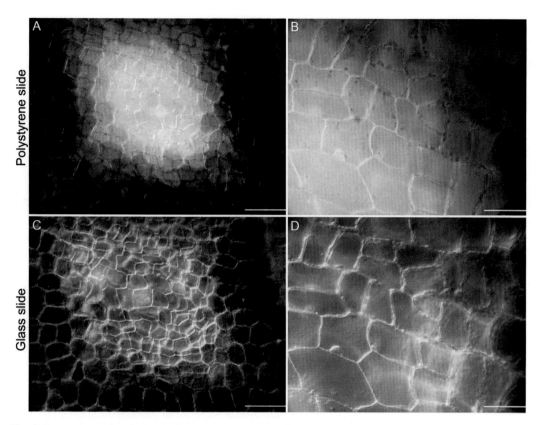

Fig. 1 Comparison of a single window stage lace plant leaf areole using custom polystyrene (**a**, **b**) and glass slides (**c**, **d**). The glass slide allows for greater resolution. All micrographs were captured using the same settings on a Nikon Eclipse 90i research microscope. Scale bars: **a**, **c** = 75 μm; **b**, **d** = 25 μm

transverse veins and a prominent central midrib. The midrib widens to its largest point at the petiole base and is smallest at the apical region of the leaf. In order to obtain adequate focal planes for long-term live cell imaging, grooves can be etched into either the polystyrene or glass material using an awl and glass etcher, respectively. The grooves should match the form of the central midrib and consequently it should become larger and deeper to one end (Fig. 2).

7. Valap sealant: Valap is a biologically inert sealant consisting of Vaseline petroleum jelly, lanolin, and paraffin wax is typically made in equivalent ratios (1:1:1 [w/w/w]) and has a melting point between 45 and 50 °C [10]. For this technique, the paraffin concentration is increased so that the final ratios are 1:1:2 for Vaseline, lanolin, and paraffin, respectively. The mixture is placed in a glass beaker, covered, and heated to 100 °C. The higher concentration of paraffin increases the hardness of the valap once it cools (thereby reducing risk of smudging objective lenses), while the higher temperature facilitates its application.

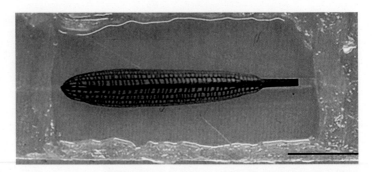

Fig. 2 Window stage lace plant leaf mounted on a polystyrene custom slide and sealed with valap. The large central midrib of the leaf is resting in a groove that allows the leaf blade to rest squarely on the slide surface. Scale bar = 1.25 cm

8. Compound light microscope, for example, the Nikon Eclipse 90i research microscope fitted with a DXM1200C digital camera (Nikon Canada Inc., Mississauga, ON, Canada).

9. Image and video acquisition software (e.g., Nikon NIS elements AR).

10. Video processing software such as Adobe Premiere Pro CC (Adobe Systems Inc., San Jose, CA, USA).

11. Other recommended materials: scissors, glass beakers with covers, glass Petri dishes distilled water (all of which are pre-sterilized and sealed prior to tissue culturing), heat source (hot plate), cotton swabs or fine paintbrush, forceps, transfer pipettes.

2.2 Growth of Ryegrass Seedlings and the Root Hair Assay for Quantification of Rates of PCD

1. Sterilizing solution: 20 % (v/v) commercial bleach (e.g., Domestos). The final concentration of sodium hypochlorite is approximately 1 % (v/v).

2. Sterile distilled water.

3. Forceps: sterilize in the autoclave.

4. Filter paper (90 mm ∅): sterilize in the autoclave.

5. Sterile plastic Petri dishes (90 mm ∅).

6. 15 ml Falcon tubes.

7. Parafilm.

8. 12-Well culture plates.

9. NaCl: Sterile 1 M stock solution. Dilute to required concentration in distilled water directly prior to use.

10. Ice.

11. Polystyrene box.

12. Access to 4 °C cold room.

13. Microscope slides and cover slips.

14. Surgical blades.

15. Fluorescein diacetate (FDA) 0.1 % w/v stock solution in acetone (store at –20 °C). Dilute to $100 \times$ (final FDA concentration 0.001 % w/v) in distilled water directly prior to use.

16. Phase contrast microscope with a FITC (fluorescein isothiocynate) filter and an attached fluorescence lamp.

17. Mechanical counters.

2.3 Growth of Arabidopsis Plants, PCD Induction, and Leaf Discs ion Leakage Measurements

1. Compact Conductivity meter, e.g., Horiba B-771 LAQUA twin from LAQUA Horiba Scientific (*see* **Note 1**).

2. Calibration buffer for the conductivity meter: 0.01 M KCl (calibration for 1.41 mS cm^{-1} of conductivity, Horiba).

3. Absorbing paper sticks: cut blue roll in 3 cm by 2 cm pieces, fold each longitudinally to fit into the conductivity meter chamber ($3 \text{ cm} \times 0.5 \text{ cm}$).

4. Cell culture multiwell plate (24-well), no need for coating.

5. A 3 mm metal cork-borer.

6. UVC induction using a DNA-cross-linker, e.g., strata linker 1800 (Stratagene LtD) or a CL-1000 UV cross-linker, (UVP LtD).

7. H_2O_2, 30 %, Sigma.

8. Sow Arabidopsis plants: sow seeds onto moist compost in plastic pots ($10 \text{ cm} \times 10 \text{ cm}$) and cover the pots with Clingfilm. To break seed dormancy, the plastic pots are placed in the cold (+4 °C) for 3 days and then transferred to short days, 8 h light, at 22 °C. After 1–2 weeks seedlings are separated and grown either individually or five to a $10 \text{ cm} \times 10 \text{ cm}$ pot or individually using Jiffy-7C coir pellet. The plants used for ion leakage test should have large leaves void of any bleaching or mottling.

3 Methods

3.1 Studying Developmental PCD in Lace Plant Leaves

The aquatic lace plant produces leaves with perforated lamina (Fig. 3a) [4]. These perforations form as superfluous cells that are removed by means of genetically regulated PCD during leaf development (Black arrow; Fig. 3a). There are five stages of leaf development relating to perforation formation: preperforation, window, perforation formation, expansion, and mature perforation [4]. Preperforation stage leaves are those that are tightly furled as they emerge from the corm and have anthocyanin pigmentation throughout the entire lamina. PCD actively occurs in window stage leaves, which exhibit a gradient of developmental PCD distinguishable by pigmentation [11]. Cells that do not undergo PCD during leaf formation are found in the region demarcated by

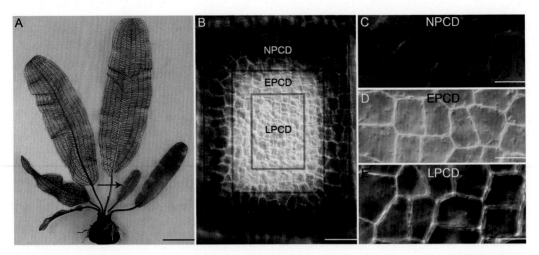

Fig. 3 Programmed cell death and the lace plant (*Aponogeton madagascariensis*). (**a**) The lace plant produces mature leaves with perforations throughout the lamina. These holes are formed via developmentally regulated programmed cell death (PCD), which actively occurs during the window stage of development (*black arrow*). (**b**) Window stage leaf areole showing a gradient of cellular death. Non-PCD (NPCD cells; **c**) persist beyond leaf morphogenesis. Cells in the early phases of PCD (EPCD; **d**) have lost anthocyanin pigmentation and those in the late phase of PCD (LPCD; **e**) are nearly transparent due to chloroplast degradation. Scale bars: **a** = 4 cm; **b** = 100 μm; **c**, **d** = 30 μm

the red anthocyanin-containing mesophyll cells (NPCD; Fig 3b, c). Adjacent to these are green cells in the early phases of PCD that have lost anthocyanin and are fated to die, but still retain abundant chlorophyll pigmentation (EPCD; Fig. 3b, d). Finally, the nearly transparent cells with little to no pigmentation remaining are nearest to death (LPCD; Fig. 3b, e). As cells are deleted a small, centralized perforation forms, which then expands as the wave of cell death continues before halting four to five cell layers from the veins by the time the leaf reaches maturity.

The accessibility and predictability of PCD, along with the plant's thin and nearly transparent leaves, that are ideal for live cell imaging, make it a tractable model system for studying developmental PCD [12–14]. The timing and intracellular dynamics of this unique form of PCD was described by Wertman et al. [14] using various stains and live cell imaging techniques including one that allowed for the continual observation of a whole leaf for several days. Long-term live cell imaging of lace plant window stage leaves (48–72 h) allows for the direct observation of the cytological features of lace plant PCD during perforation formation. The aim of this section is to provide a detailed description of that long-term live cell imaging protocol and to suggest alternative strategies that we suspect will be applicable within other systems.

1. Axenic lace plant cultures were originally provided by Dr. Michael Kane (University of Florida) and propagated according to Gunawardena et al. [12].

2. Using a UV laminar flow hood and aseptic techniques, cut away senescent leaves away using scissors and forceps.

3. In a sterile Petri dish, the mother-corm and any small cormels produced via clonal propagation are separated and dead tissues are removed using a scalpel blade and forceps. When not in use, utensils are stored in a beaker containing 95 % ethanol.

4. Plant newly cleaned corms in pre-autoclaved Magenta GA-7 vessels containing 50 ml of solid half-strength MS media.

5. Add 150 ml of the liquid media version to sealing with a PVC film (PhytoTechnology Laboratories).

6. Store the cultures are stored at 24 °C and expose them to 125 μmol m^{-2} s^{-1} fluorescent light on a 12 h light/dark cycle.

7. Alternatively, lace plant corms can be grown in filtered freshwater aquaria with a gravel substrate. The water is supplemented weekly with 1 mg/l monopotassium phosphate, 3 mg/l CSM + B plantex, and 10 mg/l potassium nitrate (Aquarium Fertilizers, Napa, California, USA). Full spectrum LED canopy lights are operated on a 12 h light/dark cycle (Fluval). Every 1–2 weeks (or as necessary), algae growths on the sides of an aquarium are removed using a brush after which 10 % of the water is changed for fresh distilled water.

8. Grow plants until they reach maturity and develop perforated leaves.

9. Remove the leaves of the window stage of development at the petiole base and then rinse thoroughly using distilled water to remove excess sucrose, debris, or possible contaminants. We recommend testing the stress limits of the tissue being used in this technique (*see* **Note 2**).

10. Carefully plate a window stage leaf into the groove of the custom slide so that the leaf blade lies flat on the slide surface.

11. Mount the specimen in distilled water, cover slipped.

12. Preheat valap to 100 °C in a covered beaker on a hot plate.

13. Apply the melted valap with a cotton swab to seal the slide.

14. After an initial coating, check for air bubbles and reapply valap where necessary to produce a complete seal.

15. Mount a window stage leaf from sterile culture or the aquarium on the custom slide and sealed as describe above.

16. Reference images are taken for comparison to the live image in order to maintain focus. Objectives with a higher numerical aperture with smaller depth of field require monitoring and fine adjustments in order to reduce alterations of focus from sources including, but not limited to: heat, vibrations, mechanical drift, as well as leaf growth and perforation formation as PCD progresses. If done manually, this requires a great deal of

labor and diligence, but technologies can be used to minimize workload (*see* **Note 3**). Tissue health must be monitored throughout the process as well (*see* **Note 2**).

17. In order to maintain tissue integrity throughout the time series, lace plant window stage leaves are rinsed with distilled water, remounted in the same orientation and sealed (as described before) every 6 h.

18. Valap can easily be removed from the custom slide. First remove the coverslip and bulk of the sealant using a razorblade then rinse the slide with hot water (>45 °C) and wipe off any remaining valap if necessary.

19. Lace plant PCD is a gradual process with initial features such as the loss of anthocyanin and a reduction of chlorophyll pigmentation occurring in early window stage leaves over 48 h prior to cell collapse (Supplementary Video 1). Although many cytological features gradually unfold over this 2 day period, the final dramatic events including nuclear displacement, tonoplast rupture, and vacuolar and plasma membrane collapse occur within the final 15–20 min of lace plant PCD [13] (Supplementary Video 2).

20. In order to capture these sudden changes, record continuous videos using Nikon NIS Elements AR software in 6 h segments which correspond to the tissue remounting steps (*see* **Note 4**).

21. Videos are then reassembled, cropped, and edited for playback speed and then compressed using Adobe Premiere Pro CC.

3.2 Quantitative Determination of Abiotic Stress Induced PCD in Ryegrass Root Hairs

The method described is adapted from Hogg et al. [15] and Kacprzyk and McCabe [16]. The root hair assay is a useful system for determination of the rates of apoptosis-like PCD (AL-PCD). AL-PCD in root hair cells is induced by moderate levels of abiotic stress and is characterized by a distinct morphology: condensation of cytoplasm and retraction of the protoplast away from the cell wall [15, 17]. In contrast, higher levels of abiotic stress induce root hairs to undergo necrotic cell death, readily distinguished from AL-PCD by the lack of protoplast retraction [15, 17]. Below we describe application of a root hair assay to score the effect of two agriculturally relevant abiotic stresses (cold and salinity) in the important pasture and forage species, perennial ryegrass (*Lolium perenne*). Previously, we described a detailed protocol for application of the root hair assay in model plants *Arabidopsis thaliana* and *Brachypodium distachyon* [15, 16]. Use of 12-well culture plates provide a convenient system for the treatment of ryegrass seedlings. Here, we present a protocol for growing ryegrass seedlings and exposing them to two cell death inducing treatments, each stress being highly relevant to agriculture: cold and salinity (NaCl). Cell death is visualized by staining with fluorescein diacetate

(FDA). FDA is a hydrophobic, cell-permeant compound that is cleaved by the cytoplasmic esterases in living cells, yielding a fluorescent product, fluorescein [18]. Viable cells stained with FDA therefore emit bright green fluorescence when illuminated with light of 490 nm wavelength.

1. Soak ryegrass seeds for 2 h in sterile distilled water and sterilize in the laminar flow hood using aseptic techniques (*see* **Note 5**).

2. Add 12 ml of sterilizing solution to the seeds in a 15 ml Falcon tube and incubate the seeds in the sterilizing solution for 5 min with mixing by inversion every minute.

3. Pour off the sterilizing solution and thoroughly washed the seeds five times with sterile distilled water.

4. Place single discs of filter paper in 9 cm ∅ Petri dishes and add 6 ml of sterile distilled water to each Petri dish to moisten the filter paper.

5. Place ryegrass seeds on the surface of filter paper using sterile forceps (approximately 20 seeds per Petri dish).

6. Seal the plates with Parafilm, and seeds are vernalized in the dark at 4 °C for 2 days.

7. Germinate the seeds at 22 °C, in the dark.

8. Use seedlings for experiments when 4 days old.

9. To provoke stress by salinity, fill wells of 12-well cultures plate with 1 ml of NaCl solution.

10. Transfer 4 day old ryegrass seedlings to individual wells (1 plant/well) using forceps.

11. Incubate the seedlings in the dark at 22 °C for 24 h.

12. Score for PCD rates, as outlined in **steps 17–20**. Typical rates of AL-PCD induced by 24 h treatment with 100 mM NaCl are approximately 50 % (Fig. 4a).

    Alternatively, cold stress can be applied. Instead of **steps 9–13**, perform **steps 13–16.**

13. Fill wells of a 12-well culture plate with 1 ml of sterile distilled water and place inside a polystyrene box half-filled with ice to prechill for 1 h.

14. Cover the well plates with more ice to fill the whole polystyrene box, then close the box and place it in a 4 °C cold room. According to our measurements, the water temperature inside the wells of 12-well plate in that system is maintained at 1 °C (±0.5 °C) for at least 24 h.

15. After the 12-well plates are prechilled, transfer the ryegrass seedlings to the individual wells using forceps (1 plant/well) and incubate the plates further at 1 °C (±0.5 °C) using the same system for 24 h.

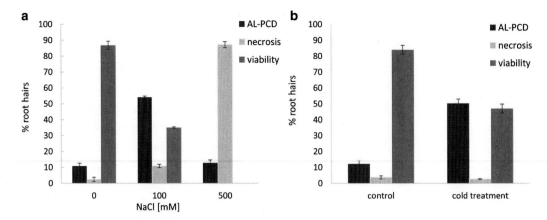

Fig. 4 Rates of AL-PCD, necrosis, and viability in ryegrass root hairs following salinity and cold treatments. Four day old seedlings of perennial ryegrass cultivar Arara were subjected to (**a**) salinity treatment (100 and 500 mM NaCl) and (**b**) cold treatment (1 °C ± 0.5 °C) over a period of 24 h. Root hairs showing green fluorescence after FDA staining were scored as alive. Root hairs showing no FDA staining and characterized by retraction and condensation of the cytoplasm were scored as AL-PCD. Root hairs showing neither FDA staining nor protoplast condensation were scored as necrotic. Means (*n* = 4) of AL-PCD, necrosis, and viability rates (±SEM) are presented

16. Score for PCD rates as outlined in **steps 17–20** (*see* **Note 6**). Typical rates of AL-PCD induced by 24 h treatment at 1 °C are approximately 50 % (Fig. 4b).

17. Dilute the FDA stock 100 times in distilled water directly prior to use.

18. For examination of root hairs' viability, cut the radicles from ryegrass seedlings with a blade and stain directly on the microscope slide with 100 μl of FDA solution.

19. Gently cover the roots with the cover slip and immediately examine under a phase contrast microscope with a FITC filter and attached fluorescence lamp.

20. The root hair assay is based on observation of dying root hairs morphology. Start examining root hairs from the root tip. The FDA stained root hairs exhibiting green fluorescence under fluorescent light are categorized as viable (Fig. 5a, b). The FDA negative root hairs are categorized as PCD if they exhibit condensation of the cytoplasm and protoplast retraction away from the cell wall (Fig. 5c, d) (*see* **Note 7**). Root hairs that are FDA negative and do not present AL-PCD morphology are scored as necrotic (Fig. 5e, f).

21. On average, score 100–150 root hairs per seedling record the results using mechanical counters.

3.3 Determination of Cellular Damage Using ion Leakage

Ion leakage is a convenient cell death assay using an inexpensive piece of equipment. The rationale is that as cells die in the treated tissue, they release ions into the external medium. Measuring the

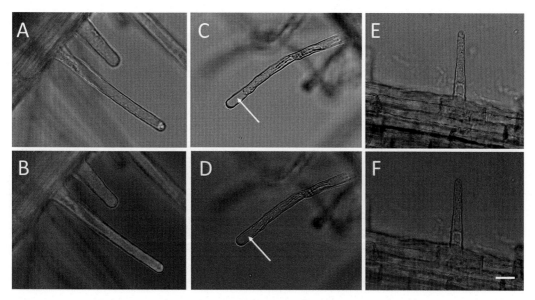

Fig. 5 Cell death morphology in root hairs of ryegrass. Morphology of a viable root hair viewed under white light (**a**). Viable root hair fluoresces green after FDA staining when viewed under fluorescent light (**b**). PCD root hair from a seedling treated with 100 mM NaCl for 24 h presents AL-PCD morphology, characterized by cytoplasmic condensation and retraction of the protoplast away from the cell wall (*arrow*) (**c**) and no fluorescence following the FDA treatment (**d**). Necrotic root hair from a seedling treated with 500 mM NaCl for 24 h presents no AL-PCD morphology (**e**) and no fluorescence after FDA staining (**f**). Scale bar = 5 μm

conductivity changes of the medium provides a quantitative measure of ion leakage from plant tissues and therefore of cell death. Depending of the experimental system used, conductivity measurements can be carried out on leaf discs [19], seedlings [20], or leaf discs punched out of a leaf at various time points after PCD induction. The protocol below is given for *Arabidopsis* leaf discs that receive a PCD-inducing treatment and are then monitored overtime. As an illustration, conditions for two PCD-inducing treatments are given. These treatments could be used as positive controls for other experimental systems.

1. Use Arabidopsis plants that are 3–4 weeks old and select leaves of similar physiological age so to minimize variability. Leaves from bolted plants have higher leakage than nonbolted plants.

2. The day before the experiment, water all the plants needed, because water status of the plant affects ion leakage results.

3. Add 1100 μl of distilled water to wells of a 24-well plate.

4. Cut suitable leaves from selected plants and place them abaxial side up, flat, on the inverted lid of a 9 cm plastic Petri dish.

5. Punch out leaf discs using a 3 mm metal borer. Avoid the midrib for the disc tissue to be homogeneous. Use forceps to carefully handle the discs. Cut discs can be left on a piece of blue roll for a short time but do not let leaf discs dry.

6. Float three leaf discs per well, abaxial side down. Handle discs with care using forceps, as tissue damage increases ion leakage. Use triplicate wells for each condition.

7. Successive time-point can be taken from the same well.

8. Calibrate the conductivity meter using a 0.01 M KCl (calibration for 1.41 mS cm^{-1}). One hour after the discs were floated on distilled water take a t_0 measure (blank) using 100 µl and a Horiba B-771 conductivity meter (*see* **Note 8**). The final volume in wells is now 1 ml.

9. To provoke cell death by irradiation, expose leaf discs with UV-C in an open 24-well cell culture plate using a UV cross-linker. For higher sensitivity than wild type, use 5 kJ/m^2; for resistance compared to wild type, use 15–20 kJ/m^2 (*see* **Note 9**). Alternatively, stimulate cell death by exposure to hydrogen peroxide by adding 30 mM H$_2$O$_2$ to wells after t_0 has been measured.

10. For sensitivity, use 20 mM with Col-0. For resistance, use 30 mM with WS and 60 mM with Ler (*see* **Notes 9** and **10**).

11. Incubate the 24-well plate in continuous light (*see* **Note 11**) and measure conductivity over time (*see* **Note 12**, for examples, of values obtained).

12. To measure conductivity, take 100 µl of water from each well and add to the reading chamber of the conductivity meter (*see* **Note 13**).

13. After each measurement, return as much of the 100 µl to the well as you can, to prevent a loss of volume.

14. As appropriate, rinse the chamber with distilled water until a reading of zero and dry using the paper sticks before the next set of measures (*see* **Notes 14** and **15**). Calculate the increase in conductivity in each well by subtracting the pretreatment conductivity reading from the posttreatment conductivity reading.

4   Notes

1. http://horiba.com; products; compact meters; B-771 LAQUA twin.

2. Detached leaf health should be tested prior to developing the experimental conditions for live cell imaging. Window stage lace plant leaves kept in Petri dishes will survive and continue growth for 4–5 days in distilled water. Additionally, after each long-term live cell imaging experiment the leaf blade should be visually scanned in order to ensure that the conditions did not have a cytotoxic effect in other parts of the organ.

3. The use a mechanical stage and Nikon's Perfect Focus (PFS) hardware would facilitate these observations. PFS allows for

drift compensation while a mechanical stage allows for a wider area to be scanned and the images can be stitched together during post-acquisition. Additionally, virtual network computing (VNC) software can allow for remote monitoring and control of the specimen and microscope. Continuous video recording yields large data sets compared to still image time series. With our hardware, stopping video recording periodically reduced instances of video corruption and facilitated downstream video processing.

4. Soaking ryegrass seeds in distilled water for 2 h prior to sterilization is not an essential part of this protocol but in our experience it enhances the subsequent seed germination.

5. Temperature controlled chamber may be used instead if available.

6. The gap between the condensed protoplast and cell wall is the hallmark of AL-PCD. Note that the extent of protoplast condensation in AL-PCD root hairs may vary. To make sure that root hairs are correctly categorized as AL-PCD carefully examine them for the presence of AL-PCD morphology by adjusting the focus, illumination, and phase contrast of the microscope. *See* Fig. 6, for examples, of different appearances of AL-PCD morphology, ranging from a less advanced cell death phenotype (Fig. 6a) to a readily recognizable gap between the condensed protoplast and cell wall present at a root tip (Fig. 6c). This gap may be also present in the middle part of the root (Fig. 6b). The condensed protoplast can also split into several units (Fig. 6d), which is more frequent in longer root hairs.

7. Untreated leaf discs leach ions due to the wounding inflicted. After 1 h the ion leakage reading is stable and constitute the blank for the leaf discs in a given well.

8. The first experiment should aim at calibrating cell death induction for the plant or ecotype used. Consider that various ecotypes have different sensitivity to a given treatment.

9. Calculation H_2O_2 doses in mM.

Final concentration (mM) required	30 % stock (µl in 1 ml)
0	0.0
5	0.6
10	1.1
20	2.3
30	3.4
40	4.5
50	5.7
60	6.8

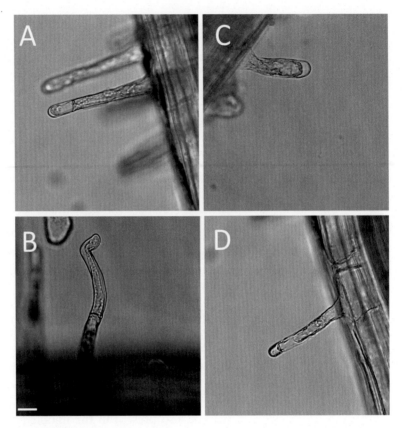

Fig. 6 Different appearances of AL-PCD root hairs from ryegrass seedling. Less advanced, but identifiable retraction of a protoplast at the root tip (**a**). Retraction of the protoplast away from the cell wall in the middle part of the root hair (**b**). Readily recognizable retraction of the protoplast away from the cell wall at the root tip (**c**). Condensed protoplast split into two units (**d**). Scale bar = 5 μm

10. Some instances of PCD have been shown to be light dependent.

11. *See* in Fig. 7, examples of conductivity values for increasing UVC dose using Col-0 with readings taken at 18 h posttreatment.

12. Pipette a few times before taking the water from each well to homogenize the liquid before measuring. Shaking the multi-well dish for a short while on a slow horizontal shaker is another possibility.

13. Measure all of the replicates for one plant or for one treatment without rinsing as values should be close, then rinse and dry the chamber using the prepared paper sticks before moving to the next set of measures.

14. If you have more than one 24-wells plate to measure, the meter will automatically switch off at some point; usually it will be near the end of the second dish. To prevent this, keep pushing the on/off switch from time to time.

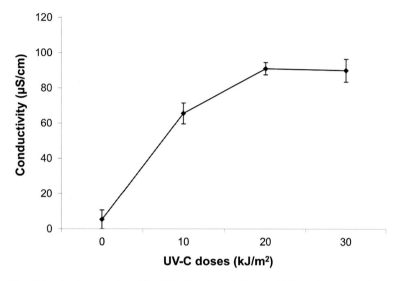

Fig. 7 Ion leakage induced by UV-C treatment. Conductivity values at 18 h for *Arabidopsis* Col-0 leaf discs treated with increasing doses of UVC delivered using a UV cross-linker. Presented values are the mean of three replicates (±SEM)

Acknowledgments

Department of Agriculture, Food & Marine (Ireland) is acknowledged for funding the work on ryegrass stress responses in the McCabe lab (VICCI grant. 14/S/819). The authors extend their thanks to the National Sciences and Engineering Research Council of Canada (NSERC) and the Killam Trusts for providing PhD funding to AND. AHLANG received funding from the Canadian Foundation for Innovation (CFI; Leaders Opportunity Fund) and the NSERC Discovery Grants Program.

References

1. Kacprzyk J, Daly CT, McCabe PF (2011) The botanical dance of death: programmed cell death in plants. In: Kader JC, Delseny M (eds) Advances in botanical research, vol 60. Academic, UK, pp 169–261

2. Lockshin RA, Zakeri Z (2004) Apotosis, autophagy, and more. Int J Biochem Cell Biol 36:2405–2419

3. Fukuda H, Watanabe Y, Kuriyama H, Aoyagi S, Sugiyama M, Yamamoto R, Demura T, Minami A (1998) Programming of cell death during xylogenesis. J Plant Res 111:253–256

4. Gunawardena AHLAN, Greenwood JS, Dengler NG (2004) Programmed cell death remodels lace plant leaf shape during development. Plant Cell 16:60–73

5. Lombardi L, Ceccarelli N, Picciarelli P, Lorenzi R (2007) Caspase-like proteases involvement in programmed cell death of *Phaseolus coccineus* suspensor. Plant Sci 172:573–578

6. Simeonova E, Sikora A, Charzyńska M, Mostowska A (2000) Aspects of programmed cell death during leaf senescence of mono- and dicotyledonous plants. Protoplasma 214:93–101

7. Rogers HJ (2006) Programmed cell death in floral organs: how and why do flowers die? Ann Bot 97:309–315

8. Heath M (2000) Hypersensitive response-related death. Plant Mol Biol 44:321–334

9. Huh GH, Damsz B, Matsumoto TK, Reddy MP, Rus AM, Ibeas JI, Narasimhan ML, Bressan RA, Hasegawa PM (2002) Salt causes ion disequilibrium-induced programmed cell death in yeast and plants. Plant J 29:649–659

10. Jerome WG, Fuseler J, Price RL (2011) Specimen preparation. In: Jerome WG, Price RL (eds) Basic confocal microscopy. Springer, New York, pp 61–77

11. Lord CEN, Wertman JN, Lane S, Gunawardena AHLAN (2011) Do mitochondria play a role in remodelling lace plant leaves during programmed cell death? BMC Plant Biol 11:102

12. Gunawardena AHLAN, Navachandrabala C, Kane M, Dengler NG (2006) Lace plant: a novel system for studying developmental programmed cell death. In: da Silva JA T (ed) Floriculture, ornamental and plant biotechnology: advances and tropical issues. Global Science Books, Middlesex, pp 157–162

13. Wright H, van Doorn WG, Gunawardena AHLAN (2009) In vivo study of developmental programmed cell death using the lace plant (*Aponogeton madagascariensis*; Aponogetonaceae) leaf model system. Am J Bot 96:865–876

14. Wertman J, Lord CEN, Dauphinee AN, Gunawardena AHLAN (2012) The pathway of cell dismantling during programmed cell death in lace plant (*Aponogeton madagascariensis*) leaves. BMC Plant Biol 12:115

15. Hogg BV, Kacprzyk J, Molony EM, O'Reilly C, Gallagher TF, Gallois P, McCabe PF (2011) An in vivo root hair assay for determining rates of apoptotic-like programmed cell death in plants. Plant Methods 7:45

16. Kacprzyk J, McCabe PF (2015) A root hair assay to expedite cell death research. In: Estevez JM (ed) Plant cell expansion, vol 1242, Methods Mol Biol. Humana Press, New York, pp 73–82

17. Reape TJ, Molony EM, McCabe PF (2008) Programmed cell death in plants: distinguishing between different modes. J Exp Bot 59:435–444

18. McCabe PF, Leaver CJ (2000) Programmed cell death in cell cultures. Plant Mol Biol 44:359–368

19. He R, Drury GE, Rotari VI, Gordon A, Willer M, Farzaneh T, Woltering EJ, Gallois P (2008) Metacaspase-8 modulates programmed cell death induced by ultraviolet light and H_2O_2 in Arabidopsis. J Biol Chem 283:774–783

20. Mishiba KI, Nagashima Y, Suzuki E, Hayashi N, Ogata Y, Shimada Y, Koizumi N (2013) Defects in IRE1 enhance cell death and fail to degrade mRNAs encoding secretory pathway proteins in the Arabidopsis unfolded protein response. Proc Natl Acad Sci U S A 110:5713–5718

Chapter 13

Modeling Metazoan Apoptotic Pathways in Yeast

David T. Bloomer, Tanja Kitevska, Ingo L. Brand, Anissa M. Jabbour, Hang Nguyen, and Christine J. Hawkins

Abstract

This chapter describes techniques for characterizing metazoan apoptotic pathways using *Saccharomyces cerevisiae*. Active forms of the major apoptotic effectors—caspases, Bax and Bak—are all lethal to yeast. Using this lethality as a readout of caspase/Bax/Bak activity, proteins and small molecules that directly or indirectly regulate the activity of these effectors can be investigated in yeast, and apoptotic inhibitors can be identified using functional yeast-based screens. Caspase activity can also be monitored in yeast by cleavage-dependent liberation of a transcription factor from the plasma membrane, enabling it to activate the lacZ reporter gene. This system can be used to define the sequences that can be efficiently cleaved by particular caspases.

Key words *Saccharomyces cerevisiae*, Yeast, Apoptosis, Caspase, Bcl-2, IAP, Bax, Programmed cell death, BH3-mimetic

1 Introduction

1.1 Researching Cell Death Pathways Using Yeast

For decades, the budding yeast *Saccharomyces cerevisiae* has been one of the most widely used genetic model organisms and a popular molecular biology workhorse. The plethora of established yeast tools and techniques enable metazoan apoptotic pathways to be modeled using this species. Yeast are well suited for researching metazoan apoptotic regulators because they are eukaryotic: in particular their subcellular organelles resemble those of mammals, insect, and nematodes (the organisms whose apoptotic pathways we have usually sought to model). Some researchers have proposed that yeast can undergo a form of cell death that in some ways resembles apoptosis [1]. However, proteins proposed to mediate yeast cell death did not affect the activity of exogenous apoptotic effector proteins from mammals, nematodes, or insects when expressed in yeast [2]; thus yeast can be considered a naïve yet eukaryotic cellular environment in which to reconstitute and study the regulation of metazoan apoptotic pathways.

Hamsa Puthalakath and Christine J. Hawkins (eds.), *Programmed Cell Death: Methods and Protocols*, Methods in Molecular Biology, vol. 1419, DOI 10.1007/978-1-4939-3581-9_13, © Springer Science+Business Media New York 2016

This chapter describes two methods we have employed to identify and characterize apoptotic regulators. The first method exploits the observation that yeast die following expression of active forms of the major apoptotic effectors: caspases, Bax and Bak (Fig. 1). Activators of these pro-apoptotic proteins can be

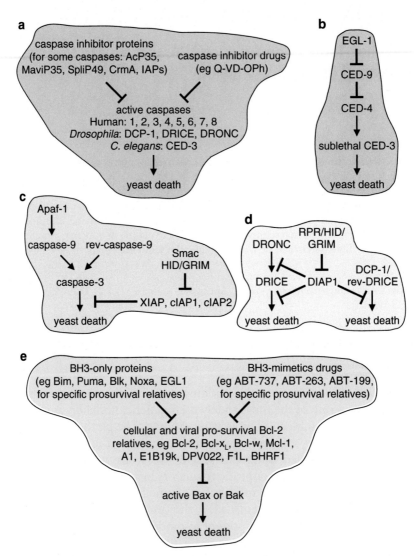

Fig. 1 Reconstitution of apoptotic pathways in yeast. Yeast can be killed by expression of many active caspases, Bax or Bak. This lethality can be exploited to assemble pathways to model the function of upstream regulators of caspase/Bax/Bak activity. (**a**) DCP-1 and human caspases-1, 5, 8 auto-activate in yeast when expressed at low to moderate levels, and these caspases have appropriate specificities to kill yeast. The native forms of other caspases are inactive or weakly active when expressed in yeast. The lethality of caspases-2, 4, DRONC, CED-3 can be enhanced by transforming yeast with multiple expression plasmids bearing different nutritional selection markers. Auto-activating forms of caspases-2, 3, 6, 9 and DRICE can be expressed by reversing the order of the subunits. Auto-activating caspase-3 can also be produced by fusing the caspase to β-galactosidase. Removing an amino-terminal sequence from caspase-7 enhances its lethality. Proteins that inhibit particular

identified by virtue of their ability to activate caspase zymogens or Bax or Bak expressed at sublethal levels. Functional screens of cDNA expression libraries can be carried out to identify inhibitors of caspases, Bax or Bak (or their activators), and antagonists of such inhibitors can also be modeled in this system. The second method was designed to define caspase specificity. Caspase cleavage within the linker of a fusion protein liberates a transcription factor from its membrane tether, so activation of a transcriptional reporter provides evidence that the caspase has the specificity required to cleave that particular linker sequence (Fig. 2).

1.2 Pathways Controlling Caspase-Dependent Yeast Death

Very high level expression of some caspases, such as that achieved using the GAL1/10 promoter, led to their autoactivation and lethality in yeast. Caspase-mediated yeast lethality manifested as loss of plasma and nuclear membrane integrity, mitochondrial function, and clonogenic potential, but did not lead to detectable DNA damage [2]. The GAL1/10 promoter has the useful feature of being induced by galactose and repressed by glucose, so plasmids encoding lethal proteins such as caspases can be transformed into yeast without provoking any toxicity, as long as glucose is supplied (Fig. 3). Raffinose can be used as a "neutral" sugar, allowing immediate induction of genes controlled by the GAL1/10 promoter after addition of galactose, rather than delayed induction following replacement of glucose with galactose. We created a truncated GAL1/10 promoter, which is also sugar-regulated but did not produce as high level transgene expression as the intact promoter [3] (Fig. 3). Other promoters which can be useful in particular contexts include copper-inducible (CUP1) [3], methionine-repressible (MET), and constitutive (ADH) promoters [4] (Fig. 3).

Galactose-inducible expression of mammalian caspases-1, 5, 8, or *Drosophila* DRONC or DCP-1, was lethal to yeast. Caspases-2, -4, -7 and *C. elegans* CED-3 were weakly toxic, but their lethality

---

Fig. 1 (continued) caspases can protect yeast from the death associated with their expression, as can incubation with the pan-caspase inhibitor drug Q-VD-OPh. (**b**) Sublethal levels of CED-3 can be activated by CED-4 in yeast, triggering death. This lethality can be inhibited by CED-9, unless EGL-1 is also expressed. (**c**) Caspase-9 can be activated in yeast by Apaf-1 or by reversing the order of its subunits ("rev-caspase-9"). A truncated version of Apaf-1 lacking the WD40 domain (Apaf-11-530) activated pro-caspase-9 in vitro without cytochrome-c or dATP [12]. This mutant cooperates efficiently with pro-caspase-9 and -3 to kill yeast; the full-length Apaf-1 protein is less active. Active caspase-9 lacks the specificity to kill yeast, but can proteolytically activate caspase-3 in yeast, leading to yeast death. Caspase-3 activity and death can be inhibited by XIAP and less potently by c-IAP1 and c-IAP2. The protection conferred by these IAPs can be antagonized by Smac/DIABLO or its insect counterparts HID or GRIM. (**d**) The corresponding core insect apoptotic machinery can be reconstituted in yeast, exploiting the observation that low levels of DRONC are insufficient to kill yeast but can activate DRICE, which is highly lethal once activated. (**e**) Yeast can also be killed by expression of Bax or Bak. Pro-survival Bcl-2 family members confer protection from this lethality, and that protection can be neutralized by co-expression of BH3-only proteins or by incubation with BH3-mimetic drugs

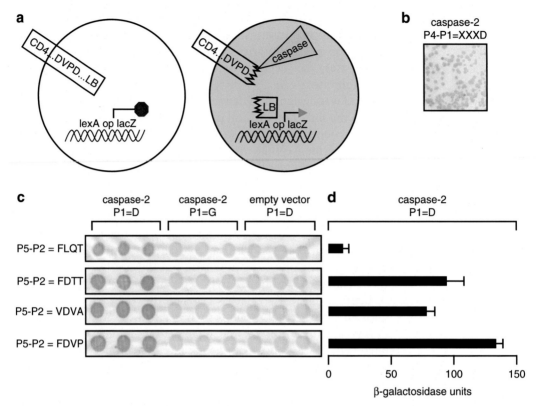

Fig. 2 Analysis of caspase specificity using the transcriptional reporter system. (**a**) A model describing the system. Yeast express a fusion protein in which the LexAB42 ("LB") transcription factor is separated from a membrane anchor by a sequence that may be susceptible to proteolysis by a caspase. If the yeast express an active caspase capable of cleaving this sequence, the transcription factor is liberated from the membrane and can induce expression of the reporter gene lacZ. (**b**) EGY48 yeast bearing the lacZ reporter gene plasmid were transformed with a caspase-2 expression plasmid and a plasmid encoding a fusion protein library with random P4-P2 residues. Induced yeast were stained with Xgal to identify colonies expressing fusion proteins containing caspase-2-sensitive cleavage sites. (**c**, **d**) EGY48 yeast bearing the lacZ reporter gene plasmid were transformed with a caspase-2 expression plasmid and a plasmid encoding a fusion proteins bearing the specified cleavage sites. Suspensions were made of three transformants expressing fusion proteins containing each of the specified P4-P2 residues. (**c**) Five µl of each suspension was spotted on repressing plates and grown for a day then filter-lifted onto inducing agar and stained with Xgal. (**d**) The same clones were grown in liquid inducing medium, lysed by freeze-thawing and β-galactosidase activity was quantitated by ONPG, as a measure of caspase-mediated cleavage at the cleavage sites contained within the fusion proteins (mean +/− SEM from three clones of each type)

could be boosted by increasing the copy number of the plasmids bearing these genes. We favor centromeric yeast expression plasmids bearing the auxotrophic markers *LEU2*, *TRP1*, *HIS3*, or *URA3*. These plasmids confer the ability to survive on media lacking leucine, tryptophan, histidine, or uracil respectively, and they are maintained in yeast at a relatively stable copy number of two to five copies per cell [5]. Transformation of yeast with two or more plasmids bearing with different selectable markers, all encoding the

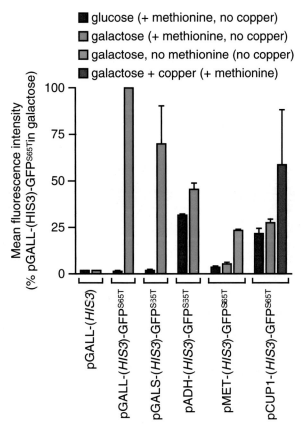

Fig. 3 Vectors for controlling transgene expression in yeast. W303α yeast were transformed with an empty vector or histidine-selectable plasmids engineered to direct expression of GFP^{S65T} under the control of various promoters: an intact GAL1/10 galactose-inducible/glucose-repressing promoter (GALL), a truncated, weaker GAL1/10 promoter (GALS), a constitutive promoter (ADH), a methionine-repressible promoter (MET), or a copper-inducible promoter (CUP1). Transformants were grown in glucose/methionine-containing medium overnight, then washed and grown for 8 h in selective media containing either 2 % glucose or 2 % galactose and containing or lacking 300 μM methionine and/or 100 μM copper sulfate. The fluorescence of each sample was then measured by flow cytometry. Data show the means and SEM from three independent transformants of each type

same weakly lethal caspase, yielded robust caspase-mediated yeast death. Alternatively, higher but more variable plasmid copy numbers (~15–34), and enhanced lethality, could be achieved by expressing caspases using vectors incorporating the "2 μ" origin of replication [6, 7].

Some executioner caspases seemed to possess specificities that enable them to kill yeast once activated, but failed to auto-activate even when expressed at extremely high levels. It was, however, possible to activate these proteases and other weakly lethal caspases, either by providing an upstream activator (e.g., sublethal levels of DRONC activated DRICE to kill yeast) or by reversing

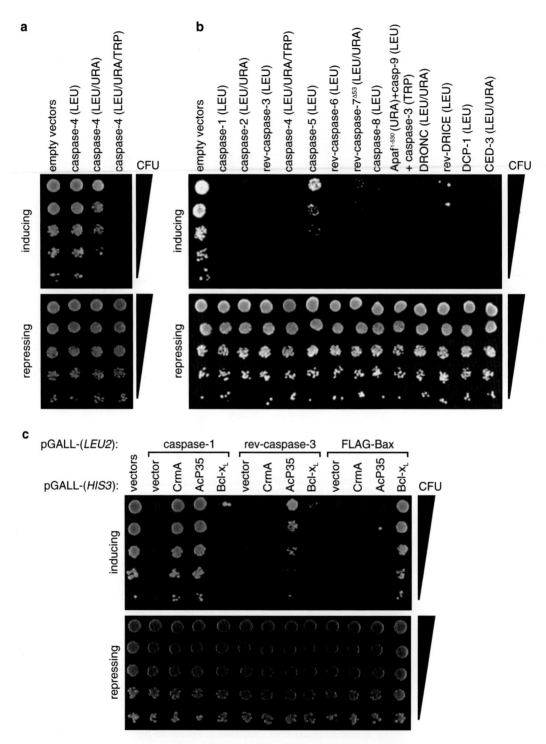

Fig. 4 Active caspases kill yeast. W303α yeast were transformed with centromeric plasmids bearing the specified selectable markers and encoding the specified genes under the control of the GAL1/10 promoter. Transformants were grown in repressing (glucose-containing) selective media, washed, and resuspended in TE at an A_{620} corresponding to 1250 CFU per μl. The suspensions were serially diluted 1:4 in TE and 5 μl of each dilution of each clone was pipetted onto repressing (glucose) and inducing (galactose) plates. Growth was

the order of the caspase subunits. Emad Alnemri pioneered this approach to caspase activation, showing that rearranged caspases in which the small subunit is amino terminal to the large subunit (separated by a cleavage site) mimicked the structure of the active enzyme [8]. We have used this trick to produce auto-activating versions of caspases-2, 3, 6, 7, 9, DRONC and DRICE in yeast [4, 9, 10]. Caspase-3 could also be expressed in an active lethal form by fusing it to β-galactosidase [4]; presumably this promotes aggregation and hence autoactivation.

Human caspase-9 is unusual. It could be forced to auto-activate in yeast by swapping the order of its subunits, or by co-expression of its natural activator Apaf-1, but active caspase-9 did not kill yeast [3, 4], presumably because its restricted P2 specificity [11] prevented it from efficiently cleaving essential yeast proteins. Active caspase-9 could, however, activate pro-caspase-3 or -7 in yeast, leading to death. In this way, it was possible to reconstitute the mammalian apoptosome in yeast. A truncation mutant of Apaf-1 (Apaf-1$^{1-530}$) that lacks the C-terminal repressive WD40 domain [12] cooperated strongly with caspases-9 and -3 to kill yeast [3, 4]. Presumably the truncation bypasses the need for derepression performed by cytosolic cytochrome-c in mammalian cells undergoing apoptosis. Surprisingly, we found that full-length Apaf-1 could also promote yeast death upon co-expression with caspases-3 and -9, although less efficiently than the truncation mutant (data not shown). *S. cerevisiae* cytochrome-c was reportedly unable to promote apoptosome activation [13], so it is unlikely that this activity is due to small amounts of cytochrome-c leaking from viable yeast mitochondria. Instead, we suspect that a small but evidently sufficient proportion of overexpressed full-length Apaf-1 adopts a conformation that facilitates activation of enough caspase-9 molecules to cleave and activate executioner caspases. The approaches outlined above can be exploited to establish yeast-based assays for monitoring the activity of a large number of caspases from mammals, nematodes, and insects (Fig. 4).

Fig. 4 (continued) photographed after 3 days (galactose plates) or 2 days (glucose plates). Growth on inducing plates indicates survival and proliferation of yeast expressing the transgenes. (**a**) Caspase-4-mediated toxicity is more potent when the protease is expressed from multiple centromeric plasmids. (**b**) Many caspases can be used to kill yeast. The indicated caspases were expressed either in their natural configuration or with the order of the subunits swapped ("rev"). Caspases (and Apaf-1) were expressed from one or more plasmids bearing leucine (LEU), uracil (URA), or tryptophan (TRP) selectable markers. Empty vectors were included in transformations as required to enable all clones to grow on media lacking all three nutrients. (**c**) Yeast death induced by caspases or Bax can be prevented by co-expression of inhibitors. Yeast bearing leucine-selectable plasmids directing galactose-inducible expression of either empty vector, caspase-1, rev-caspase-3, or FLAG-Bax (or empty vector) were transformed with histidine-selectable, galactose-inducible plasmids encoding either the caspase-1/8 inhibitor CrmA, the broad spectrum caspase inhibitor AcP35, or the Bax/Bak inhibitor Bcl-x$_L$

Active lethal caspases could be prevented from killing yeast by co-expression of cellular or viral proteins that can inhibit their enzymatic activity. We have used this approach to test the activity and specificity of candidate caspase inhibitors, including p35 family members from insect viruses [14–16] (Fig. 4). This technique can be adapted to screen cDNA expression libraries to identify proteins that inhibit particular caspases, allowing the yeast to survive and form colonies despite expressing the caspase ([3], Fig. 5). It can also be used to visualize caspase inhibition by small molecules such as Q-VD-OPh [17].

a

Plasmid(s) encoding pro-apoptotic proteins	RNA source for library	Inhibitor identity, accession number (verified in vitro?)
pGALL-(*LEU2*)-DCP-1	*Drosophila* embryo	DIAP1 (yes)
pYX143 KAS caspase-3-lacZ	human glioma	c-IAP1 (yes)
pGALL-(*URA3*)-DRONC + pGALS-(*LEU2*)-DRONC + pGALL-(*TRP1*)-DRICE	*Ac*MNPV-infected Sf21 cells	AcP35 (yes) 63% identical to XP_004926257.1 (no) EIF3p40, BAM17757.1 (no)
pGALL-(*URA3*)-DRONC + pGALS-(*LEU2*)-DRONC + pGALL-(*TRP1*)-DRICE	*Ld*MNPV-infected Sf21 cells	none (only 45% library screened)
pGALL-(*LEU2*)-CED-3 validated versus pGALS-(*LEU2*)-CED-3 + pMET1-(*URA3*)-CED-4	*C. elegans*	F49E8.2, NP_001255303.1 (no) BRP-1, NP_001255152.1 (no)

b

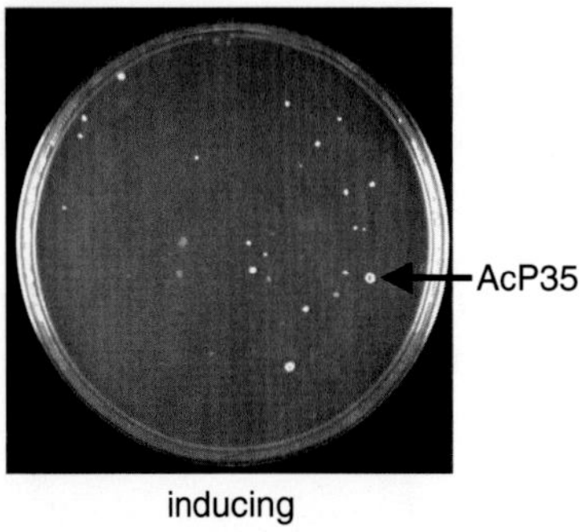

inducing

Fig. 5 Isolation of caspase inhibitors by functional screening of cDNA expression libraries in yeast. (a) RNA extracted from the specified sources was used to create galactose-inducible yeast expression libraries. W303α yeast bearing the specified caspase expression plasmids were transformed with these libraries and plated onto inducing media. Only yeast that acquired a library plasmid that encoded a caspase inhibitor would be expected to survive and form colonies. Library plasmids were isolated from these colonies, amplified in bacteria, and re-transformed into yeast to verify that they protected yeast from caspase-dependent death. The right column shows the identities of these "yeast-validated" candidate inhibitors. Those that have been confirmed to inhibit caspases in vitro are noted. (b) The AcMNPV/Sf21 library was screened for DRONC/DRICE inhibitors. A transformation plate is shown: the indicated colony bore a library plasmid encoding the pan-caspase inhibitor AcP35. The authors thank Rollie Clem for the RNA that was used to make the AcMNPV and LdMNPV libraries and Michael Hengartner for the RNA to make the *C. elegans* library.

1.3 Pathways Controlling Bax- or Bak-Dependent Yeast Death

We have not investigated Bax/Bak-mediated yeast death in as much detail as caspase-dependent killing, but others have reported that expression in yeast of active forms of Bax and Bak leads to loss of mitochondrial potential and release of cytochrome-*c* (reviewed by [18]). We observed that addition of amino-terminal tags to Bax and Bak, particularly the human orthologues, significantly boosted their lethality (Fig. 6). This probably reflects the tag's ability to encourage externalization of the amino-terminal α-helix, bypassing an early Bax/Bak activation step [19]. Co-expression of pro-survival proteins inhibited Bax or Bak from killing yeast (Figs. 4 and 7), and this protection could be alleviated by co-expression of BH3-only proteins, or exposure to BH3-mimetic drugs [17] (Fig. 7). Direct activation of Bax or Bak by BH3-only proteins could also be modeled, by expressing sublethal levels of Bax or Bak in the presence or absence of the potential direct activator. In this way, we observed that Puma enhanced killing by sublethal levels of Bax expressed from the methionine-repressible promoter (data not shown).

1.4 Transcriptional Reporter System for Testing Caspase Specificity

We developed a yeast-based system for detecting protease-mediated cleavage of an engineered protein substrate [3]. The substrate has three parts: a portion of the human CD4 protein encompassing the transmembrane domain, a linker region containing potential cleavage site(s), and a transcription factor (LexA-B42). When intact, the transcription factor is tethered to the plasma membrane, and cannot stimulate expression of a reporter gene like lacZ.

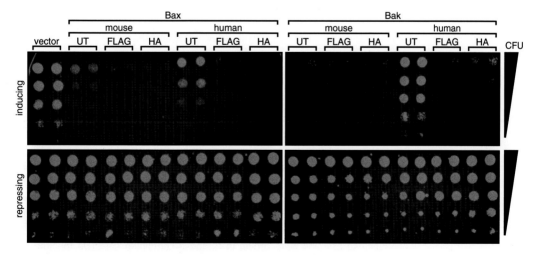

Fig. 6 Yeast growth is impaired by expression of Bax or Bak, and lethality is enhanced by N-terminal tags. W303α yeast were transformed with an empty vector or galactose-inducible plasmids encoding murine or human Bax or Bak; either untagged (UT) or tagged at the amino terminus with FLAG or HA epitopes. Spotting was performed as detailed in the legend to Fig. 4

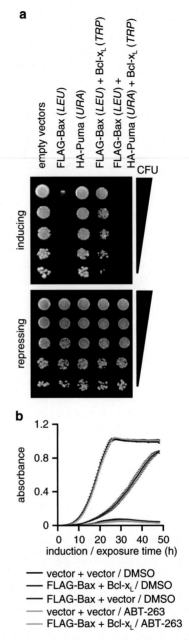

Fig. 7 Antagonism of pro-survival Bcl-2 proteins by BH3-only proteins and BH3-mimetics can be modeled in yeast. (**a**) FY1679-28C yeast were transformed with plasmids encoding the listed Bcl-2 family members. Spotting onto inducing and repressing media was performed as described in the legend to Fig. 4. (**b**) Three independent transformants bearing galactose-inducible plasmids encoding the specified proteins were grown in raffinose-containing selective medium then transferred into medium containing galactose and either 10 μM ABT-263 in DMSO or the equivalent concentration of DMSO. The absorbance of each culture was measured every 0.5 h for 48 h. Error bars indicate standard errors of the means from analyses of three independent clones of each type

However, if a caspase is co-expressed, and the linker region of the chimeric protein contains a sequence amenable to proteolysis by the caspase, cleavage liberates the transcription factor from the membrane, allowing it to access the nucleus and trigger expression of the reporter gene (Fig. 2). This system could be employed to screen expression libraries for proteases with particular specificities, but we have mainly used it to define the minimal specificity of selected caspases [20, 21]. For this application, we generated fusion protein libraries bearing random residues in the linker separating the transmembrane domain from the transcription factor. Yeast bearing an expression plasmid encoding the caspase of interest are transformed with such a library, and the resulting transformants are assayed (using X-gal or ONPG) for expression of the reporter gene (lacZ), which indicates that the fusion protein encoded by the library plasmid can be efficiently cleaved by the caspase. Sequencing the linker region of genes encoding the library fusion proteins from blue ("positive") and white ("negative") clones defines cleavable and noncleavable sites for that caspase.

As explained above, many caspases are toxic to yeast, so in this system the caspase is expressed in a galactose-dependent manner using the GAL1/10 promoter described above. Transformants are initially grown on media containing glucose to repress expression, then transferred to a filter (recording the orientation of the filter relative to the agar plate). The filter is then incubated on a plate containing galactose, to induce expression, then stained with Xgal to identify the clones that contain a caspase and fusion protein with a cleavage-sensitive site. The corresponding transformants can be recovered from the glucose-containing plate for further characterization. To more quantitatively analyze reporter gene expression, indicating cleavage efficiency, the β-galactosidase substrate ONPG can be used.

2 Materials

2.1 Yeast Strains

1. Experiments involving caspase or Bax/Bak-dependent death have predominantly used the yeast strain W303α (MATα; can1-100; leu2-3,-112; his3-11,-15; trp1-1; ura3-1; ade2-1).

2. The following strains and their derivatives have also been used for particular studies exploring the impact of mutations in genes associated with yeast cell death or drug transport on caspase/Bax/Bak lethality: BY4741 (MATa; his3Δ1; leu2Δ0; met15Δ0; ura3 Δ0); FY1679-28C (MATa, ura3-52, trp1Δ63, leu2Δ1, his3Δ200, GAL2+).

3. The transcriptional reporter gene assays for detecting caspase cleavage employ EGY48 (MATa, ura3, trp1, his3, LexAop6-LEU2).

2.2 Plasmids

1. pSH18-34 (Invitrogen). This is a uracil-selectable plasmid in which the lacZ gene is located downstream of eight lexA operator sequences, enabling lexA-dependent β-galactosidase expression.

2. The methods described below involve expressing apoptotic regulators in yeast using centromeric, galactose-inducible yeast plasmids bearing nutritional markers: pGALL-(*LEU2*), pGALL-(*TRP1*), pGALL-(*HIS3*), and pGALL-(*URA3*). These constructs were derived from the pRS31X series of plasmids [5] by introduction of the GAL1/10 promoter, a polylinker, and the actin terminator [3]. As mentioned above, similar vectors have also been created containing different promoters. All plasmids are available upon request.

2.3 Media

1. 10× glucose: 20 % (w/v) glucose. Dissolve 100 g of glucose in 300 ml of water, add water to 500 ml. Filter sterilize through a 0.2 μm filter into a sterile bottle. Store at room temperature.

2. 10× galactose: 20 % (w/v) galactose. Dissolve 100 g of galactose in 300 ml of water, add water to 500 ml. Filter sterilize through a 0.2 μm filter into a sterile bottle. Store at room temperature.

3. 10× raffinose: 20 % (w/v) raffinose. Dissolve 100 g of raffinose in 300 ml of water, add water to 500 ml. Filter sterilize through a 0.2 μm filter into a sterile bottle. Store at room temperature.

4. YPglc/gal (complete liquid media): Dissolve 10 g peptone and 5 g yeast extract in 450 ml water. Autoclave for 15 min on wet cycle. Add 50 ml of 10× glucose or galactose (above) while still hot. Store at room temperature.

5. YPglc/gal agar (complete solid media): Dissolve 10 g peptone and 5 g yeast extract in 450 ml water. Add 10 g agar. Autoclave for 15 min on wet cycle. Store at room temperature. Melt in a microwave on low power, remove the required volume, and add 1/10th volume of 10× glucose or galactose before pouring into plates.

6. 10× dropout solution. Dissolve 3.5 g dropout supplement lacking leucine, tryptophan, histidine, and uracil (Sigma #Y2001) in 500 ml water. Filter sterilize through a 0.2 μm filter into a sterile bottle. Store at 4 °C.

7. 100× nutrient supplements: Dissolve 300 mg of leucine or 200 mg of tryptophan, histidine, or uracil in 100 ml of water. Filter sterilize through a 0.2 μm filter into a sterile bottle. Store at 4 °C.

8. Minimal liquid media. Dissolve 3.4 g of yeast nitrogen base without amino acids (BD#291940) in 400 ml of water. Add 50 ml of 10× dropout solution. Filter sterilize through a 0.2 μm filter into a sterile bottle. Store at 4 °C. Add 1/100th volume of 100× nutrient supplement(s) and 1/10th volume of 10× glucose, galactose, or raffinose as required.

9. Minimal agar media. Dissolve 3.4 g of yeast nitrogen base without amino acids (BD#291940) in 400 ml of water. Add 10 g of agar. Autoclave for 15 min on wet cycle. Store at room temperature. Melt in a microwave on low power and remove the required volume. Add 1/100th volume of 100× nutrient supplement(s), 1/10th volume of 10× dropout solution, and 1/10th volume of 10× glucose or galactose as required, before pouring into plates.

2.4 Other Reagents

1. Single stranded (carrier) DNA. Purchase fish sperm DNA (e.g., Roche #11467140001) and aliquot 5 ml per tube into multiple 10 ml tubes and freeze. Thaw one tube and denature DNA by heating at 98 °C for 15 min. Chill on ice for 10 min and aliquot 500 μl per tube. Store at –20 °C.

2. 10× LiAc: 1 M Lithium acetate. Dissolve 10.2 g Lithium acetate dihydrate in 90 ml of water. Add water to 100 ml. Autoclave 15 min on wet cycle.

3. 10× TE: 100 mM Tris–HCl pH 7.5, 10 mM EDTA. Dissolve the following in 90 ml of water: 1.2 g Tris, 0.37 g ETDA. Adjust the pH to 7.5 with HCl. Add water to 100 ml. Autoclave 15 min on wet cycle.

4. 1× LiAc/TE: 0.1 M lithium acetate, 10 mM Tris–HCl pH 7.5, 1 mM EDTA. Mix 50 ml 10× LiAc, 50 ml of 10× TE with 400 ml water. Autoclave 15 min on wet cycle.

5. TE: 10 mM Tris–HCl pH 7.5, 1 mM EDTA. Mix 50 ml 10× TE with 450 ml water. Autoclave 15 min on wet cycle.

6. 50 % PEG3350: Mix 250 g Poly(ethylene glycol)3350 in 500 ml of water until dissolved (this takes a few hours). Autoclave 15 min on wet cycle. Store at room temperature.

7. PEG/LiAc/TE: 40 % PEG3350, 0.1 M lithium acetate, 10 mM Tris–HCl pH 7.5, 1 mM EDTA. Into a sterile 50 ml tube, pipette 5 ml of 10× LiAc and 5 ml of 10× TE, and pour (sterilely) 40 ml of 50 % PEG. Mix (e.g., on rocker) for ~1 h.

8. Hybond-N+ nylon membrane: either purchase circular filters or cut to match the dimensions of the petri dishes

9. Xgal (5-Bromo-4-chloro-3-indolyl β-d-galactopyranoside): 20 mg/ml Xgal in dimethylformamide. Dissolve 100 mg of Xgal in 5 ml of dimethylformamide. Store at –20 °C protected from light.

10. 4 mg/ml ONPG (o-Nitrophenyl β-d-galactopyranoside). Dissolve 100 mg in 25 ml of Z buffer (below). Filter sterilize through a 0.2 μm filter into a sterile bottle. Store at –20 °C.

11. Z buffer: 60 mM Na_2HPO_4, 40 mM NaH_2PO_4, 10 mM KCl, 1 mM $MgSO_4$, pH 7.0. Dissolve 4.3 g of Na_2HPO_4, 4.8 g of NaH_2PO_4, 0.74 g of KCl, and 0.12 g of $MgSO_4$ in 900 ml of water. Adjust the pH to 7.0 and add water to 1 L. Autoclave 15 min on wet cycle. Store at room temperature. Just before performing assays, add 2.7 μl β-mercaptoethanol and *either* 15 μl of 20 mg/ml Xgal *or* 225 μl of 4 g/ml ONPG per ml of Z buffer.

12. 1 M Na_2CO_3.

13. Lysis buffer for plasmid DNA extraction: 2 % (v/v) Triton X-100, 1 % SDS (w/v), 100 mM NaCl, 10 mM Tris–HCl pH 8, 1 mM EDTA pH 8. Store at room temperature.

14. Phenol/chloroform/isoamyl alcohol, 25:24:1 (v:v:v). Store at 4 °C protected from light.

15. Glass beads (~500 μm diameter).

16. 3 M NaOAc pH 5.2. Store at room temperature.

2.5 Equipment

1. These methods require a plate-reading spectrophotometer. For Subheading 3.4, it must be able to be programmed to perform cycles of shaking for ~2 min, followed by reading absorbance at 600–620 nm, followed by incubation at 30 °C. This (or some equivalent arrangement) is necessary for obtaining absorbance measurements of resuspended yeast every 30 min for up to 48 h.

2. Shaking incubator that can be set at 30 °C, 230 rpm.

3. Plate incubator that can be set at 30 °C.

4. Microfuge that can be set to 14,000×g, 4 °C.

3 Methods

3.1 Lithium Acetate Transformation

This is a common protocol for transforming *S. cerevisiae*, which is based on a method published by Ito et al. [22]. The quantities specified below are sufficient for ten transformations, but can be scaled-up if required.

1. Streak the yeast strain from a glycerol stock frozen at –80 °C onto a YPglc plate (if using a parental strain) or the appropriate selective agar plate containing glucose (if the strain already harbors a vector, such as a caspase or Bax/Bak expression plasmid).

2. After 2 or 3 days, when good-sized colonies have grown, inoculate a colony into 5 ml of liquid media containing glucose and grow, shaking at around 230 rpm, at 30 °C for ~18 h (*see* **Note 1**).

3. Expand the yeast into 50 ml of YPglc and grow, shaking for another 4–5 h.

4. Pellet the yeast by centrifuging for 5 min at $4000 \times g$.

5. Resuspend pellet in 25 ml 1×LiAc/TE, spin again (5 min, $4000 \times g$).

6. Resuspend pellet in 1 ml 1×LiAc/TE.

7. In sterile 1.5 ml tubes, prepare the DNA mixtures to be transformed. Pipette 10 µl of ssDNA (of a 10 mg/ml stock) into each tube, and add ~1 µg of each plasmid (*see* **Note 2**). To screen a cDNA library for inhibitors, transform a clone bearing the caspase/Bax/Bak plasmid with 8 µg of the library DNA spread across eight separate transformations. Also perform control transformations with empty vector and (if available) a plasmid encoding a known inhibitor (e.g., AcP35 for most caspases, Bcl-x_L for Bax/Bak).

8. Add 100 µl of yeast suspension to each tube.

9. Add 600 µl of pre-mixed PEG/LiAc/TE to each tube.

10. Invert ~5 times to mix.

11. Incubate at 30 °C for 30 min.

12. Add 70 µl of DMSO to each tube.

13. Invert ~5 times to mix.

14. Incubate at 42 °C for 15 min.

15. Pellet the yeast by centrifuging for 15 s at $14,000 \times g$. If transforming a single plasmid with a defined insert, resuspend the pellet in 70 µl of TE and spread the yeast onto a 10 cm selectable repressive plate (*see* **Note 3**). If transforming a cDNA library, merge the resuspended yeast transformed with the library, plate 1 % and 10 % on 10 cm repressing selective plates (to determine the transformation frequency) and the remainder onto three 15 cm inducing selective plates. Plate the control transformations onto 10 cm inducing selective plates.

16. Incubate the plates at 30 °C for 2–3 days (glucose plates) or 4–7 days (galactose plates).

17. Use Subheading 3.2 to extract library plasmid DNA from library screening plates to validate and identify candidate inhibitors, or use Subheading 3.3 to visualize the impact of the transgene(s) on yeast survival and growth.

3.2 Extracting Library Plasmid DNA from Yeast

1. Grow each of the transformants whose library plasmids are to be extracted for 12–18 h in 1 ml of selective liquid media containing glucose (or galactose, if isolating caspase/Bax/Bak inhibitors), shaking, at 30 °C.

2. Pellet the yeast by centrifuging for 15 s at $14,000 \times g$.

3. Resuspend in 1 ml of YPglc and grow, shaking, for 4–5 h (this rapid growth leads to weaker cell walls and better lysis).

4. Pellet the yeast by centrifuging for 15 s at 14,000×g, discard supernatant.

5. Add a roughly equal volume of glass beads to the tubes and give them a quick spin to ensure all the beads are at the bottom. Samples can be frozen at this point if necessary.

6. Add 500 μl of lysis buffer to the tubes and thoroughly resuspend pellets.

7. Add 500 μl of phenol/chloroform/isoamyl alcohol to each tube.

8. Make sure the lids are securely closed and wear gloves. Vortex at top speed for 2 min.

9. Centrifuge for 3 min at 14,000×g.

10. Take top layer to a fresh tube, add 50 μl of 3 M NaOAc and 500 μl of isopropanol. Mix by inversion a few times.

11. Precipitate DNA by spinning for 15 min, 14,000×g at 4 °C

12. Gently discard supernatant.

13. Gently pipette 1 ml of ice-cold 70 % ethanol into each tube.

14. Centrifuge for 1 min at 14,000×g then gently discard supernatant.

15. Spin again for 15 s, use a yellow tip to remove remaining liquid.

16. Place tubes in 42 °C heatblock for 2 min with lids open, to evaporate residual ethanol (don't leave longer than this as overdried DNA is difficult to dissolve).

17. Resuspend the pellet in 200 μl of TE, leave at 42 °C for 1–2 min (lids closed) to completely dissolve DNA.

18. Electroporate 2 μl into electrocompetent bacteria (*see* **Note 4**), plate on ampicillin-containing media.

19. Perform a standard "miniprep" procedure to extract DNA from the bacteria for transformation into yeast to confirm the inhibitor activity, and for sequencing.

3.3 Visualizing Caspase/Bax/Bak Activity and Inhibition in Yeast, by Semi-Quantitative Spotting onto Agar Plates

1. Grow each of the transformants to be tested for 12–18 h in 1 ml of selective repressive liquid media, shaking, at 30 °C. These transformants may, for example, bear a leucine-selectable plasmid encoding caspase/Bax/Bak plus a second histidine-selectable plasmid either lacking an insert or encoding a candidate inhibitor. In all assays, be sure to also include a transformant containing empty vectors bearing the appropriate selectable markers: these controls will indicate the amount of yeast growth expected on galactose-containing media if the caspase/Bax/Bak is completely inhibited.

2. Pellet the yeast by centrifuging for 15 s at $14,000 \times g$.

3. Resuspend in 500 µl TE and re-pellet.

4. Resuspend in 500 µl TE, mix 50 µl of each yeast suspension with 150 µl of water in separate wells of flat-bottomed clear 96-well plates, and measure absorbance at 620 nm (A_{620}) using a plate-reading spectrophotometer. (The yeast cell walls prevent lysis in water in the time required to read absorbance).

5. Use a sterile 96-well plate to prepare serial dilutions of each yeast clone to be analyzed: each clone in a separate column. In the top row, add the appropriate volume of each yeast suspension to achieve 250,000 colony forming units (CFU) in 200 µl of TE (*see* **Note 5**).

6. Pipette 160 µl of TE into rows 2–5 of the plate. Prepare serial fivefold dilutions of the yeast suspensions by taking 40 µl from the first row into the second, mixing, and then taking 40 µl from the second row into the third, and so on. Multichannel pipettors make this step a lot easier and quicker.

7. While the yeast are still in suspension, carefully pipette 5 µl from each well onto both a selective repressing plate (containing glucose) and also onto a selective inducing plate (containing galactose). This will result in 6250 CFU of each clone being "spotted" in the first row, and fivefold fewer being spotted in each subsequent row (*see* **Note 6**).

8. Leave the plates puddle-side facing up, with lids ajar, on the bench for ~15 min, to allow much of the liquid to soak into the plate and evaporate. Alternatively, if a laminar flow hood is available, leave the plates in it for ~5 min with the lids completely off.

9. Incubate the plates at 30 °C for 2 days (repressing glucose plate) or 3–4 days (inducing galactose plate). Figure 4 shows typical results from this kind of assay.

3.4 Visualizing Activity of BH3-Mimetics or Small Molecule Caspase Inhibitors in Yeast

This method uses optical density as a measure of yeast survival and propagation. **Steps 1–4** are the same as those for the spotting assay described above.

1. Grow each of the transformants to be tested (including the empty vector control) for 12–18 h in 1 ml of selective repressive liquid media. It is useful to assay at least three transformants bearing each plasmid combination, to determine the extent of clone-to-clone variability.

2. Pellet the yeast by centrifuging for 15 s at $14,000 \times g$.

3. Resuspend in 500 µl TE and re-pellet.

4. Resuspend in 500 µl TE, mix 50 µl of each yeast suspension with 150 µl of water in separate wells of flat-bottomed clear 96-well plates, and measure absorbance at 620 nm (A_{620}) using a plate-reading spectrophotometer.

5. Pipette 150 µl of minimal media containing 2 % raffinose (no glucose, no galactose) per well into as many wells of a sterile 96-well plate as the clones to be assayed.

6. Pipette 250,000 CFU of each yeast suspension (*see* **Note 5**) into the corresponding well of the 96-well plate.

7. Incubate, shaking at 30 °C for 5 h.

8. Towards the end of the incubation, prepare a second sterile clear, flat-bottomed, 96-well plate with the media +/− drugs to be tested. Into each well, pipette 150 µl of minimal media containing galactose plus either the BH3-mimetic drug (at one or more concentrations) or equivalent amount of solvent. It is important to measure the growth of each clone in media containing and lacking the drug. One convenient option is to arrange the plate so each clone occupies a column, and each row contains either just solvent or defined concentrations of the drug(s) in a constant concentration of solvent. Do not use the outer wells of the plate: instead pipette 150 µl of TE into the wells surrounding the samples to minimize evaporation. Equilibrate this plate at 30 °C.

9. Use a multichannel pipettor to resuspend the yeast growing in the first plate. Transfer 10 µl of each clone into the appropriate wells of the second plate containing media +/− drugs.

10. Place the plate into the plate-reading spectrophotometer programmed to maintain 30 °C and shake for 2 min then take readings at A_{620} every 0.5 h for up to 48 h.

11. Plot the absorbance of each sample relative to time, to visualize the effect of the transgenes and drugs on yeast growth.

12. Figure 7 shows examples of this kind of data.

3.5 Defining the Minimal Specificity of a Protease Using Yeast

This method uses Xgal staining to visualize reporter gene activity, as a readout of caspase-mediated fusion protein cleavage. It can be used as described below to identify proteins containing a caspase-sensitive cleavage site, and can easily be adapted to quickly test caspase-mediated cleavage of particular fusion proteins engineered to bear particular sequences between the CD4 and lexAB42 domains. Some indication of cleavage efficiency can be gleaned from the intensity of the blue color produced due to Xgal cleavage by β-galactosidase, and/or the speed with which that blue color is observed. For a more quantitative assessment of β-galactosidase activity, and hence caspase cleavage efficiency, the liquid ONPG assay outlined in Subheading 3.6 can be used.

1. Create or obtain (e.g., from us) a plasmid encoding a fusion protein containing the transmembrane portion of CD4 and the lexAB42 transcription factor, separated by a linker containing nucleotides encoding the sequence XXXD, under the con-

trol of the GAL1/10 promoter. We create these libraries using redundant oligonucleotides bearing NNS codons.

2. Create or obtain a second galactose-inducible plasmid encoding an active form of the caspase or other protease of interest.

3. Transform the EGY48 yeast strain with a caspase expression plasmid and pSH18-34 (a uracil-selectable plasmid encoding lexA-inducible β-galactosidase), using Subheading 3.1 (*see* **Note 7**).

4. Plate onto selectable repressing media and incubate at 30 °C for 2–3 days.

5. Transform the "XXXD" library into a transformant clone. Plate 90 % of the transformation onto one 15 cm selectable repressing plate and 10 % onto a second selective repressing plate, and incubate both for 2–3 days.

6. Prepare Xgal-drenched 3MM paper: cut two pieces of 3MM paper to fit inside a 15 cm petri dish. Pour 10 ml of Z buffer containing β-mercaptoethanol and Xgal into an empty 15 cm petri dish. Put the 3MM papers into the plate, avoiding bubbles. Remove the excess liquid by pipetting.

7. Carefully lay a nylon filter over a transformation plate bearing nicely separated colonies. Try to avoid dragging the filter, as this will smear the yeast.

8. Dip a 22 gauge needle in a colored dye (e.g., agarose gel loading dye, Coomassie blue stain) and punch asymmetrically located holes through filter into the agar. These colored holes will enable colonies on the Xgal-stained filter to be subsequently aligned with the corresponding colonies on the plate.

9. Use two pairs of forceps on opposite sides of the filter to carefully lift it off the plate without smearing.

10. Place the filter into a bath of liquid nitrogen to freeze, then remove and lie on a petri dish lid to thaw.

11. Transfer the filter onto the Xgal-soaked 3MM papers.

12. Wrap the petri dish with plastic wrap and incubate at 37 °C for 1–8 h. If/when colonies stain blue, align the filter with the transformation plate to identify the corresponding colonies.

13. Streak out positive colonies and repeat staining, to process a single clone. Extract library plasmid DNA (Subheading 3.2) for further analysis and to define the caspase-cleavable sequence.

3.6 Quantitative Assessment of β-Galactosidase Activity, as a Readout of Caspase Cleavage Efficiency

Although more quantitative than the Xgal-based method described above, this approach is less sensitive.

1. Grow each of the transformants to be tested for ~18 h in 1 ml of selective repressive liquid media. It is useful to assay at least three transformants bearing each plasmid combination, to determine the extent of clone-to-clone variability.

2. Pellet the yeast by centrifuging for 15 s at $14,000 \times g$.

3. Resuspend in 500 µl of TE and re-pellet.

4. Resuspend in 500 µl of TE, mix 50 µl of each yeast suspension with 150 µl of water in separate wells of flat-bottomed clear 96-well plates and measure A_{620} using a plate-reading spectrophotometer.

5. Based on the A_{620} readings, pellet all of the most dilute suspension and smaller volumes of the more concentrated suspensions, so roughly equal numbers of yeast are collected for each clone.

6. Resuspend each pellet in 1 ml of inducing selective medium.

7. Incubate, shaking at 30 °C for 10–24 h.

8. Pellet the yeast by centrifuging for 15 s at $14,000 \times g$.

9. Resuspend each sample in 400 µl Z buffer and measure and record A_{620}.

10. Pipette 100 µl into a fresh 1.5 ml tube and freeze at –80 °C from 15 min to overnight.

11. Prepare Z buffer containing β-mercaptoethanol and ONPG and equilibrate to 30 °C.

12. Thaw tubes to be analyzed in a 37 °C waterbath or heatblock (*see* **Note 8**).

13. Add 700 µl Z/β-ME/ONPG buffer to each tube. Note the time.

14. Incubate each tube at 30 °C until yellow color develops, up to 3 h.

15. For each tube, when distinct yellow color is visible, add 400 µl of 1 M Na_2CO_3 to stop the reaction and record the time that elapsed between resuspending in ONPG buffer (i.e., **step 12**) and adding Na_2CO_3.

16. Pellet the yeast by centrifuging for 5 min at $14,000 \times g$.

17. Measure A_{414} of each sample's supernatant.

18. Express β-galactosidase activity according to the formula:

$$\frac{2500 \times A_{414} \, (\text{step } 16)}{\text{time till yellow (seconds)} \times A_{620} \, (\text{step } 8)}$$

4   Notes

1. Good aeration is important, so use vented flasks or only fill closed containers (e.g., tubes) to 10 % or less of their capacity.

2. It is possible to simultaneously transform multiple plasmids with different selectable markers. The transformation efficiency does decrease, however: this method reproducibly yields

hundreds of colonies when two plasmids are co-transformed, and usually tens of colonies when three plasmids are transformed. Quadruple transformation are often unsuccessful, however. If three or four plasmids must be present, it would be advisable to perform sequential transformations.

3. If one or two plasmids are being transformed, it is often not necessary to plate the entire transformation mixture (unless it is crucial to maximize the numbers of transformants, as when screening libraries). To save time in this context, after the heat-shock step be careful not to mix the contents of the tube and simply plate 70 µl of the mixture from the bottom of the tube (most of the yeast will have settled there) directly onto the plate.

4. "Hit" clones from library screens would usually contain at least two plasmids, for example a leucine-selectable plasmid encoding a caspase and a histidine-selectable library-derived plasmid encoding a putative caspase inhibitor. We have used three techniques to specifically isolate the library plasmids: (A) DNA extracted from the yeast can be transformed into the KC8 strain of *E. coli* (Clontech), which harbors auxotrophic leuB, trpC, and hisB mutations that can be complemented by the *LEU2, TRP1*, and *HIS3* genes present on the yeast expression plasmids described above. KC8 cells transformed with DNA from yeast clones can be plated onto ampicillin-containing M9 minimal media lacking the appropriate amino acid, to select for bacterial transformants that acquired only the library plasmids (and not those encoding the caspase, for example). (B) DNA isolated from yeast can be transformed into any competent *E. coli* strain and plated onto complete media containing ampicillin, then colony-PCR can be performed using primers that anneal to sequences uniquely present in the library plasmid (e.g., within the *HIS3* gene). (C) We introduced the kanamycin-resistance cassette from pDORR221 (Invitrogen) into a restriction site located within the β-lactamase gene of pGALL-(*LEU2*) plasmid, thereby creating a plasmid that was leucine-selectable in yeast, and kanamycin-selectable (but not ampicillin-selectable) in bacteria. To screen for caspase inhibitors, caspase genes were subcloned into this vector, and cDNA libraries were constructed in the histidine/ampicillin-selectable pGALL-(*HIS3*) plasmid. This strategy enables simple isolation of library plasmids encoding potential caspase inhibitors, by transforming bacteria with yeast DNA then plating on media containing ampicillin.

5. Determine empirically the relationship between CFU and absorbance, use the plate-reading spectrophotometer you plan to use for these assays. Measure the A_{620} of a suspension of yeast bearing one or more plasmids, create around six serial 1:4

dilutions in TE and plate 70 μl of each onto selective repressing plates. Count the colonies on plates where this is possible, to determine the relationship between CFU and A_{620}.

6. To produce neat "spottings," position the tip just above the surface of the plate and slowly depress the pipettor to release the liquid: try not to gouge the agar. It can be helpful to place a template (we use a piece of paper marked with dots corresponding to the positions of the wells), under the agar plate. If this step is not done quickly, the yeast will settle to the bottom of the wells. They can be easily resuspended using a multichannel pipette if necessary. Spotting assays work best when the plates are relatively well dried, as the puddles soak in better and are less likely to dribble when the plates are inverted. To ensure the plates are suitably dry, after pouring the plates leave them in a laminar flow hood with the lids off for ~15–20 min before using.

7. The EGY48 strain contains a lexA-inducible LEU2 gene whose expression allows growth on media lacking leucine to be used as a readout of transcription factor activity (and hence caspase-mediated substrate cleavage). To avoid confounding selective pressure to retain the caspase plasmid with reporter gene activity, we avoid using leucine-selectable caspase expression plasmids in this system—instead we tend to express caspases from histidine and/or tryptophan-selectable plasmids.

8. Each sample may turn yellow at a different time, and the time taken to develop the yellow color is critical to accurately estimate reporter gene activity. To ensure the color development in each tube can be monitored properly, and each reaction stopped when they reach equivalent color intensity, it is advisable to process only a few samples at a time.

References

1. Madeo F, Herker E, Wissing S, Jungwirth H, Eisenberg T, Frohlich KU (2004) Apoptosis in yeast. Curr Opin Microbiol 7(6):655–660

2. Puryer MA, Hawkins CJ (2006) Human, insect and nematode caspases kill Saccharomyces cerevisiae independently of YCA1 and Aif1p. Apoptosis 11(4):509–517

3. Hawkins CJ, Wang SL, Hay BA (1999) A cloning method to identify caspases and their regulators in yeast: identification of Drosophila IAP1 as an inhibitor of the Drosophila caspase DCP-1. Proc Natl Acad Sci 96(6):2885–2890

4. Hawkins CJ, Silke J, Verhagen AM, Foster R, Ekert PG, Ashley DM (2001) Analysis of candidate antagonists of IAP-mediated caspase inhibition using yeast reconstituted with the mammalian Apaf-1-activated apoptosis mechanism. Apoptosis 6(5):331–338

5. Sikorski R, Hieter P (1989) A system of shuttle vectors and yeast host strains designed for efficient manipulation of DNA in Saccharomyces cerevisiae. Genetics 122:19–27

6. Christianson TW, Sikorski RS, Dante M, Shero JH, Hieter P (1992) Multifunctional yeast high-copy-number shuttle vectors. Gene 110(1):119–122

7. Karim AS, Curran KA, Alper HS (2013) Characterization of plasmid burden and copy number in Saccharomyces cerevisiae for optimization of metabolic engineering applications. FEMS Yeast Res 13(1):107–116, 110/1111/1567-1364.12016. Epub 12012 Nov 12020

8. Srinivasula S, Ahmad M, MacFarlane M, Luo Z, Huang Z, Fernandes-Alnemri T, Alnemri E (1998) Generation of constitutively active recombinant caspases-3 and -6 by rearrangement of their subunits. J Biol Chem 273(17):10107–10111

9. Ho PK, Jabbour AM, Ekert PG, Hawkins CJ (2005) Caspase-2 is resistant to inhibition by inhibitor of apoptosis proteins (IAPs) and can activate caspase-7. FEBS J 272(6):1401–1414

10. Wang SL, Hawkins CJ, Yoo SJ, Muller HA, Hay BA (1999) The *Drosophila* caspase inhibitor DIAP1 is essential for cell survival and is negatively regulated by REAPER, HID and GRIM, which disrupt DIAP1-caspase interactions. Cell 98(4):453–463

11. Thornberry N, Rano T, Peterson E, Rasper D, Timkey T, Garcia-Calvo M, Houtzager V, Nordstrom P, Roy S, Vaillancourt J, Chapman K, Nicholson D (1997) A combinatorial approach defines specificities of members of the caspase family and granzyme B. Functional relationships established for key mediators of apoptosis. J Biol Chem 272(29):17907–17911

12. Srinivasula S, Ahmad M, Fernandes-Alnemri T, Alnemri E (1998) Autoactivation of procaspase-9 by Apaf-1-mediated oligomerization. Mol Cell 1(7):949–957

13. Kluck RM, Martin SJ, Hoffman BM, Zhou JS, Green DR, Newmeyer DD (1997) Cytochrome c activation of CPP32-like proteolysis plays a critical role in a Xenopus cell-free apoptosis system. EMBO J 16(15):4639–4649

14. Brand IL, Civciristov S, Taylor NL, Talbo GH, Pantaki-Eimany D, Levina V, Clem RJ, Perugini MA, Kvansakul M, Hawkins CJ (2012) Caspase inhibitors of the P35 family are more active when purified from yeast than bacteria. PLoS One 7(6):39248. doi:10.1371/journal.pone.0039248

15. Brand IL, Green MM, Civciristov S, Pantaki-Eimany D, George C, Gort TR, Huang N, Clem RJ, Hawkins CJ (2011) Functional and biochemical characterization of the baculovirus caspase inhibitor MaviP35. Cell Death Dis 2, e242

16. Jabbour AM, Ekert PG, Coulson EJ, Knight MJ, Ashley DM, Hawkins CJ (2002) The p35 relative, p49, inhibits mammalian and *Drosophila* caspases including DRONC and protects against apoptosis. Cell Death Differ 9(12):1311–1320

17. Beaumont TE, Shekhar TM, Kaur L, Pantaki-Eimany D, Kvansakul M, Hawkins CJ (2013) Yeast techniques for modeling drugs targeting Bcl-2 and caspase family members. Cell Death Dis 4:e619, 10.1038/cddis.2013.143

18. Khoury CM, Greenwood MT (2008) The pleiotropic effects of heterologous Bax expression in yeast. Biochim Biophys Acta 1783(7):1449–1465

19. Walensky LD, Gavathiotis E (2011) BAX unleashed: the biochemical transformation of an inactive cytosolic monomer into a toxic mitochondrial pore. Trends Biochem Sci 36(12):642–652. doi:10.1016/j.tibs.2011.08.009

20. Kitevska T, Roberts SJ, Pantaki-Eimany D, Boyd SE, Scott FL, Hawkins CJ (2014) Analysis of the minimal specificity of caspase-2 and identification of Ac-VDTTD-AFC as a caspase-2-selective peptide substrate. Biosci Rep 17:17

21. Westein SJ, Scott FL, Hawkins CJ (2008) Analysis of the minimal specificity of CED-3 using a yeast transcriptional reporter system. Biochim Biophys Acta 1783(3):448–454

22. Ito H, Fukuda Y, Murata K, Kimura A (1983) Transformation of intact yeast cells treated with alkali cations. J Bacteriol 153(1):163–168

Chapter 14

Characterizing Bcl-2 Family Protein Conformation and Oligomerization Using Cross-Linking and Antibody Gel-Shift in Conjunction with Native PAGE

Grant Dewson

Abstract

The Bcl-2 family of proteins tightly controls the intrinsic or mitochondrial pathway of apoptosis. This family is subdivided based on function into pro-survival proteins (Bcl-2, Bcl-x$_L$, Bcl-w, Mcl-1, Bfl-1/A1) and pro-apoptotic proteins. The pro-apoptotic subset is further divided into those proteins that initiate the pathway, the BH3-only proteins (including Bim, Puma, Noxa, and Bid), and those that execute the pathway, Bak and Bax. Whether a cell lives or dies in response to apoptotic stress is determined by the interactions of the Bcl-2 family, which is in turn influenced by their conformation. We describe here a protocol to interrogate the interactions and conformation of the Bcl-2 family of proteins under native conditions.

Key words Apoptosis, Bak, Bak, Bcl-2, Cross-linking, Mitochondria, Native polyacrylamide gel electrophoresis, Oligomerization

1 Introduction

The Bcl-2 family participates in a dynamic network of interactions at the mitochondrial outer membrane to regulate apoptosis. This myriad of interactions is influenced by the conformation of these proteins and ultimately controls whether Bak and Bax become activated to damage the mitochondrial outer membrane. This breach in the integrity of mitochondria is usually sufficient to bring about a cells' demise. However, death is ensured by the release of apoptogenic factors from the mitochondrial inter-membrane space including cytochrome c that activate apoptotic proteases termed caspases that demolish the cell.

Both Bak and Bax participate in multiple interactions at mitochondria in healthy cells. These include interactions that govern mitochondrial localization or pro-apoptotic potential such as with VDAC2 [1–3] and the pro-survival Bcl-2 proteins [4, 5]. They have also been proposed to interact with components of the

Hamsa Puthalakath and Christine J. Hawkins (eds.), *Programmed Cell Death: Methods and Protocols*, Methods in Molecular Biology, vol. 1419, DOI 10.1007/978-1-4939-3581-9_14, © Springer Science+Business Media New York 2016

mitochondrial fission/fusion machinery [6]. Once apoptosis is induced, Bak and Bax dissociate from these regulatory interactions and instead self-associate to form the putative apoptotic pore. The structure of this pore is currently unknown, although evidence suggests that both Bak and Bax form stable homodimers that multimerize by an as yet undetermined mechanism to form the pore that damages the mitochondrial outer membrane [7–11]. Understanding these interactions is critical to decipher how the apoptotic activity of Bax and Bak is so exquisitely controlled. Interrogating the complex interactions under native conditions is paramount, as interactions between the Bcl-2 family members are profoundly influenced by subcellular localization and certain detergents [12]. Blue native polyacrylamide gel electrophoresis (BN-PAGE) and clear native PAGE (CN-PAGE) are powerful approaches to characterize the interactions of membrane proteins (*see* Fig. 1). BN-PAGE has been informative in interrogating the molecular interactions and protein conformation change of Bcl-2 family proteins [2, 13–16]. Using native PAGE in combination

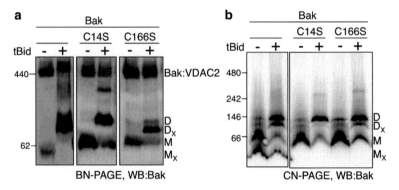

Fig. 1 BN-PAGE and CN-PAGE resolve different complexes of Bak. Mitochondria-enriched fractions from MEFs expressing human Bak or mutants of Bak with either of the endogenous cysteines (C14 or C166) mutated to serine were solubilized in 1 % digitonin without DTT. Membrane fractions were then analyzed on (**a**) BN-PAGE or (**b**) CN-PAGE and immunoblotted for Bak. CN-PAGE can provide superior resolution of the low molecular weight forms of Bak, but fails to resolve the larger complex involving VDAC2 complex. Native PAGE in the absence of reducing agent (DTT) reports disulfide linkage of Bak that is diagnostic of conformation and intermolecular interfaces. M_x, intramolecularly disulfide-linked monomer (diagnostic of activated Bak); M, non-disulfide-linked monomer (diagnostic of activated Bak); D, non-disulfide-linked homodimer; D_x, intermolecularly disulfide-linked homodimer. Note that samples were electrophoresed on a single gel but intervening lanes have been removed for clarity. This research (**b**) was originally published in the *Journal of Biological Chemistry*. Ma S, Hockings C, Anwari K, Kratina T, Fennell S, Lazarou M, Ryan MT, Kluck RM, Dewson G. Assembly of the Bak apoptotic pore: a critical role for the Bak protein α6 helix in the multimerization of homodimers during apoptosis. J Biol Chem. 2013; 288(36):26027–38. © the American Society for Biochemistry and Molecular Biology

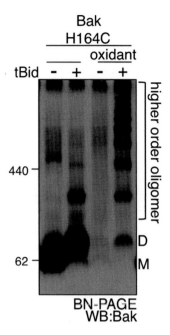

Fig. 2 Cross-linking native PAGE can inform protein interfaces. Mitochondria from MEFs expressing a variant of human Bak with a single cysteine engineered in the α6 helix (H164C) were treated with apoptotic stimulus (recombinant tBid) followed by induction of cysteine cross-linking (in this example cysteines were cross-linked by oxidant-induced disulfide bonds). Linkage of the single cysteine in α6 stabilized higher order oligomeric complexes of Bak on native PAGE, thereby implicating the α6 as a potential interface for multimerization of homodimers

with cysteine cross-linking (*see* Fig. 2) and antibody gel-shift (*see* Fig. 3) can provide additional information regarding the conformations of proteins in specific complexes and inform the interfaces involved.

2 Materials

2.1 Cell Fractionation Reagents

1. Phosphate buffered saline (PBS): 137 mM NaCl, 2.7 mM KCl, 10 mM Na_2HPO_4, 1.8 mM KH_2PO_4.

2. Complete™ protease inhibitors without EDTA: Make a 50× stock by dissolving one tablet in 1 ml of deionized water and store at –20 °C until use.

3. Digitonin (BioSynth): Make a 10 % (w/v) in DMSO and store at room temperature.

4. Permeabilization buffer: 20 mM Hepes/KOH, pH 7.5, 250 mM sucrose, 50 mM KCl, 2.5 mM $MgCl_2$.

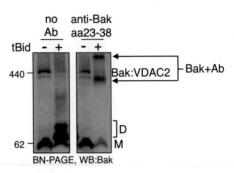

Fig. 3 Antibody gel-shift informs protein conformation in native complexes. Mitochondria from MEFs were incubated with an apoptotic stimulus (recombinant tBid) followed by conformation-specific antibody targeting the N-terminus of Bak (aa23–28) prior to BN-PAGE and immunoblotting for Bak. The conformation-specific antibody did not gel-shift either monomeric Bak (M) or that complexed with VDAC2 indicating that Bak was inactive. Following Bak activation with tBid, the antibody bound and gel-shifted the homodimer (D) population confirming that the dimer comprised Bak in an activated conformation. This research was originally published in the *Journal of Biological Chemistry*. Ma S, Hockings C, Anwari K, Kratina T, Fennell S, Lazarou M, Ryan MT, Kluck RM, Dewson G. Assembly of the Bak apoptotic pore: a critical role for the Bak protein α6 helix in the multimerization of homodimers during apoptosis. J Biol Chem. 2013; 288(36):26027–38. © the American Society for Biochemistry and Molecular Biology

5. Q-VD.oph: Dissolve in sterile DMSO to a concentration of 25 mM and store at –20 °C.

6. Dithiothreitol (DTT): Make a 2 M stock in water and store at –20 °C.

7. *N*-ethylmaleimide (NEM): Make a 1 M stock in ethanol on the day of the experiment.

8. Trypan blue: Make a 0.4 % w/v in PBS. Boil to dissolve and cool to room temperature.

2.2 Disulfide and Cross-linking Reagents

1. Bismaleimidoethane (BMOE, 8 Å linker arm): Dissolve in DMSO to a concentration of 10 mM on the day of the experiment.

2. Bismaleimidohexane (BMH, 12 Å linker arm): Dissolve in DMSO to a concentration of 10 mM on the day of the experiment.

3. Copper (II)(1,10-phenanthroline)3: Dissolve 1,10-phenanthroline to a concentration of 20 mM in 20 % ethanol. Dissolve copper sulfate to concentration of 300 mM in water. To make a 10 mM stock solution (concentration refers to the phenanthroline concentration), the phenanthroline and copper sulfate solutions are diluted to 10 mM and 30 mM respectively in 20 % ethanol.

4. Chemical cross-linking buffer: 20 mM Hepes/KOH, pH 7.5, 250 mM sucrose, 1 mM EDTA, 50 mM KCl, 2.5 mM MgCl$_2$.

5. *tris*(2-carboxyethyl)phosphine (TCEP): Make a 1 mM stock in water on the day of use.

6. Dithiothreitol (DTT): Make a 2 M stock in water and stored at –20 °C.

7. *N*-ethylmaleimide (NEM): Make a 1 M stock in ethanol on the day of the experiment.

2.3 Antibody Reagents for Gel-Shift

1. Conformation-specific antibodies: Several conformation-specific antibodies that only recognize the activated form of Bak and Bax have been characterized and are commercially available (*see* **Note 1** and Fig. 4) [12, 17, 18].

2. Positive control antibodies: Antibodies that recognize all conformations. Bax 2D2 (Santa Cruz Biotechnology, CA) recognizes both inactive and activated human Bax [19] and Bak 2 14 (Millipore) recognizes both inactive and activated human Bak (albeit the inactive form slightly less efficiently) [18] (*see* **Note 9**).

3. Negative control antibodies: Isotype-matched antibody raised against an irrelevant protein.

2.4 Blue Native PAGE Reagents

1. Anode buffer: 25 mM imidazole/HCl pH 7.0.

2. BN-PAGE Loading buffer: 5 % Coomassie Blue G-250, 500 mM aminocaproic acid, filtered through a 0.45 µm filter.

3. Cathode buffer (*blue*) (10×): 500 mM Tricine, 75 mM imidazole pH 7, 0.2 % w/v Coomassie Blue G-250, filtered through a 0.45 µm filter.

4. Cathode buffer (*clear*) (10×): 500 mM Tricine, 75 mM imidazole, pH 7.0.

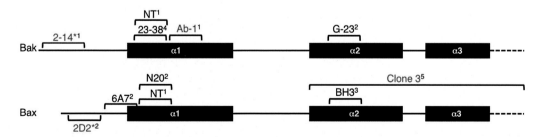

Fig. 4 Bax and Bak conformation-specific antibodies. Numerous commercial antibodies are available that specifically, or preferentially, recognize the activated forms of Bax and Bak. Some commonly used antibodies are indicated showing their recognition sites spanning the first three α-helices (including the BH3 domain) of Bak and Bax. Antibodies that only recognize human protein are indicated (*red*). *Antibody recognizes both the activated and inactive forms and is a useful positive control. [1]Millipore, [2]Santa Cruz Biotechnology, [3]Abgent, [4]Sigma-Aldrich, [5]BD Biosciences

5. Solubilization buffer: 50 mM NaCl, 5 mM aminocaproic acid, 1 mM EDTA, 50 mM imidazole/HCl pH 7.

6. Digitonin: Make a 10 % (w/v) solution in solubilization buffer and heat at 95 °C for 5 min to dissolve.

7. NativeMark™ prestained molecular weight markers (Life Technologies or equivalent).

2.5   Clear Native PAGE Reagents

1. Solubilization buffer: 50 mM NaCl, 5 mM aminocaproic acid, 1 mM EDTA, 50 mM imidazole/HCl pH 7.

2. Digitonin: Make a 10 % (w/v) solution in solubilization buffer and heat at 95 °C for 5 min to dissolve.

3. Ponceau S solution.

4. Glycerol: 50 % v/v in water.

5. Anode buffer: 25 mM imidazole/HCl pH 7.0.

6. Cathode buffer (1×): 50 mM Tricine, 7.5 mM imidazole, 0.05 % w/v deoxycholate, 0.01 % w/v n-dodecyl-D-maltopyranoside, pH 7.0.

7. NativeMark™ prestained molecular weight markers (Life Technologies or equivalent).

2.6   Electrotransfer Reagents

1. Native PAGE transfer buffer: Tris-Glycine/MeOH, 0.037 % SDS.

2. Coomassie R-250 staining solution.

3. Destaining solution: 50 % v/v methanol, 25 % v/v acetic acid (*see* **Note 2**).

4. Polyvinylidene fluoride (PVDF) or nitrocellulose transfer membrane.

3   Methods

3.1   Cell Fractionation

1. Treat cells with apoptotic stimuli as required (*see* **Note 3**).

2. Harvest cells, and pellet at $2500 \times g$ for 5 min.

3. Remove media and resuspend cell pellet in ice-cold PBS and pellet cells again at $2500 \times g$, for 5 min.

4. Remove PBS and resuspend the cell pellet in permeabilization buffer supplemented with digitonin (0.025 % w/v) and 1× protease inhibitors at 1×10^7 cells/ml (*see* **Notes 3** and **4**).

5. Incubate on ice for 10 min.

6. Verify permeabilization by trypan blue uptake. Remove 5 μl of permeabilized cells and add an equal volume of trypan blue. Assess trypan blue uptake by light microscopy. More than 90 % cells should be permeabilized (i.e., blue).

7. Spin at $13,000 \times g$ for 5 min at 4 °C to separate supernatant (cytosol) from pellet (heavy membrane—includes mitochondria, but also plasma membrane and nuclei).

8. If combining with disulfide linkage, proceed to Subheading 3.2. If combining with antibody gel-shift, proceed to Subheading 3.3. If proceeding straight to native PAGE, proceed to Subheading 3.4.

3.2 Induction of Cysteine Linkage

1. Membrane fractions are resuspended at 1×10^7 cells/ml in permeabilization buffer without digitonin for disulfide linkage (proceed to **step 2**) or cross-linking buffer for chemical cross-linking either BMOE or BMH (proceed to **step 6**) (*see* Fig. 1). Combining cysteine linkage with BN-PAGE stabilizes higher order oligomers of Bak. Linkage of engineered single cysteine mutants of Bak informs regions that are in an interface that links Bak homodimers to form the higher order apoptotic pore [15]. *Induce disulfide linkage*

2. Add 10 mM copper (II)(1,10-phenanthroline)$_3$ to a final concentration of 1 mM and incubate for 30 min on ice.

3. Quench disulfide linkage with 10 mM NEM.

4. Spin to pellet the membranes at $13,000 \times g$ for 5 min at 4 °C and remove supernatant.

5. Proceed to Subheading 3.4 to analyze samples by native PAGE (*see* **Note 6**).
   Cross-link with chemical cross-linker

6. Add 10 mM BMH or BMOE to a final concentration of 0.5 mM (keep final DMSO concentration to <10 %) and incubate at room temperature for 30 min in the dark.

7. Quench disulfide linkage with 10 mM NEM or 2 mM DTT.

8. Spin to pellet the membranes at $13,000 \times g$ for 5 min at 4 °C and remove supernatant.

9. Proceed to Subheading 3.4 to analyze samples by native PAGE (*see* **Note 6**).

3.3 Antibody Gel-Shift

Combining native PAGE with antibody gel-shift with conformation-specific antibodies can inform the activation status of Bak and Bax in particular complexes (*see* Fig. 3). Additionally, if the epitope recognized by the antibody is known its ability to gel-shift the complex under native conditions can inform the protein interfaces involved.

1. Add 1 μg of conformation-specific antibody (*see* Fig. 4) to permeabilized cells and incubate on ice for 30 min (*see* **Notes 7–9**).

2. Spin to pellet the membranes at $13,000 \times g$ for 5 min at 4 °C and remove supernatant.

3. Proceed to Subheading 3.4 to analyze samples by native PAGE (*see* **Note 5**).

3.4 Preparation of Samples for Native PAGE

1. Resuspend membrane fractions in solubilization buffer including 1 % digitonin by pipetting rapidly with a P200 pipette and incubate for 30 min on ice (*see* **Notes 10** and **11**).

2. Pellet insoluble debris at $13,000 \times g$ for 5 min and retrieve the supernatant.

3. Add 1/10 volume of 10× BN-PAGE loading buffer and proceed to BN-PAGE (*see* **Note 12**).

3.5 Blue Native PAGE [20] (See Fig. 1a)

1. Native PAGE gradient gels (standardly 3–12 % or 4–16 %) can be poured using a gradient mixer. Alternatively, pre-cast minigels can be purchased from Life Technologies.

2. Before loading your samples, add sufficient Cathode buffer (*blue*) to fill the *wells only*. Do not fill the cathode chamber entirely. This will allow you to see the samples as they are loaded and help prevent displacement of the sample. Once the samples are loaded, fill cathode chamber with Cathode buffer (*blue*) carefully avoiding the wells.

3. Load sample as standardly sufficient to detect the protein of interest by immunoblotting. This will vary depending on cell type so needs to be established empirically. It is important not to overload the gel as this will reduce the definition of the complexes.

4. Load samples alongside native molecular weight markers.

5. Add Anode buffer to the anode chamber to ½ way up the gel.

6. Run minigels at constant current at 8 mA per gel. Run large (15 cm) gels at 8 mA until Coomassie dye front enters the separation gel then 14 mA for 3–5 h.

7. When the Coomassie dye front has migrated approximately 1/3 of the way through the separation gel, replace Cathode buffer (*blue*) with Cathode buffer (*clear*). This will allow visualization of the proteins and also aid subsequent transfer.

8. Restart the electrophoresis until the Coomassie dye front nears the bottom of the gel.

3.6 High Resolution Clear Native PAGE (See Note 13) [21] (See Fig. 1b)

1. Add 0.01 % v/v Ponceau S and 5 % v/v glycerol to the solubilized membrane fractions (*see* **Note 14**).

2. For clear native PAGE, the cathode buffer is supplemented with 0.05 % w/v deoxycholate and 0.01 % w/v *n*-dodecyl-D-maltopyranoside instead of Coomassie Blue G-250.

3. Electrophorese as for blue native PAGE except there is no need to change the cathode buffer.

3.7 Electrotransfer and Immunoblotting

1. Carefully dismantle the gel cassette and equilibrate the gel briefly in native PAGE transfer buffer (*see* **Note 15**).

2. Electrotransfer proteins to PVDF membrane according to the standard procedure (*see* **Note 2**). We find that 30 V for 2.5 h for minigels and 400 mA for 2 h for large gels efficiently transfers supramolecular protein complexes.

3. After electrotransfer, incubate the blot in Coomassie R-250 stain for 10 min with gentle agitation.

4. To visualize the molecular weight markers (*see* **Note 16**), destain the gel by incubating in detain solution until the majority of the dye has been removed. The blot will still appear light blue, but this will not impact on subsequent immunoblotting.

5. Wash the blot several times in deionized water to remove all traces of the destain solution.

6. Block the membranes in 5 % w/v skimmed milk in TBS-T and immunoblot membranes under conditions optimized for the antibody/antigen (*see* **Note 17**).

4 Notes

1. Numerous conformation-specific antibodies that recognize only the activated forms of Bax and Bak are available (*see* Fig. 4). Although pro-survival proteins are also argued to change conformation [22], antibodies that distinguish between the pro-survival conformers are not well characterized.

2. Commonly blue native gels are transferred to PVDF membrane as the blot is destained in 50 % methanol prior to immunoblotting. However, if necessary nitrocellulose can be used, but the destaining solution must contain <20 % methanol.

3. It is important to block activated caspases during apoptosis as their activity may lead to loss of protein. We routinely treat cells with the broad range caspase inhibitor Q-VD.oph (25 μM for 45 min) prior to treatment with apoptotic stimulus.

4. If gel-shift analysis or cysteine linkage is not being performed, one should consider adding a reducing agent at this permeabilization step to prevent the formation of disulfide bonds and so mimic the generally reducing environment of the cytosol to which most of the Bcl-2 family proteins are exposed. This artifactual disulfide linkage can provide important information regarding protein conformation and/or protein interactions (*see* Fig. 1); however if this is not a requirement, disulfide linkage can be prevented with the addition of DTT (2 mM) or NEM (10 mM) is sufficient to prevent disulfide linkage. If gel-shift analysis is to be performed, DTT should be omitted to

prevent disruption of structural disulfides in the antibody. If disulfide linkage is to be performed, omit reducing agents and NEM. When using the maleimide-based cysteine cross-linkers BMH or BMOE, the addition of 0.05 mM TCEP aids linkage by keeping the cysteines reduced, but will not inhibit cysteine cross-linking.

5. At this step the membrane fraction is more difficult to resuspend. Pipette up and down several times with a P200 until no clumps are visible. Disaggregation of cell clumps aids the efficiency of cell permeabilization.

6. When running oxidized disulfide-linked samples alongside reduced (BMH, BMOE) samples on the same native gel, leave at least one (preferably two) empty lane between the samples as the reducing agent will leach across the gel during electrophoresis and disrupt disulfide-linked complexes.

7. The amount of antibody needs to be optimized and will depend on the antibody used, the volume of sample to be treated, and the expression of the target protein.

8. An irrelevant antibody of the same isotype should be used as a negative control. An antibody that is not conformation-specific should be used as a positive control. If unavailable for the endogenous protein, this could be an antibody against an epitope tag.

9. When immunoblotting a gel-shift experiment, much like an immunoprecipitation, it is preferable to use an antibody raised in a different species than the antibody used for gel-shift. Otherwise the secondary antibody will bind avidly to the high concentration of antibody in the gel and will overwhelm the signal and complicate interpretation.

10. Cytosolic fractions can also be run under native conditions. If comparing directly to membrane fractions (for example to monitor Bax translocation during apoptosis), we recommend adding 1 % digitonin to the cytosolic fractions. This ensures that the proteins are treated and migrate similarly on native PAGE.

11. This step is critical for the efficient solubilization of the membrane complexes and their resolution on BN-PAGE.

12. *Critical*: Do not heat the samples as this will denature the proteins and their complexes. As complexes tend to disassociate on storage, only prepare enough sample as required to be run on the day of the experiment.

13. We observe that high resolution clear native PAGE provides better resolution of Bak and Bax homo-oligomers than blue native PAGE, but tends to disrupt the large complex with VDAC2 [15] (*see* Fig. 1).

14. Ponceau S is not to stain proteins, but to aid in sample loading and to provide a dye front.

15. The abundant mitochondrial respiratory complexes can commonly be directly visualized on the gel post-electrophoresis.

16. Molecular weight markers are a good reference to compare between gels. However, it should be noted that the migration of protein complexes on blue native PAGE is not only determined by molecular weight but also influenced by its shape and bound Coomassie dye and detergent. Thus molecular weight markers cannot give an accurate estimation of the mass of a protein complex.

17. Detection of proteins on native blots by immunoblotting is dependent on the antibody recognizing the native protein and the level of expression of the protein of interest. Antibodies that work well for SDS-PAGE may not necessarily work well on native gels. If necessary, to aid immunodetection the blot can be incubated at 65 °C for 30 min in Tris-Glycine buffer supplemented with 2 % (w/v) SDS and 5 % (v/v) β-mercaptoethanol to denature the proteins and reveal occluded epitopes [13].

Acknowledgments

This work was supported by the National Health and Medical Research Council of Australia (637335), Australian Research Council (FT100100791), and was made possible through Victorian State Government Operational Infrastructure Support and Australian Government NHMRC IRIISS 9000220.

References

1. Ma SB, Nguyen TN, Tan I et al (2014) Bax targets mitochondria by distinct mechanisms before or during apoptotic cell death: a requirement for VDAC2 or Bak for efficient Bax apoptotic function. Cell Death Differ 21(12):1925–1935

2. Lazarou M, Stojanovski D, Frazier AE et al (2010) Inhibition of Bak activation by VDAC2 is dependent on the Bak transmembrane anchor. J Biol Chem 285(47):36876–36883

3. Cheng EH, Sheiko TV, Fisher JK et al (2003) VDAC2 inhibits BAK activation and mitochondrial apoptosis. Science 301(5632):513–517

4. Edlich F, Banerjee S, Suzuki M et al (2011) Bcl-x(L) retrotranslocates Bax from the mitochondria into the cytosol. Cell 145(1):104–116

5. Llambi F, Moldoveanu T, Tait SWG et al (2011) A unified model of mammalian BCL-2 protein family interactions at the mitochondria. Mol Cell 44(4):517–531

6. Karbowski M, Lee YJ, Gaume B et al (2002) Spatial and temporal association of Bax with mitochondrial fission sites, Drp1, and Mfn2 during apoptosis. J Cell Biol 159(6):931–938

7. Brouwer JM, Westphal D, Dewson G et al (2014) Bak core and latch domains separate during activation, and freed core domains form symmetric homodimers. Mol Cell 55(6):938–946

8. Czabotar PE, Westphal D, Dewson G et al (2013) Bax crystal structures reveal how BH3 domains activate Bax and nucleate its oligomerization to induce apoptosis. Cell 152(3):519–531

9. Dewson G, Ma S, Frederick P et al (2012) Bax dimerizes via a symmetric BH3:groove interface during apoptosis. Cell Death Differ 19(4):661–670

10. Dewson G, Kratina T, Czabotar P et al (2009) Bak activation for apoptosis involves oligomerization of dimers via their alpha6 helices. Mol Cell 36(4):696–703

11. Dewson G, Kratina T, Sim HW et al (2008) To trigger apoptosis Bak exposes its BH3 domain and homo-dimerizes via BH3:groove interactions. Mol Cell 30(3):369–380

12. Hsu YT, Youle RJ (1997) Nonionic detergents induce dimerization among members of the Bcl-2 family. J Biol Chem 272(21):13829–13834

13. Valentijn AJ, Upton JP, Gilmore AP (2008) Analysis of endogenous Bax complexes during apoptosis using blue native PAGE: implications for Bax activation and oligomerization. Biochem J 412(2):347–357

14. Er E, Lalier L, Cartron PF et al (2007) Control of Bax homodimerization by its carboxyl terminus. J Biol Chem 282(34):24938–24947

15. Ma S, Hockings C, Anwari K et al (2013) Assembly of the Bak apoptotic pore: a critical role for the Bak alpha6 helix in the multimerization of homodimers during apoptosis. J Biol Chem 288(36):26027–26038

16. Ospina A, Lagunas-Martinez A, Pardo J et al (2011) Protein oligomerization mediated by the transmembrane carboxyl terminal domain of Bcl-XL. FEBS Lett 585(19):2935–2942

17. Griffiths GJ, Corfe BM, Savory P et al (2001) Cellular damage signals promote sequential changes at the N-terminus and BH-1 domain of the pro-apoptotic protein Bak. Oncogene 20(52):7668–7676

18. Alsop AE, Fennell SC, Bartolo RC et al (2015) Dissociation of Bak alpha1 helix from the core and latch domains is required for apoptosis. Nat Commun 6:6841

19. Nechushtan A, Smith CL, Hsu YT et al (1999) Conformation of the Bax C-terminus regulates subcellular location and cell death. EMBO J 18(9):2330–2341

20. Wittig I, Braun HP, Schagger H (2006) Blue native PAGE. Nat Protoc 1(1):418–428

21. Wittig I, Karas M, Schagger H (2007) High resolution clear native electrophoresis for in-gel functional assays and fluorescence studies of membrane protein complexes. Mol Cell Proteomics 6(7):1215–1225

22. Dlugosz PJ, Billen LP, Annis MG et al (2006) Bcl-2 changes conformation to inhibit Bax oligomerization. EMBO J 25(11):2287–2296

Using Förster-Resonance Energy Transfer to Measure Protein Interactions Between Bcl-2 Family Proteins on Mitochondrial Membranes

Justin P. Pogmore, James M. Pemberton, Xiaoke Chi, and David W. Andrews

Abstract

The Bcl-2 family of proteins regulates the process of mitochondrial outer membrane permeabilization, causing the release of cytochrome c and committing a cell to apoptosis. The majority of the functional interactions between these proteins occur at, on, or within the mitochondrial outer membrane, complicating structural studies of the proteins and complexes. As a result most in vitro studies of these protein-protein interactions use truncated proteins and/or detergents which can cause artificial interactions. Herein, we describe a detergent-free, fluorescence-based, in vitro technique to study binding between full-length recombinant Bcl-2 family proteins, particularly cleaved BID (cBID) and BCL-X_L, on the membranes of purified mitochondria.

Key words Apoptosis, Bcl-2 family, Detergent free, Förster-resonance energy transfer, FRET, Protein interaction, Mitochondria

1 Introduction

Mitochondrial outer membrane permeabilization (MOMP) generally commits cells to apoptosis through release of pro-apoptotic molecules such as cytochrome c and Smac/Diablo from the mitochondrial inner membrane space via. MOMP is regulated by interactions between Bcl-2 family proteins at, on, and within the mitochondrial outer membrane (MOM). The Bcl-2 family consists of anti-apoptotic and pro-apoptotic proteins that inhibit and promote MOMP, respectively. The anti-apoptotic proteins (Bcl-XL, Mcl-1, Bcl-2, A1) inhibit pro-apoptotic proteins by transient and stable direct binding interactions at mitochondria. The pro-apoptotic proteins have been subdivided into those that oligomerize to form pores within the MOM (e.g., BAX and BAK), the direct

Hamsa Puthalakath and Christine J. Hawkins (eds.), *Programmed Cell Death: Methods and Protocols*, Methods in Molecular Biology, vol. 1419, DOI 10.1007/978-1-4939-3581-9_15, © Springer Science+Business Media New York 2016

activators of these proteins (e.g., BID, BIM), and those that sensitize cells to MOMP (e.g., BAD, NOXA) by inhibiting anti-apoptotic proteins through mutual sequestration [1].

In vitro techniques such as isothermal calorimetry or fluorescence anisotropy can be used to study protein-protein interactions in solution. However, it's been demonstrated that the binding between several Bcl-2 family members, such as cBID and BAX, requires the presence of a membrane [2] that serves both as an active platform for the interactions of Bcl-2 family members and as the substrate permeabilized. Other common techniques such as co-immunoprecipitation are used to analyze interactions between proteins in solubilized cells; however the detergents needed can artifactually promote or prevent authentic Bcl-2 protein interactions [2, 3].

We have successfully used the fluorescence-based technique Förster-Resonance Energy Transfer (FRET) to study interactions between full-length recombinant Bcl-2 family proteins [2]. To make these measurements single-cysteine versions of the proteins are labeled with a pair of fluorophores in which the wavelengths of the light emitted by one (the donor) overlaps the excitation spectrum of the other (the acceptor). For the FRET experiments shown here, the donor is cBID labeled with Alexa 568 and the acceptor is BCL-X_L labeled with Alexa 647. When the distance between the two fluorophores is small (less than 100 Å), the excited donor can transfer energy nonradiatively to the acceptor via dipole-dipole interactions. This energy transfer results in a decrease of the donor emission which we measure in our FRET experiments. FRET is a very sensitive method for detecting protein-protein interactions because the efficiency of FRET (E) decreases to the sixth power of distance (r):

$$E = \frac{R_0^6}{R_0^6 + r^6}$$

where R_0 is the distance between a donor-acceptor pair at which the FRET efficiency is 50 %, also referred to as the Förster distance. For the donor-acceptor dye pair of Alexa 568 and Alexa 647, the reported R_0 is 82 Å [4]. To relate the observed FRET efficiency to binding between donor- and acceptor-labeled proteins requires a series of measurements in which the donor concentration is constant and the acceptor is titrated across a range of concentrations. Using such data it is possible to calculate a binding curve and a dissociation constant (K_D) for the interaction between cBID and BCL-X_L (Fig. 2).

For this assay single-cysteine mutants of cBID and BCL-X_L were created and labeled with thiol-reactive maleimide derivatives of the dyes (ALEXA 568 and 647). Given the limited number of cysteine residues in the Bcl-2 family, we chose to create single-cysteine mutants of our proteins so we can label them with maleimide attached fluorophores. However, it is also possible to label free amino groups such as that of the Lysine side chain using succinimide esters and other commercially available reactive dyes. To measure FRET it is critical that only one position on the protein be labeled due to the distance dependence of the measurements. Single-cysteine mutants can be generated by site-directed mutagenesis. There are a myriad of fluorescent probes available that can react with the thiol group of the cysteine amino acid. The specific fluorophores used will depend on the system in question and the equipment available. The fluorescent experiments described here include mitochondria, which have intrinsic fluorescence in the higher energy visible/UV spectrum due to cytochromes [5], FADH, and other endogenous fluorophores from the cell [6]. Thus, we chose dyes with longer excitation and emission wavelengths for minimal interference. When labeling proteins one should set up a parallel reaction with a cysteine-less mutant to check for nonspecific labeling. We recommend following the labeling protocol supplied by the manufacturer of the fluorescent probe initially; however it is often the case that the protocol must be altered to optimize labeling efficiency and specificity for your protein of interest.

In cells, the pro-apoptotic proteins, BAX (Cytoplasmic and peripherally bound to mitochondria) and BAK (embedded in the MOM), bind to both BCL-X_L and cBID; thus the endogenous BAX and BAK will compete for the fluorescent proteins we add to the reaction and will permeabilize the mitochondria complicating interpreting the data from our experiment. Thus, we use mitochondria purified from the liver of BAK deficient mice (Bak$^{-/-}$) for the FRET assay described here. Liver cells from BAK$^{-/-}$ animals contain BAX; however BAX is cytosolic and is washed from the mitochondrial pellet [7]. As Bcl-XL and Bid are mostly cytosolic and endogenous levels of Bcl-2 and Mcl-1 are low, there is little competition for these proteins when using mouse liver mitochondria. A series of purification steps are described below for the purification of mitochondria from dissected mouse liver.

To ensure the purified proteins are functional, we trigger the release of cytochrome c from mitochondria by using purified Bax and cBid and then measure how efficiently BCL-X_L inhibits the process. This is done by adding purified proteins to mitochondria and immunoblotting for cytochrome c in the supernatant and pellet fractions. Retention of cytochrome c prior to MOMP is also an indication of the integrity of the mitochondria that were prepared. Proteins released from mitochondria are separated by pelleting the mitochondria by centrifugation. As a loading control for the immunoblots, we used heat shock protein 60 (HSP60). The HSP60 protein is retained in the pellet, even if the outer mitochondrial membrane is permeabilized, since it is located within the mitochondrial matrix [8]. Active labeled proteins can then be used to obtain a FRET binding curve for the interaction between cBID and BCL-X$_L$ on mitochondrial membranes as described below.

2 Materials

2.1 Purification of Bcl-2 Family Recombinant Proteins

1. *BAX Lysis Buffer* (pH 7): 10 mM HEPES, 100 mM NaCl, 0.2 % (m/v) CHAPS, 1 mM PMSF, DNase 10 µg/mL, 10 % (w/v) Glycerol. *BAX Wash Buffer* (pH 7): 10 mM HEPES, 500 mM NaCl, 0.5 % (w/v) CHAPS, 10 % (w/v) Glycerol. *BAX Cleavage Buffer* (pH 7): 10 mM HEPES, 200 mM NaCl, 0.2 mM EDTA, 0.1 % (w/v) CHAPS, 200 mM Hydroxylamine, 10 % (w/v) Glycerol (*see* **Note 1**). *BAX Dialysis Buffer* (pH 7): 10 mM HEPES, 200 mM NaCl, 0.2 mM EDTA, 10 % (w/v) Glycerol.

2. *BCL-XL Lysis Buffer* (pH 8): 20 mM Tris (*see* **Note 2**), 500 mM NaCl, 0.5 mM EDTA, 1 % (w/v) CHAPS, 1 mM PMSF, DNase 10 µg/mL. *BCL-XL Wash Buffer* (pH 8): 20 mM TRIS, 200 mM NaCl, 0.2 % (w/v) CHAPS, 20 % (w/v) Glycerol, 1 mM PMSF. *BCL-XL Cleavage Buffer* (pH 8): BCL-X$_L$ Wash Buffer supplemented with 200 mM Hydroxylamine. *BCL-XL Elution Buffer* (pH 8): 20 mM TRIS, 0.2 % (w/v) CHAPS, 20 % (w/v) Glycerol. *BCL-XL Dialysis Buffer* (pH 8): 20 mM TRIS, 20 % (w/v) Glycerol.

3. BID Lysis Buffer (pH 7): 10 mM HEPES, 100 mM NaCl, 10 mM Imidazole, DNase 10 µg/mL, 1 mM PMSF. *BID Wash Buffer* (pH 7): 10 mM HEPES, 300 mM NaCl, 1 % (w/v) CHAPS, 10 mM Imidazole. *BID Elution Buffer* (pH 7): 10 mM HEPES, 100 mM NaCl, 0.1 % (w/v) CHAPS, 200 mM Imidazole, 10 % Glycerol. *BID Cleavage Buffer* (pH 7): 50 mM HEPES, 100 mM NaCl, 0.1 % (w/v) CHAPS, 10 % (w/v) Glycerol, 1 mM EDTA, 10 mM DTT. *BID Dialysis Buffer* (pH 7): 10 mM HEPES, 100 mM NaCl, 0.1 mM EDTA, 10 % (w/v) Glycerol.

4. LB-ampicillin (to make 1.5 L): 10 g Yeast Extract, 15 g NaCl, 150 mg Ampicillin Salt, 1.5 L Milli-Q dH$_2$O.

5. Human Caspase 8: such as from Enzo Life Sciences (Catalog number: ALX-804-447-C100).

2.2 Labeling Proteins with Donor and Acceptor Maleimide ALEXA Fluorophores

1. 100 % DMSO, anhydrous.

2. Maleimide fluorophores from Molecular probes, Life Technologies: Alexa 647 C2-maleimide (Catalog number: A-20347), Alexa 568 C5-maleimide (Catalog number: A-20341).

3. Dialysis Buffers: BAX—10 mM HEPES pH 7.0, 100 mM NaCl, 0.1 mM EDTA. BCL-X_L—20 mM TRIS pH 8.0.

4. 10 % (w/v) CHAPS.

5. Solution of 8 M Urea pH 7.0 in distilled water.

6. 1 M solution of Dithiothreitol (DTT) in distilled water (*see* **Note 3**).

7. 10 mM solution of the reducing agent tris(2-carboxyethyl) phosphine (TCEP).

8. Sephadex G-25 fine beads from GE Healthcare Life Sciences (*see* **Note 4**).

9. Ni-NTA agarose beads from Qiagen.

2.3 Mitochondria Preparation

1. Mouse liver from BAK knockout mice (The Jackson Laboratory, stock number 004183).

2. AT Buffer: 300 mM Trehalose, 10 mM HEPES-KOH pH 7.7, 10 mM KCl, 1 mM EGTA, 1 mM EDTA, 0.1 % BSA.

3. AT-KCl Buffer: 300 mM Trehalose, 10 mM HEPES-KOH pH 7.7, 80 mM KCl, 1 mM EGTA, 1 mM EDTA, 0.1 % BSA.

4. Regenerating Buffer: 300 mM Trehalose, 10 mM HEPES-KOH pH 7.7, 80 mM KCl, 1 mM EGTA, 1 mM EDTA, 0.1 % BSA, 5 mM Succinate, 2 mM ATP, 10 μM phosphocreatine, 10 μg/mL creatine kinase (*see* **Note 5**).

5. Liquid nitrogen and dry ice (solid CO_2).

6. Potter-Elvehjem homogenizer (30 mL size).

2.4 Cytochrome c Release

1. Obtain energized mitochondria from Subheading 3.3.

2. Black, nonbinding surface, Corning 96 well plate (reference number 3881).

3. Alexa 568 126C cBID and Alexa 647 152C BCL-X_L from Subheading 3, and wild-type BAX from Subheading 3.1.

4. Cytochrome *c* antibody (we make ours in-house, but it is commonly available for purchase through many biotechnology companies).

2.5 cBID–BCL-X_L FRET Interaction in the Presence of Purified Mitochondria

1. Purified cBID, and BCL-X_L labeled with Alexa 586 and Alexa 647, respectively, from Subheading 3.2.

2. Purified mitochondria from Subheading 3.3.

3. Instrument capable of end-point fluorescence measurements. We used a TECAN Infinite M1000 PRO.

4. Corning 96 well, clear bottom, half area, black, nonbinding surface plate.

3   Methods

3.1   Purification of Bcl-2 Family Recombinant Proteins

Unless otherwise stated, all protocols should be carried out on ice or 4 °C.

1. Both BAX and BCL-X_L are expressed with a carboxyl-terminal intein-chitin binding domain (IMPACT expression systems, New England Biolabs), while BID is expressed with an amino-terminal 6× Histidine Tag. Plasmids for BAX, BID, and BCL-X_L are transformed into *Escherichia coli* (BL21-AI for BAX and BID, DH5α for BCL-X_L, New England Biolabs), plated on LB-ampicillin agar, and incubated overnight at 37 °C. The next day, a single colony is used to inoculate a small culture of LB-ampicillin (~200 mL) and grown again overnight at 30–37 °C.

2. The next day, take 30–40 mL of the confluent overnight cultures and inoculate 1.5 L cultures of LB-ampicillin. Grow cultures at 37 °C with shaking until they reach an optical density (OD_{600}) between 0.6 and 0.8.

3. Induce the 1.5 L cultures with 0.2 % (w/v) Arabinose (BAX and BID) or 1 mM IPTG (BCL-X_L). Reduce the temperature to 30 °C and continue to grow for an additional 4–5 h (*see* **Note 6**).

4. Harvest the bacteria by centrifugation. Bacterial Pellets can then be stored at –20 °C.

5. Resuspend the bacterial pellets in BAX, BCL-X_L, or BID lysis buffer (~4 mL per bacterial gram) by vortexing and using an 18-gauge needle to remove large chunks. The bacterial lysates are then mechanically lysed via French Press or Homogenizer (AVASTIN) (*see* **Note 7**). Centrifuge the lysed samples and recover the supernatant.

6. Add 1–2 mL of Chitin Bead Resin (New England Biolabs) to the recovered bacterial lysates of BAX and BCL-X_L, and 1 mL of Ni-NTA Resin (Thermo Fisher Scientific) to BID. Let incubate for 1–2 h at 4 °C.

7. Load the incubated lysates into Econo-Pac chromatography columns (BioRad), and allow the lysate to pass through several times. This ensures even packing and saturation of affinity beads.

8. Wash each column with 50 mL of BAX, BCL-X_L, or BID wash buffer. Elute BID from the column by adding 10 mL of BID elution buffer and collecting the first five 1 mL fractions. Perform a Bradford assay to quickly determine which fractions contain the highest concentration of BID. BID must first be labeled before continuing. Please refer to Subheading 3.2 for the labeling of BID, then return to **step 10c** for the cleavage of BID protocol.

9. After washing BAX and BCL-X$_L$ columns, add 10 mL of freshly prepared BAX and BCL-X$_L$ cleavage buffer. Allow the first 8 mL to pass through, and then cap the column to retain ~2 mL of cleavage buffer. Let sit for 48 h at 4 °C. Afterwards, remove the column cap and elute BAX or BCL-X$_L$ (*see* **Note 8**).

10. (a) BAX: A 0.2 mL bed volume of DEAE-Sepharose beads is equilibrated with 2–3 mL of BAX cleavage buffer (without hydroxylamine added). The 2 mL BAX elution is passed through the column three to four times in order to remove nucleic acids that tend to elute with BAX.

    (b) BCL-X$_L$: Equilibrate a 0.3 mL bed volume high-performance phenyl sepharose column with ~3 mL of BCL-X$_L$ wash buffer. Apply BCL-X$_L$ fractions from the chitin column, and wash again with ~5 mL of BCL-X$_L$ wash buffer. Elute BCL-X$_L$ with BCL-X$_L$ elution buffer (~3 mL). Collect three fractions and assess via Bradford to determine which has the highest concentration of BCL-X$_L$.

    (c) BID: To make cBID (cleaved BID) take your labeled BID with the highest concentration (~1 mL) and add the appropriate amounts of buffer components to achieve cleavage buffer concentrations. I.e., BID Dialysis Buffer is at 10 mM HEPES, so add 40 mM HEPES to achieve a final concentration of 50 mM HEPES as stated in BID Cleavage Buffer. Add 500 U of human Caspase-8 (Enzo Life Sciences) to your BID aliquot, and leave for 48 h at room temperature to cleave.

11. BAX, BID, and BCL-X$_L$ aliquots are then put into 16–18 kDa dialysis tubing, ensuring both ends of the tubing are securely clipped. These proteins are dialyzed with BAX, BID, and BCL-X$_L$ Dialysis buffer respectively, 1 L for 4 h, 2 L overnight, then 1 L for 4 h the next morning.

12. Proceed to Subheading 3.2 Labeling of Proteins before completing **step 13** below.

13. Repeat dialysis **step 11** for BAX and BCL-X$_L$ after labeling the proteins.

14. Proteins can then be aliquoted into non-stick hydrophobic tubes (Fisher Scientific), flash-frozen with liquid nitrogen, and stored at −80 °C.

3.2 Labeling of Proteins with Donor and Acceptor Maleimide ALEXA Fluorophores

Steps are carried out at room temperature unless otherwise indicated.

1. For BID, BAX, and BCL-X$_L$, the labeling reaction is carried out after dialysis of the protein (*see* **Note 9**). Ensure the pH of the protein sample to be labeled is between 7.0 and 7.5, which

is suitable to allow the cysteines to be most reactive, while lowering the reactivity of primary amines that could potentially be labeled (*see* **Note 10**).

2. Adjust the buffer conditions of the protein sample such that CHAPS is increased to 0.5 % of the final volume. The dye is typically dissolved in DMSO and the amount of dye added to the protein sample should be 10–15 times the mole amount of protein in the reaction.

3. Add the reducing agent, TCEP, four times the mole amount of protein to prevent disulfide bond formation (*see* **Note 11**).

4. The reaction should rotate in a microfuge tube wrapped in foil at room temperature for 2–3 h (*see* **Note 12**). After the specified time has elapsed, quench the reaction with 1 mM DTT.

5. (a) Apply BAX or BCL-X_L labeled protein solution to a pre-equilibrated (with the appropriate dialysis buffer) G25 column with a bed volume ten times the reaction volume (e.g., 1.5 mL reaction volume = 15 mL). Discard the flow through and add dialysis buffer to the top of the column whilst collecting 500 µL fractions (*see* **Note 13**).

   (b) Apply BID labeled protein solution to the Ni-NTA column (0.2 mL bed volume of beads) and pass through the column three times to ensure all labeled protein binds. The column is washed with 50 mL of BID wash buffer, and eluted with 5 mL of BID elution buffer (*see* **Note 14**).

6. Pool the fractions with the most protein and check the percent labeling efficiency by dividing the concentration of the dye by the concentration of the protein and multiplying by 100 (*see* **Note 15**).

$$\% \text{ Labeling Efficiency} = \left(\frac{[\text{Dye}]}{[\text{Protein}]} \right) \times 100$$

7. Dialyze the final labeled protein sample once again using dialysis buffers with 20 % glycerol.

8. After dialysis is complete, measure the protein and dye concentration again to get a final % labeling efficiency. Aliquot the labeled, dialyzed protein into one-time use samples. Proteins should be flash-frozen with liquid nitrogen, should be kept in solution with at least 10 % glycerol, and should never be refrozen after use.

The labeled protein needs to be assayed for functionality before use in a FRET assay. Labeling could potentially alter protein folding, orientation, or steric inactivation of binding sites. The activities of these proteins are tested for release of cytochrome *c* in the presence of mitochondria as described in the next section.

3.3   Mitochondria Preparation

Steps are carried out at room temperature unless otherwise indicated. Procedure adapted from Yamaguchi et al. [9].

1. Euthanize mice (BAK$^{-/-}$) using CO_2 and remove liver. Place in 20 mL AT buffer. You can process up to four livers at once.

2. Wash liver sample five times in 5–10 mL of AT buffer to remove erythrocytes (*see* **Note 16**).

3. Mince liver using surgical scissors in 5 mL/liver of ice-cold AT. Adjust volume to 30 min.

4. Homogenize the minced tissue using four strokes (one stroke is up and down) of a motor-driven Potter-Elvehjem homogenizer (30 mL size).

5. Centrifuge homogenate for 10 min at $600 \times g$ to remove tissue components.

6. Remove the supernatant to a fresh tube and centrifuge for 15 min at $3500 \times g$.

7. Decant the supernatant carefully (*see* **Note 17**).

8. Transfer the supernatant to a fresh tube and centrifuge at $5500 \times g$ for 10 min.

9. Resuspend the final mitochondrial pellet to the required protein concentration in AT buffer and freeze (described below). For four livers, add 50 µL of AT buffer initially, measure the protein concentration by Bradford Assay, then adjust the volume so that the final protein concentration is no more than 50 mg/mL (*see* **Note 18**). Quickly move to the freezing step.

Freezing

1. Obtain dry ice and crush into small pellets in an ice bucket (*see* **Note 19**).

2. Set up prechilled, labeled microfuge tubes by setting them in the dry ice.

3. Dispense 10 µL into the prechilled tubs without touching the tip to the side of the tube (*see* **Note 20**). The mitochondria will freeze as soon as they touch the cold tube.

4. Place each tube immediately into liquid nitrogen after each aliquot is pipetted, and store aliquots at –80 °C until ready to use.

5. When mitochondria are needed for FRET assays, use the mitochondria regeneration protocol described below. After aliquots are frozen, they are one-time use samples and should not be re-frozen.

Mitochondria regeneration

1. Rapidly thaw a mitochondria aliquot by holding tube between your fingers and then place on ice.

2. Wash mitochondrial pellet by washing once in AT-KCl buffer, spinning the pellet down at $10,000 \times g$ for 10 min, removing the supernatant, and resuspending the pellet in regenerating buffer.

3. The regenerated mitochondria can be used directly for fluorescence assays.

3.4 Cytochrome c Release

Here we need to ensure that our proteins are functional by either releasing cytochrome c (BID + BAX) or inhibiting release of cytochrome c (BID + BAX + BCL-X_L). This is also an indication of the integrity of the mitochondria that were prepared. With immunoblot analysis, we should see that the mitochondria pellet retains cytochrome c while the addition of BID and BAX should release cytochrome c into the supernatant.

1. Using the regenerated mitochondria from Subheading 3.3, add 98–100 µL (depending on how much protein you are adding) to each well desired in a 96 well plate (*see* **Note 21**).

2. Add proteins to the desired wells. We will need one with only mitochondria, one well with Alexa 568 cBID 4 and 25 nM BAX to release cytochrome c from the mitochondria, and the rest containing the same concentration of cBID and BAX plus a titration of Alexa 647 BCL-X_L from 4 to 20 nM (*see* **Note 22**).

3. Mix well and incubate the plate at 37 °C for an hour.

4. Spin down the plate, or transfer each reaction into non-stick tubes and spin at $5000 \times g$ for 5 min. Separate the pellet and supernatant. Resuspend the mitochondrial pellet in an equal volume (100 µL) of regeneration buffer.

5. Add 2× loading buffer to each sample and run standard polyacrylamide gel electrophoresis to resolve proteins. Transfer proteins to PVDF or nitrocellulose western blotting paper and carry out standard immunoblotting with the cytochrome c antibody (1:5000 dilution). Figure 1 shows cytochrome c release of mitochondria in either the pellet or supernatant fractions.

3.5 Measuring the Interaction Between cBID and BCL-X_L in Mitochondria Using FRET

1. Mitochondria are obtained (Subheading 3.3) at a concentration of 1 mg/mL.

2. The fluorimeter (TECAN Infinite M1000 PRO) is set to record the fluorescence of Alexa 647 (567 nm excitation, 5.0 nm slit width; 600 nm emission, 10 nm slit width) at 37 °C reading once per minute. Frequency set to 400 Hz and 50 flashes with an integration time of 2 µs. Instrument reads from the bottom of the plate (wells have clear bottoms).

3. 98–100 µL of regenerated mitochondria from Subheading 3.3 is added into 16 wells of a 96 well plate. The signal is read for 15 min or until the signal remains stable. Two reactions are required for the detection of a FRET interaction; one that

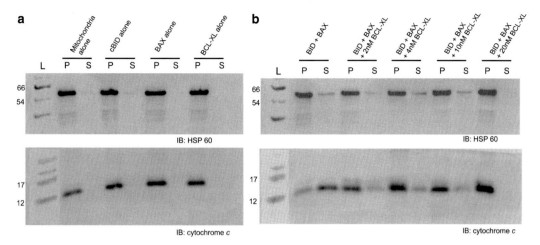

Fig. 1 Cytochrome *c* release from purified mitochondria using BID and BAX, which is inhibited by BCL-X$_L$. (**a**) The pellet (P) and supernatant (S) fractions are shown for the mitochondria alone, 2 nM cBID protein alone, 10 nM Bax protein alone, and 20 nM BCL-X$_L$ protein alone. Immunoblotting for cytochrome *c*, and the loading control HSP60. (**b**) The pellet and supernatant fractions are shown for 2 nM BID + 10 nM BAX, and a titration of BCL-X$_L$ from 4 to 20 nM. Blots were stained for HSP60 loading control and cytochrome *c* as labeled

contains both the donor-labeled cBID and acceptor-labeled BCL-X$_L$ (F_{DA}) and a control that contains the donor-labeled cBID and unlabeled acceptor BCL-X$_L$ (F_D) to account for environment changes that may affect the signal of the donor fluorophore (*see* **Note 23**).

4. After the mitochondria background read, obtain an initial fluorescence reading (F_0) by adding 4 nM Alexa 568 cBID to each well and collecting data again for 10–15 min so the reaction reaches 37 °C. Also, it is important to ensure that the addition of the donor alone reaches a stable plateaued signal at the desired temperature.

5. To obtain the final fluorescence read (F), add either labeled Alexa 647 BCL-X$_L$ or unlabeled BCL-X$_L$ to the wells as described and read for 1–2 h at 1 min intervals.

6. When 1–2 h have passed or the kinetic data reaches a plateau, average the values for the last 10–15 min of the assay. Divide the averaged points for each well from F and F_0 (F/F_0) for each well to normalize data collected (*see* **Note 24**).

7. The % FRET efficiency (E) can be obtained by comparing the relative fluorescence intensities of the donor in the presence and absence of the acceptor with the following calculation:

$$\% E = \left(1 - \frac{F_{DA}}{F_D}\right) \times 100$$

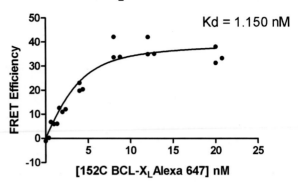

Fig. 2 cBID interacts with BCL-X$_L$ in the presence of purified mitochondria. The donor concentration was kept constant at 4 nM while the concentration of the acceptor was titrated from 0 to 20 nM as indicated. Each data point represents an experiment done in triplicate with $n=3$

where F_{DA} is the F/F_0 value for the donor-labeled protein in the presence of the acceptor-labeled protein, and F_D is the F/F_0 value for the donor-labeled protein in the presence of the unlabeled acceptor.

8. By plotting the FRET efficiency against the concentration of the acceptor used in the plate, one can obtain a binding curve as shown in Fig. 2 for the FRET interaction between donor-labeled cBID and acceptor-labeled BCL-X$_L$.

9. Fit the data to the equation below to find the dissociation constant (K_D) for this interaction (*see* **Note 25**):

$$f\left(\left[B_0\right]\right) = \frac{[AB]}{[A_0]} = \frac{[A_0]+[B_0]+K_D - \sqrt{\left([A_0]+[B_0]+K_D\right)^2 - 4[A_0][B_0]}}{2[A_0]}$$

where $[A_0]$ is the initial concentration of the donor protein, $[B_0]$ is the concentration of the donor protein when system reaches equilibrium.

4   Notes

1. The addition of hydroxylamine lowers the pH of the cleavage buffers. Be sure to adjust the pH of the buffers using NaOH. Hydroxylamine is used instead of other reducing agents since cleavage of the intein with hydroxylamine will not leave a free thiol group as is the case for cleavage with DTT or BME. This free thiol is labeled with maleimide dyes and therefore using hydroxylamine is required to prevent unwanted labeling of the protein.

2. The pH of TRIS buffers is temperature dependent and will change with decreasing temperature. Make sure to adjust the pH again after cooling the buffers to 4 °C.

3. 1 mL of 1 M DTT will last for many labeling reactions. When not using DTT, keep frozen at –20 °C.

4. Beads can be reused by washing with four column volumes of dialysis buffer and stored at 4 °C (can also refer to the manufacturers specifications for bead storage and regeneration).

5. This buffer was designed to regenerate the mitochondrial inner membrane transmembrane potential. Add 5 μL creatine kinase (5 mg/mL) from a stock stored at –20 °C into the 500 μL mitochondrial sample.

6. For increased protein yield and reduced degradation, the induced bacterial cultures can be grown at 17 °C overnight.

7. Mechanical lysis produces heat; take frequent breaks to ensure samples remain at 4 °C. Keep all samples chilled at 4 °C for the remainder of the protocol.

8. Hydroxylamine in the cleavage buffer causes cleavage of the intein-chitin binding domain, allowing full-length BAX or BCL-X_L to be eluted 48 h later.

9. Labeling of BID can be performed prior to cleavage and dialysis. Imidazole in the buffer does not affect the labeling reaction. For proteins fused with intein-chitin binding domain (BAX, BCL-X_L), hydroxylamine instead of other common reagents such as beta-mercaptoethanol or DTT is used to induce intein self-cleavage. Hydroxylamine-induced self-cleavage does not leave a thiol group at the C-terminus of the protein, preventing potential labeling at the c-terminus of the protein. Hydroxylamine needs to be removed by dialysis prior to labeling.

10. The total reaction volume for BID, BAX, and BCL-X_L would typically be 1–1.5 mL with a protein concentration of 10–100 μM depending on the yield from protein purification.

11. If labeling efficiency is low, you can try adding urea to the reaction at a final concentration of 1 M to help expose cysteine residues.

12. The reaction can be incubated at 4 °C overnight. Also, you can increase the room temperature incubation for an hour or two if labeling efficiencies are low, but if you are increasing the incubation time, ensure that your cysteine-less labeling control is not being labeled.

13. For this step, use the protein's dialysis buffer without glycerol to elute from the G25 column. With the bright Alexa Fluor dyes, you should be able to see the labeled protein fraction travel down the column faster than the free, unreacted dye.

Typically the labeled protein will elute in the 12th–15th fractions. If you use multiple dyes on the column be sure to thoroughly wash the column before adding the next labeling reaction to prevent contamination between dyes. Rather than collecting fractions of a specific volume, it is easier to collect the protein as you see it coming out of the column since the labeled protein is intensely colored. Use the same wash and elution buffers from the BID purification method.

14. Typically we use protein samples that are >80 % labeled for our fluorescence-based assays. Also, dialysis of the protein sample will remove any free dye that might still be in the final elution from the G25 column. Protein concentration is typically measured by Bradford assay (BID), protein A280 (BAX), or BCA assay (BCL-X_L) while the dye concentration is measured by the optical density of the sample at the peak absorbance wavelength and molar extinction coefficient of the dye provided by the product information.

15. We find that using nonbinding surface, hydrophobic microtubes are best to store these proteins.

16. You can set up two glass jars with a small volume of AT buffer. Pick up the mouse liver with forceps and wash the livers in the first jar, and do a final rinse in the second jar. You want to do this until the jar of buffer the livers are washed in remains clean.

17. Occasionally, there is an adipose layer that floats on top of the desired supernatant that you should avoid collecting.

18. Mitochondria at a concentration greater than 50 mg/mL may not contain enough trehalose to stabilize them, thus higher concentrations of mitochondria are not recommended [9].

19. Liquid nitrogen also works for flash freezing mitochondria.

20. Only pipette 10 µL to facilitate quick freezing. Do not touch the pipette tip to the tube or the mitochondria will freeze in the tip. If this does occur, discard the aliquot and the frozen mitochondria in the tip.

21. With this assay, we fill eight wells. One well with only mitochondria, one well with just cBID and BAX, and the rest of the wells with cBID, BAX, and a titration of BCL-X_L from 1 to 20 nM. Also, the in vitro proteins used will be Alexa 568 126C cBID, Alexa 647 152C BCL-X_L, and wild-type BAX.

22. To try and optimize the distance for better FRET efficiencies for a protein-protein interaction with a given donor-acceptor pair, you can label different single-cysteine mutants. You may get a better idea of the orientation of these labels by looking at the structures of the proteins in question, if available.

23. Similar plate setup to Subheading 3.4. One well will contain only mitochondria, while the rest will contain Alexa 568

cBID and a titration with Alexa 647 BCL-X$_L$ from 0.1 to 20 nM. Set up row A (fluorescence with the donor and acceptor, F_{DA}) with eight wells, and row B (fluorescence with the donor and the unlabeled acceptor protein, F_D) with eight wells. The wells in row A will contain the donor and the acceptor for the main fluorescence measurement. The wells in row B will contain a control for row A, the donor and the unlabeled acceptor.

24. For example, for one well in row A, take the average of the last 10 points in the F measurement and divide by the average of the last 10 points in the F_0 measurement. Because row A contains fluorescently labeled acceptor protein, the value you obtain for this F/F_0 is the F_{DA} value for that well.

25. For Bcl-2 family proteins, it is common that the concentration of donor used in the experiment is greater than the dissociation constant in the nM range. If this is the case, we cannot use the conventional Hill slope equation to fit our curve. Instead we need to use a modified equation that takes into account the protein concentration as explained by Shamas-Din et al. [10]. In short, for the binding interaction between cBID and BCL-X$_L$ we can create the equation $A + B \rightarrow AB$, where A is cBID and B is BCL-X$_L$, with the dissociation constant equal to:

$$K_D = \frac{[A][B]}{[AB]}$$

If $[A] << K_D$, then we can exclude $[A]$ from the equation (Hill slope) and it becomes much simpler to fit the curve:

$$f([B_0]) = \frac{[AB]}{[A_0]} = \frac{[B_0]}{K_D + [B_0]}$$

where f is the fraction of bound molecule.

However, when the K_D is below 100 nM, as frequently seen within the Bcl-2 family $[A]$ is not $<< K_D$, so we need to take $[A]$ into account. Thus, using the form of the quadratic equation and the appropriate substitutions given our new equation, the expression for the equation to fit the binding curve becomes:

$$f([B_0]) = \frac{[AB]}{[A_0]} = \frac{[A_0] + [B_0] + K_D - \sqrt{([A_0] + [B_0] + K_D)^2 - 4[A_0][B_0]}}{2[A_0]}$$

Lastly, since we are using FRET to produce our binding curve, we assume our FRET signal is proportional to the fraction bound. When the FRET efficiency is half of the maximum fret efficiency of the curve, then $[B] = K_D$.

Acknowledgments

Funding was provided by the Canadian Institutes of Health Research (CIHR) grant FRN12517 to DWA.

References

1. Shamas-Din A, Kale J, Leber B, Andrews DW (2013) Mechanisms of action of Bcl-2 family proteins. Cold Spring Harb Perspect Biol 5(4):a008714

2. Lovell J, Billen L, Binder S, Shamas-Din A, Fradin C, Leber B, Andrews D (2008) Membrane binding by tBid initiates a series of events culminating in membrane permeabilization by Bax. Cell 135:1074–1084

3. Hsu YT, Youle RJ (1997) Nonionic detergents induce dimerization among members of the Bcl-2 family. J Biol Chem 272(21): 13829–13834

4. Johnson I, Spence MTZ (2010) The molecular probes handbook. Life Technologies Corporation, Eugene, OR

5. Esteve-Núñez A, Sosnik J, Visconti P, Lovley DR (2008) Fluorescent properties of c-type cytochromes reveal their potential role as an extracytoplasmic electron sink in Geobacter sulfurreducens. Environ Microbiol 10: 497–505

6. Tohmi M, Takahashi K, Kubota Y, Hishida R, Shibuki K (2009) Transcranial flavoprotein fluorescence imaging of mouse cortical activity and plasticity. J Neurochem 109(s1):3–9

7. Shibasaki F, Kondo E, Akagi T, McKeon F (1997) Suppression of signalling through transcription factor NF-AT by interactions between calcineurin and Bcl-2. Nature 386(6626):728–731

8. Cechetto JD, Soltys BJ, Gupta RS (2000) Localization of mitochondrial 60-kD heat shock chaperonin protein (Hsp60) in pituitary growth hormone secretory granules and pancreatic zymogen granules. J Histochem Cytochem 48(1):45–56

9. Yamaguchi R, Andreyev A, Murphy AN, Perkins GA, Ellisman MH, Newmeyer DD (2007) Mitochondria frozen with trehalose retain a number of biological functions and preserve outer membrane integrity. Cell Death Differ 14(3):616–624

10. Shamas-Din A, Satsoura D, Khan O, Zhu W, Leber B, Fradin C, Andrews DW (2014) Multiple partners can kiss-and-run: Bax transfers between multiple membranes and permeabilizes those primed by tBid. Cell Death Dis 5:e1277

Chapter 16

Preparing Samples for Crystallization of Bcl-2 Family Complexes

Marc Kvansakul and Peter E. Czabotar

Abstract

High-resolution protein structures determined by X-ray crystallography or NMR have proven invaluable for deciphering the molecular mechanisms underlying the function of a vast range of proteins. Here, we describe methods to generate complexes of proteins belonging to the Bcl-2 family of proteins with either biological ligands or small molecule antagonists.

Key words Bcl-2 family proteins, Structural biology, X-ray crystallography, Drug discovery

1 Introduction

Members of the Bcl-2 family of proteins fall into two opposing -factions, the prosurvival group and the proapoptotic group (Fig. 1). The interplay between members of these rival family factions ultimately determines cellular fate, and structural insights into these interactions have led to a wealth of information into Bcl-2 mediated signaling and its role in disease (see, e.g., [1]). Despite substantial unresolved challenges in the preparation of complexes of full-length Bcl-2 constructs, mechanisms of action governing the biology of these proteins are increasingly well understood. These advances have relied heavily on the structural analysis of protein complexes of the various family members bound to relevant partners.

The first structural analysis of a Bcl-2 family protein complex was achieved using NMR and revealed in detail the interactions between Bcl-x_L and a short 16-mer peptide spanning the BH3 domain of the proapoptotic executioner molecule Bak [2] (Table 1, Fig. 2). The interaction was mediated through hydrophobic interactions between the amphipathic BH3 helix and a groove on the surface of the prosurvival protein, a salt bridge between a conserved Aspartate on the BH3 peptide and a conserved Arginine on the

Hamsa Puthalakath and Christine J. Hawkins (eds.), *Programmed Cell Death: Methods and Protocols*, Methods in Molecular Biology, vol. 1419, DOI 10.1007/978-1-4939-3581-9_16, © Springer Science+Business Media New York 2016

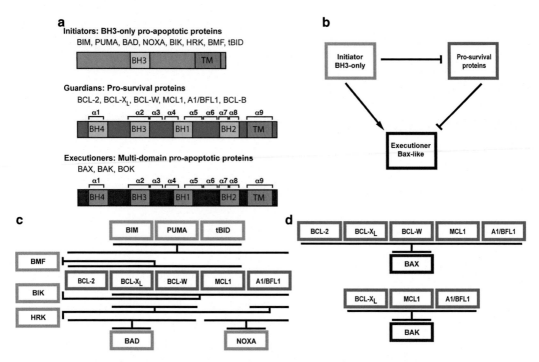

Fig. 1 Bcl-2 family members and interactions. (a) The Bcl-2 protein family consists of two opposing groups, the prosurvival proteins and the proapoptotic proteins. The proapoptotic members can be further subdivided into the BH3-only proteins, whose role is to initiate death signaling, and the executioner proteins Bax and Bak (and possibly Bok) that are responsible for mitochondrial outer membrane permeabilization (MOMP), a point of no return in the death signaling pathway. (b) Apoptotic signaling is initiated through the upregulation of BH3-only proteins. These inhibit the activity of the prosurvival proteins and can directly interact with and activate the executioner proteins. Prosurvival proteins inhibit activated executioners by binding to their BH3 domains, and possibly other regions, to prevent oligomerization. An excess of BH3-only proteins competes for this interaction, releasing activated Bax-like proteins so that they can oligomerize and initiate MOMP. (c) Some BH3-only proteins, such as Bim, Puma, and Bid, interact with the full suite of prosurvival proteins whereas others, such as Bad and Noxa, interact with only a subset [3, 32, 33]. (d) Bak is primarily inhibited by Bcl-x_L, Mcl-1, and A1 [34] and Bax is most likely inhibited by the full range of prosurvival proteins [35]

prosurvival protein was also observed. Subsequent structural analyses were informed by the realization that 26-mer peptides of BH3 domains of proapoptotic BH3-only proteins faithfully recapitulate key aspects of these interactions [3]. This work also provided the first insights into the specificity of interactions occurring between different family members (Fig. 1). Structures for a large number of various complexes have now been solved (Table 1).

A number of complexes have also now been solved for prosurvival proteins in complex with peptides corresponding to the BH3 regions of the Bax-like executioner proteins (Table 1). However, the absence of a structure of a full-length mammalian prosurvival Bcl-2 protein bound to a Bax-like protein is hampering a complete understanding of the intricacies of prosurvival Bcl-2-mediated regulation

Table 1
PDB entries of Bcl-2 family member complexes (*see* Note 6)

Protein complex	Species	PDB code	Method	Reference
Bcl-x$_L$:Bak BH3	Human	1BXL	NMR	[2]
Bcl-x$_L$:Bad BH3	Human	1G5J	NMR	[39]
Bcl-x$_L$:Bim	Mouse	1PQ1	X-ray	[36]
CED9:EGL-1 BH3	*C. elegans*	1TY4	X-ray	[40]
CED4:CED9	*C. elegans*	2A5Y	X-ray	[4]
Bcl-w:Bid BH3	Human	1ZY3	NMR	[41]
Bcl-x$_L$:Beclin-1 BH3	Human	2P1L	X-ray	[24]
M11L:Bak BH3	Myxoma virus/human	2JBY	X-ray	[42]
Mcl-1:Bim BH3	Human/rat	2NL9	X-ray	[12]
Mcl-1:mNoxaB BH3	Human/rat	2NLA	X-ray	[12]
Mcl-1:mNoxaB BH3	Mouse	2JM6	NMR	[12]
Bax:BimSAHB	Human	2K7W	NMR	[43]
Bcl-x$_L$:Beclin-1 BH3	Human	2PON	NMR	[44]
A1/Bfl-1:Bim BH3	Human	2VM6	X-ray	[45]
Bcl-x$_L$:Bad BH3	Mouse	2BZW	X-ray	[46]
M11:Beclin-1 BH3	mγHV68/mouse	3BL2	X-ray	[46]
A1/Bfl-1:Puma BH3	Mouse	2VOF	X-ray	[47]
A1/Bfl-1:Bmf BH3	Mouse	2VOG	X-ray	[47]
A1/Bfl-1:Bak BH3	Mouse	2VOH	X-ray	[47]
A1/Bfl-1:Bid BH3	Mouse	2VOI	X-ray	[47]
Mcl-1:Puma BH3	Mouse	2ROC	NMR	[48]
Mcl-1:mNoxaA BH3	Mouse	2ROD	NMR	[48]
Mcl-1:mutBim BH3	Human/mouse	3D7V	X-ray	[49]
BHRF1:Bim BH3	EBV/human	2V6Q	X-ray	[37]
BHRF1:Bak BH3	EBV/human	2XPX	X-ray	[37]
M11:Beclin-1 BH3	mγHV68/mouse	3DVU	X-ray	[50]
Bcl-x$_L$:Foldamer	Human	3FDM	X-ray	[51]
Bcl-x$_L$:BimBH3	Human	3FDL	X-ray	[51]
Bcl-x$_L$:BimBH3L12F	Human	3IO8	X-ray	[26]
Mcl-1:BimL12Y	Human/rat	3IO9	X-ray	[26]
Mcl-1:Bim BH3	Human	2PQK	X-ray	[52]
Mcl-1:Bim BH3	Human	2PQK	X-ray	[52]

(continued)

**Table 1
(continued)**

Protein complex	Species	PDB code	Method	Reference
Mcl-1:BimI2dY BH3	Human	3KJ0	X-ray	[52]
Mcl-1:BimI2dA BH3	Human	3KJ1	X-ray	[52]
Mcl-1:BimF4aE BH3	Human	3KJ2	X-ray	[52]
Mcl-1:Bid BH3	Human	2KBW	NMR	[53]
Mcl-1:Mcl-1 BH3	Human	3MK8	X-ray	[54]
Mcl-1:MB7	Human	3KZ0	X-ray	[55]
Bcl-2:Bak BH3	Human	2XA0	X-ray	[56]
Mcl-1:Bax BH3	Human	3PK1	X-ray	[57]
Bcl-x$_L$:Bax BH3	Human	3PL7	X-ray	[57]
sJA:Bak BH3	Schistosome/human	3QBR	X-ray	[58]
Bcl-x$_L$:Soul BH3	Human	3R85	X-ray	[59]
Bcl-x$_L$:Puma Foldamer	Human	2YJ1	X-ray	[60]
Bax:vMIA	Human/CMV	2LR1	NMR	[61]
Bcl-x$_L$:αβ foldamer 4C	Human	4A1W	X-ray	[62]
Bcl-x$_L$:αβ foldamer 2C	Human	4A1U	X-ray	[62]
Bcl-b:Bim BH3	Human	4B4S	X-ray	[31]
A1/Bfl-1:Bid BH3	Human	4ZEQ	X-ray	Not published
A1/Bfl-1:Bak BH3	Human	3I1H	X-ray	Not published
A1/Bfl-1:Noxa BH3	Human	3MQP	X-ray	Not published
Mcl-1:αβPuma2	Human	4BPI	X-ray	[63]
Mcl-1:αβPuma3	Human	4BPJ	X-ray	[63]
Mcl-1:αβPuma5	Human	4BPK	X-ray	[63]
Bcl-x$_L$:BimLOCK BH3	Human	2YQ7	X-ray	[64]
Bcl-x$_L$:BimSAHB BH3	Human	2YQ6	X-ray	[64]
Bcl-x$_L$:Puma BH3	Human	4HNJ	X-ray	[65]
Bcl-x$_L$:Puma BH3	Human	2M04	NMR	[65]
Bax:Bid BH3	Human	4BD2	X-ray	[15]
Bax:Bax BH3	Human	4BD6	X-ray	[15]
Bax BH3-in-Groove dimer	Human	4BDU	X-ray	[15]
Bak:Bid SAHB BH3	Human	2M5B	NMR	[66]

(continued)

Table 1
(continued)

Protein complex	Species	PDB code	Method	Reference
Mcl-1:Mcl-1BH3	Human	4HW4	X-ray	[67]
Bcl-w:Bcl-w BH3	Human	4CIM	X-ray	[68]
Bcl-x$_L$:Bcl-xL BH3	Human	4CIN	X-ray	[68]
BHRF1:BINDI	EBV	4OYD	X-ray	[69]
Bak BH3-in-Groove dimer	Human	4U2V	X-ray	[14]
F1L:Bim BH3	Vaccinia virus/human	4D2M	X-ray	[11]
F1L:Bak BH3	Vaccinia virus/human	4D2L	X-ray	[11]
Bcl-x$_L$:p53	Human	2MEJ	NMR	[70]
DPV022:Bim BH3	Deerpoxvirus/human	4UF3	X-ray	[10]
DPV022:Bak BH3	Deerpoxvirus/human	4UF1	X-ray	[10]
DPV022:Bax BH3	Deerpoxvirus/human	4UF2	X-ray	[10]
F1L:Bid BH3	Variola virus/human	5AJJ	X-ray	[71]
F1L:Bak BH3	Variola virus/human	5AJK	X-ray	[71]
Bcl-x$_L$:Bid BH3	Human	4QVE	X-ray	[72]
Bcl-x$_L$:Bim BH3	Human	4QVF	X-ray	[72]
Bax:Bim BH3mini	Human	4ZIF	X-ray	[30]
Bax:Bim BH3mini	Human	4ZIH	X-ray	[30]
Bax:Bim BH3	Human	4ZIE	X-ray	[30]
Bax:Bid BH3mini	Human	4ZIG	X-ray	[30]
BaxI66A:Bid BH3	Human	4ZII	X-ray	[30]
Bcl-x$_L$:BimBH3 with AKT site	Human	4YJ4	X-ray	[73]

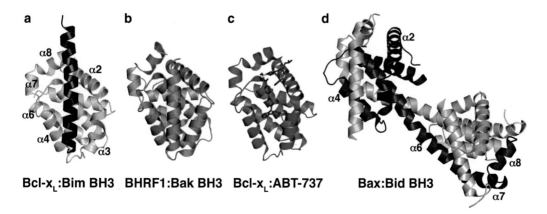

Bcl-x$_L$:Bim BH3   BHRF1:Bak BH3   Bcl-x$_L$:ABT-737   Bax:Bid BH3

Fig. 2 Structures of Bcl-2 relatives complexed with BH3 domains or BH3 mimetics. (**a**) Structure of Bcl-x$_L$ with Bim [36]. (**b**) Structure of BHRF1 with Bak BH3 [37]. (**c**) Crystal structure of ABT-737 bound to Bcl-x$_L$ [38]. (**d**) Structure of Bid BH3 bound to Bax domain swapped dimer [15]

of Bax and Bak. Nonetheless, the structure determination of a full-length complex of CED9 bound to CED4, two key regulators of intrinsic apoptosis in the worm *C. elegans* [4], suggests that these challenges are not insurmountable. More recently structures have also been solved for complexes of Bax bound to activating BH3-only proteins, providing insight into the initiation of Bax conformational change, and of Bax and Bak dimers, providing insight into the ensuing oligomerization of these proteins (Table 1).

Here, we will focus on methods and strategies related to the analysis of Bcl-2 family protein complexes with crystallography. However, it should be noted that other structural and biophysical techniques have contributed greatly to our understanding of Bcl-2 family protein structure, function, and drug discovery including NMR (e.g., [2, 5]), Fluorescence Resonance Energy Transfer (FRET; e.g., [6]), Double Electron-Electron Resonance spectroscopy (DEER; e.g., [7]), and chemical cross-linking (e.g., [8]).

As with all attempts at protein crystallization there are a variety of different strategies to obtain diffracting crystals of target proteins [9]. Routinely, initial crystallization trials are performed with a desired construct in a large number of crystallization conditions, and sometimes at a range of protein concentrations, in order to find conditions in which the protein is enticed toward formation of a crystal rather than precipitation. However, often crystallization conditions for target constructs are not forthcoming despite extensive screening and in these situations alternative construct strategies are often tried. In the case of the Bcl-2 family of proteins, a range of different construct design strategies have been successful as follows (*see* **Notes 1–3**).

2 Materials

1. Recombinant prosurvival protein (e.g., vaccinia virus F1L protein, and Bcl-x_L) purified to homogeneity in final sample buffer (e.g., 25 mM Hepes pH 7.5, 150 mM NaCl).

2. Synthetic BH3 domain peptide (e.g., Bim BH3, Uniprot accession code O43521-3, residues 51–77, Genscript) dissolved in H_2O.

3. Centrifugal concentrator (MWCO 10 kDa, Merck Millipore).

3 Methods

Preparation of complexes of prosurvival proteins bound to peptides of their proapoptotic counterparts has led to important insights into Bcl-2 mediated signaling and its role in disease. In this example, we demonstrate how to prepare a complex of vaccinia virus

F1L with the human Bim BH3 domain peptide (*see* **Note 4**). This method has been successfully used to prepare complexes for crystallization trials of prosurvival Bcl-2 proteins bound to BH3 domain peptides with affinities ranging from 1 nM to 7 µM [10, 11]. Similar approaches can be used to prepare complexes between Bcl-2 family proteins and small molecules (*see* **Note 5**). Final concentrations for crystallization experiments may vary depending on the sample.

1. Wash a 5 mL centrifugal concentrator with 5 mL of final sample buffer by centrifugation.

2. Add 1 mg of prosurvival protein in final sample buffer and top up with additional buffer to a final volume of 4 mL.

3. Aspirate a 1.25 molar excess of BH3 domain peptide.

4. Slowly add peptide to centrifugal concentrator while stirring with pipette to avoid local precipitation of sample.

5. Concentrate sample to a final concentration of 5 mg/mL of prosurvival protein.

6. Top up sample with additional buffer to a final volume of 4 mL.

7. Concentrate sample to a final concentration of 5 mg/mL of prosurvival protein. Final concentrations for crystallization experiments may vary with each sample.

4   Notes

1. A common strategy for obtaining diffracting crystals of difficult targets is to attempt to crystallize the protein of interest from different species. Structures of Bcl-2 family proteins from a variety of different species have been crystallized (Tables 1 and 2) and in some cases chimeric constructs consisting of sequence from two different species have proved useful [12]. Naturally for drug discovery programs, it is usually desirable to use human constructs and so for these projects alternative strategies for enabling crystallization may be pursued.

2. One method by which crystallization can be enhanced is through the use of protein fusion partners. These can act to both aid with protein solubility and may also provide extra opportunities for the formation of crystal contacts upon which a crystals lattice can build. One recent notable success has been achieved with a maltose binding protein fusion with Mcl-1 [13]. This construct provided the first crystal structure for apo Mcl-1 and enabled ligand bound Mcl-1 structures to be obtained through both soaking of compounds into the apo crystals and through cocrystallization of compound and protein. Fusion partners have also enabled the crystallization

Table 2
PDB entries of Bcl-2 family members in complex with compounds

Protein:drug complex	Species	PDB code	Method	Reference
Bcl-x$_L$:N3B	Human	1YSI	NMR	[5]
Bcl-x$_L$:4FC/TN1	Human	1YSG	NMR	[5]
Bcl-x$_L$:43B	Human	1YSN	NMR	[5]
Bcl-2:43B	Human	1YSW	NMR	[5]
Bcl-x$_L$:43B	Human	2O1Y	NMR	[74]
Bcl-2:43B	Human	2O21	NMR	[74]
Bcl-2:LIU	Human	2O22	NMR	[74]
Bcl-2:LI0	Human	2O2F	NMR	[74]
Bcl-x$_L$:LI0	Human	2O2M	NMR	[74]
Bcl-x$_L$:LIW	Human	2O2N	NMR	[74]
Bcl-x$_L$:ABT-737	Human	2YXJ	X-ray	[38]
Bcl-2:DRO	Human	2W3L	X-ray	[75]
Bcl-x$_L$:W1191542	Human	3INQ	X-ray	[26]
Bcl-x$_L$:HI0	Human	3QKD	X-ray	[76]
Bcl-2:398	Human	4AQ3	X-ray	[77]
Bcl-x$_L$:0Q5	Human	4EHR	X-ray	[78]
Bcl-x$_L$:B50	Human	3SPF	X-ray	[79]
Bcl-x$_L$:03B	Human	3SP7	X-ray	[79]
Bcl-x$_L$:33B	Human	2LP8	NMR	[80]
Mcl-1:PRD_000921	Human	4G35	X-ray	[81]
Bcl-x$_L$:WEHI-539	Human	3ZLR	X-ray	[27]
Bcl-x$_L$:1HI	Human	3ZK6	X-ray	[27]
Bcl-x$_L$:H0Y	Human	3ZLN	X-ray	[27]
Bcl-x$_L$:X8U	Human	3ZLO	X-ray	[27]
Bcl-2:1E9	Human	4IEH	X-ray	[82]
Mcl-1:19H	Human	4HW2	X-ray	[67]
Mcl-1:19G	Human	4HW3	X-ray	[67]
Bcl-2:ABT-263	Human	4LVT	X-ray	[83]
Bcl-2:1XV	Human	4LXD	X-ray	[83]
Bcl-2:1Y1	Human	4MAN	X-ray	[83]
Mcl-1:LC3	Human	3WIX	X-ray	[84]
Mcl-1:LC6	Human	3WIY	X-ray	[84]

(continued)

Table 2
(continued)

Protein:drug complex	Species	PDB code	Method	Reference
Bcl-x$_L$:LC6	Human	3WIZ	X-ray	[84]
Bcl-x$_L$:X0R	Human	4C5D	X-ray	[28]
Bcl-x$_L$:X0D	Human	4C52	X-ray	[28]
Mcl-1:2UU	Human	4OQ5	X-ray	[85]
Mcl-1:2UV	Human	4OQ6	X-ray	[85]
Bcl-w:013_D12	Bos Taurus	4K5A	X-ray	[86]
Bcl-w:UNP	Bos Taurus	4K5B	X-ray	[86]
Bcl-x$_L$:38H	Human	4TUH	X-ray	[87]
Mcl-1:3M6	Human	4WGI	X-ray	[88]
Bcl-x$_L$:3CQ	Human	1QVX	X-ray	[29]
Mcl-1:4M7	Human	4ZBF	X-ray	[89]
Mcl-1:4M6	Human	4ZBI	X-ray	[89]
Mcl-1:BRDI1	Human	4WMR	X-ray	[13]
Mcl-1:865	Human	4WMT	X-ray	[13]
Mcl-1:19H	Human	4WMU	X-ray	[13]
Mcl-1:BRDI3	Human	4WMV	X-ray	[13]
Mcl-1:BRDI4	Human	4WMW	X-ray	[13]
Mcl-1:BRDI5	Human	4WMX	X-ray	[13]
Mcl-1:BRDI6	Human	4WMY	X-ray	[13]

Three letter codes from the PDB entries are used to describe ligands unless a specific name for the compound has been published

of truncated constructs of Bax and Bak that reveal details for the initial steps of dimerization. For example, it was recently discovered that one of the conformational changes occurring to these proteins upon activation includes separation into "core" (α2–α5 and possibly including 1) and "latch" (α6–α7) domains [14, 15]. Fusion of GFP to the "core" domains of these proteins [16] enabled their expression and crystallization and revealed the atomic details of the dimer units upon which the larger Bax and Bak oligomers build [8, 17].

3. Often it proves useful to make truncations or modifications to constructs in order to enable proteins to be expressed, purified, and/or crystallized. The vast majority of Bcl-2 constructs used for structural studies have lacked the C-terminal transmembrane domain (α9 helix), primarily because it is difficult to

produce sufficient quantities of soluble protein containing this highly hydrophobic region. Bax, however, is a notable exception as it can be expressed as a full-length protein in relatively high quantities [18]. Expression and purification of full-length constructs for Bak [19], Bcl-x_L [20], and Bcl-w [21] have also been reported; however, these have not been used in structural studies. Another region of the Bcl-2 family fold that is often modified is the loop between the $\alpha 1$ and $\alpha 2$ helices. This segment is large and unstructured in most family members and is thus often either shortened (e.g., Bcl-x_L $\Delta 45$–84 [22]), or replaced with the shorter loop from another family member (e.g., the Bcl-2 loop being replaced with sequence from Bcl-x_L [23]). A particularly useful construct for crystallization has been Bcl-x_L in which the $\alpha 1$–$\alpha 2$ loop is dramatically shortened (lacking residues 27–82) such that the $\alpha 1$ cannot fold correctly with the remainder of the protein. Instead this constructs forms a domain swap dimer, with the $\alpha 1$ of one monomer folding into its neighbor to complete the Bcl-2 fold [24, 25]. These dimers readily produce crystals in a number of different crystal forms and thus have proven extremely fruitful for drug discovery (e.g., [26–29]). Similarly, a domain swapped dimer version of Bax, in which the $\alpha 6$–$\alpha 8$ "latch" region swaps with a neighbor, has been useful for solving structures of Bax bound to activator BH3 domains (Fig. 2) [15, 30]. One possible reason for enhanced crystallization of these dimer constructs is that the dimerization interface provides a point of symmetry on which the crystal can build. In a similar manner, in the first structure solved of Bcl-x_L bound to a compound within the benzothiazole series (Bcl-x_L:1HI from PDB code 3ZK6 [27]), the compound itself dimerizes between two proteins across a twofold axis within the crystal, this may have similarly enhanced the crystallization of this low affinity inhibitor complex. Notably, however, the compound did not dimerize Bcl-x_L in gel filtration experiments and so may only act within the crystal or at the high concentrations of protein found within the crystallization drop.

4. An alternative method of producing complexes of prosurvival protein bound to BH3 domain peptides is to express both as a single chain construct with a protease cleavable linker [31]. It has been found in some cases that this aids the expression of the prosurvival protein and ensures complete saturation of all available binding sites. The constructs consisted of a C-terminally truncated form of the prosurvival protein linked to human Bim$_s$ BH3 peptide via a (GS) linker. This enables the Bcl-2 hydrophobic groove to be fully occupied with the native ligand. The final expression construct thus consists of: 6His-x-Bcl-2ΔC-x-$(GS)_9$-x-Bim-BH3 (where -x- represents a TEV

cleavage site ENLYFQGS). Following initial affinity purification TEV-cleaveable linkers are cleaved via incubation with TEV protease, followed by reapplication of cleaved sample to affinity resin to remove uncleaved protein and purification tag. The final sample can then be concentrated for crystallization.

5. Preparation of complexes of prosurvival proteins with small molecules for crystallization can often be achieved using similar methods to those described above for prosurvival:BH3 domain peptide complexes (Table 2). However, an added difficulty with small molecules is that the ligands are usually dissolved in DMSO which can sometimes hinder crystallization. Furthermore, small molecules often have significantly reduced affinity for their target proteins as compared to wildtype BH3-only proteins. In the preparation of such samples, DMSO is most efficiently removed from sample mixtures of protein and ligand through buffer exchange, but for low affinity targets this might also result in loss of compound. One approach to minimize such loss is to add a molar excess of compound to protein at high concentrations in small volumes and then to dilute these samples to a final DMSO concentration of 1 % (or lower), followed by concentration using low molecular weight centrifugal filters back to the desired final molarity. Using this strategy, the solubility of the compound in solution is reduced during the dilution step thereby minimizing the rate of ligand dissociation during the purification step.

6. Table 1 demonstrates that an enormous collection of structures of Bcl-2 family protein complexes has now been accumulated. These structures have informed our understanding of the molecular mechanisms controlling apoptosis and guided the development of inhibitors targeting these proteins. However, the family portrait is by no means complete. We are yet to determine a structure of a prosurvival protein in complex with a full-length Bax-like executioner protein and there are a large number of viral derived family members for which structures have not yet been solved. Such structures are likely to offer further insights into the molecular interactions governing these pathways and may provide new strategies for targeting them for novel therapeutic outcomes.

Acknowledgements

This work was supported by an Australian Research Council Future Fellowship (FT130101349) to MK and a National Health and Medical Research Council Senior Research Fellowship to PEC as well as project grants 1079706, 1059331, and 1023055 (to PEC) and 1082383 and 1007918 (to MK). We also thank Amanda Voudouris for assistance in preparing the manuscript.

References

1. Kvansakul M, Hinds MG (2013) Structural biology of the Bcl-2 family and its mimicry by viral proteins. Cell Death Dis 4:e909. doi:10.1038/cddis.2013.436

2. Sattler M, Liang H, Nettesheim D, Meadows RP, Harlan JE, Eberstadt M, Yoon HS, Shuker SB, Chang BS, Minn AJ, Thompson CB, Fesik SW (1997) Structure of Bcl-x_L-Bak peptide complex: recognition between regulators of apoptosis. Science 275(5302):983–986

3. Chen L, Willis SN, Wei A, Smith BJ, Fletcher JI, Hinds MG, Colman PM, Day CL, Adams JM, Huang DC (2005) Differential targeting of prosurvival Bcl-2 proteins by their BH3-only ligands allows complementary apoptotic function. Mol Cell 17(3):393–403. doi:10.1016/j.molcel.2004.12.030

4. Yan N, Chai J, Lee ES, Gu L, Liu Q, He J, Wu JW, Kokel D, Li H, Hao Q, Xue D, Shi Y (2005) Structure of the CED-4-CED-9 complex provides insights into programmed cell death in Caenorhabditis elegans. Nature 437(7060):831–837

5. Oltersdorf T, Elmore SW, Shoemaker AR, Armstrong RC, Augeri DJ, Belli BA, Bruncko M, Deckwerth TL, Dinges J, Hajduk PJ, Joseph MK, Kitada S, Korsmeyer SJ, Kunzer AR, Letai A, Li C, Mitten MJ, Nettesheim DG, Ng S, Nimmer PM, O'Connor JM, Oleksijew A, Petros AM, Reed JC, Shen W, Tahir SK, Thompson CB, Tomaselli KJ, Wang B, Wendt MD, Zhang H, Fesik SW, Rosenberg SH (2005) An inhibitor of Bcl-2 family proteins induces regression of solid tumours. Nature 435(7042):677–681

6. Lovell JF, Billen LP, Bindner S, Shamas-Din A, Fradin C, Leber B, Andrews DW (2008) Membrane binding by tBid initiates an ordered series of events culminating in membrane permeabilization by Bax. Cell 135(6):1074–1084. doi:10.1016/j.cell.2008.11.010, S0092-8674(08)01439-6 [pii]

7. Bleicken S, Jeschke G, Stegmueller C, Salvador-Gallego R, Garcia-Saez AJ, Bordignon E (2014) Structural model of active Bax at the membrane. Mol Cell 56(4):496–505. doi:10.1016/j.molcel.2014.09.022

8. Dewson G, Kratina T, Sim HW, Puthalakath H, Adams JM, Colman PM, Kluck RM (2008) To trigger apoptosis, Bak exposes its BH3 domain and homodimerizes via BH3:groove interactions. Mol Cell 30(3):369–380. doi:10.1016/j.molcel.2008.04.005, S1097-2765(08)00265-7 [pii]

9. Luft JR, Newman J, Snell EH (2014) Crystallization screening: the influence of history on current practice. Acta Crystallogr F Struct Biol Commun 70(Pt 7):835–853. doi:10.1107/S2053230X1401262X

10. Burton DR, Caria S, Marshall B, Barry M, Kvansakul M (2015) Structural basis of Deerpox virus-mediated inhibition of apoptosis. Acta Crystallogr D Biol Crystallogr 71(Pt 8):1593–1603. doi:10.1107/S1399004715009402

11. Campbell S, Thibault J, Mehta N, Colman PM, Barry M, Kvansakul M (2014) Structural insight into BH3 domain binding of vaccinia virus antiapoptotic F1L. J Virol 88(15):8667–8677. doi:10.1128/JVI.01092-14

12. Czabotar PE, Lee EF, van Delft MF, Day CL, Smith BJ, Huang DC, Fairlie WD, Hinds MG, Colman PM (2007) Structural insights into the degradation of Mcl-1 induced by BH3 domains. Proc Natl Acad Sci U S A 104(15):6217–6222. doi:10.1073/pnas.0701297104

13. Clifton MC, Dranow DM, Leed A, Fulroth B, Fairman JW, Abendroth J, Atkins KA, Wallace E, Fan D, Xu G, Ni ZJ, Daniels D, Van Drie J, Wei G, Burgin AB, Golub TR, Hubbard BK, Serrano-Wu MH (2015) A maltose-binding protein fusion construct yields a robust crystallography platform for MCL1. PLoS One 10(4):e0125010. doi:10.1371/journal.pone.0125010

14. Brouwer JM, Westphal D, Dewson G, Robin AY, Uren RT, Bartolo R, Thompson GV, Colman PM, Kluck RM, Czabotar PE (2014) Bak core and latch domains separate during activation, and freed core domains form symmetric homodimers. Mol Cell 55(6):938–946. doi:10.1016/j.molcel.2014.07.016

15. Czabotar PE, Westphal D, Dewson G, Ma S, Hockings C, Fairlie WD, Lee EF, Yao S, Robin AY, Smith BJ, Huang DC, Kluck RM, Adams JM, Colman PM (2013) Bax crystal structures reveal how BH3 domains activate Bax and nucleate its oligomerization to induce apoptosis. Cell 152(3):519–531. doi:10.1016/j.cell.2012.12.031

16. Suzuki N, Hiraki M, Yamada Y, Matsugaki N, Igarashi N, Kato R, Dikic I, Drew D, Iwata S, Wakatsuki S, Kawasaki M (2010) Crystallization of small proteins assisted by green fluorescent protein. Acta Crystallogr D Biol Crystallogr 66(Pt 10):1059–1066. doi:10.1107/S0907444910032944

17. Dewson G, Kratina T, Czabotar P, Day CL, Adams JM, Kluck RM (2009) Bak activation for apoptosis involves oligomerization of dimers via their alpha6 helices. Mol Cell 36(4):696–703. doi:10.1016/j.molcel.2009.11.008, S1097-2765(09)00821-1 [pii]

18. Suzuki M, Youle RJ, Tjandra N (2000) Structure of Bax: coregulation of dimer formation and intracellular localization. Cell 103(4):645–654, doi:S0092-8674(00)00167-7 [pii]

19. Leshchiner ES, Braun CR, Bird GH, Walensky LD (2013) Direct activation of full-length pro-apoptotic BAK. Proc Natl Acad Sci U S A 110(11):E986–E995. doi:10.1073/pnas.1214313110

20. Yethon JA, Epand RF, Leber B, Epand RM, Andrews DW (2003) Interaction with a membrane surface triggers a reversible conformational change in Bax normally associated with induction of apoptosis. J Biol Chem 278(49):48935–48941. doi:10.1074/jbc.M306289200

21. Hinds MG, Lackmann M, Skea GL, Harrison PJ, Huang DC, Day CL (2003) The structure of Bcl-w reveals a role for the C-terminal residues in modulating biological activity. EMBO J 22(7):1497–1507. doi:10.1093/emboj/cdg144

22. Muchmore SW, Sattler M, Liang H, Meadows RP, Harlan JE, Yoon HS, Nettesheim D, Chang BS, Thompson CB, Wong SL, Ng SL, Fesik SW (1996) X-ray and NMR structure of human Bcl-xL, an inhibitor of programmed cell death. Nature 381(6580):335–341. doi:10.1038/381335a0

23. Petros AM, Medek A, Nettesheim DG, Kim DH, Yoon HS, Swift K, Matayoshi ED, Oltersdorf T, Fesik SW (2001) Solution structure of the antiapoptotic protein bcl-2. Proc Natl Acad Sci U S A 98(6):3012–3017. doi:10.1073/pnas.041619798

24. Oberstein A, Jeffrey PD, Shi Y (2007) Crystal structure of the Bcl-XL-Beclin 1 peptide complex: Beclin 1 is a novel BH3-only protein. J Biol Chem 282(17):13123–13132. doi:10.1074/jbc.M700492200

25. Kvansakul M, Yang H, Fairlie WD, Czabotar PE, Fischer SF, Perugini MA, Huang DC, Colman PM (2008) Vaccinia virus antiapoptotic F1L is a novel Bcl-2-like domain-swapped dimer that binds a highly selective subset of BH3-containing death ligands. Cell Death Differ 15(10):1564–1571

26. Lee EF, Czabotar PE, Yang H, Sleebs BE, Lessene G, Colman PM, Smith BJ, Fairlie WD (2009) Conformational changes in Bcl-2 pro-survival proteins determine their capacity to bind ligands. J Biol Chem 284(44):30508–30517. doi:10.1074/jbc.M109.040725, M109.040725 [pii]

27. Lessene G, Czabotar PE, Sleebs BE, Zobel K, Lowes KN, Adams JM, Baell JB, Colman PM, Deshayes K, Fairbrother WJ, Flygare JA, Gibbons P, Kersten WJ, Kulasegaram S, Moss RM, Parisot JP, Smith BJ, Street IP, Yang H, Huang DC, Watson KG (2013) Structure-guided design of a selective BCL-X(L) inhibitor. Nat Chem Biol 9(6):390–397. doi:10.1038/nchembio.1246, nchembio.1246 [pii]

28. Brady RM, Vom A, Roy MJ, Toovey N, Smith BJ, Moss RM, Hatzis E, Huang DC, Parisot JP, Yang H, Street IP, Colman PM, Czabotar PE, Baell JB, Lessene G (2014) De-novo designed library of benzoylureas as inhibitors of BCL-XL: synthesis, structural and biochemical characterization. J Med Chem 57(4):1323–1343. doi:10.1021/jm401948b

29. Tao ZF, Hasvold L, Wang L, Wang X, Petros AM, Park CH, Boghaert ER, Catron ND, Chen J, Colman PM, Czabotar PE, Deshayes K, Fairbrother WJ, Flygare JA, Hymowitz SG, Jin S, Judge RA, Koehler MF, Kovar PJ, Lessene G, Mitten MJ, Ndubaku CO, Nimmer P, Purkey HE, Oleksijew A, Phillips DC, Sleebs BE, Smith BJ, Smith ML, Tahir SK, Watson KG, Xiao Y, Xue J, Zhang H, Zobel K, Rosenberg SH, Tse C, Leverson JD, Elmore SW, Souers AJ (2014) Discovery of a potent and selective BCL-XL inhibitor with in vivo activity. ACS Med Chem Lett 5(10):1088–1093. doi:10.1021/ml5001867

30. Robin AY, Krishna Kumar K, Westphal D, Wardak AZ, Thompson GV, Dewson G, Colman PM, Czabotar PE (2015) Crystal structure of Bax bound to the BH3 peptide of Bim identifies important contacts for interaction. Cell Death Dis 6:e1809. doi:10.1038/cddis.2015.141

31. Rautureau GJ, Yabal M, Yang H, Huang DC, Kvansakul M, Hinds MG (2012) The restricted binding repertoire of Bcl-B leaves Bim as the universal BH3-only prosurvival Bcl-2 protein antagonist. Cell Death Dis 3:e443. doi:10.1038/cddis.2012.178, cddis 2012178 [pii]

32. Kuwana T, Bouchier-Hayes L, Chipuk JE, Bonzon C, Sullivan BA, Green DR, Newmeyer DD (2005) BH3 domains of BH3-only proteins differentially regulate Bax-mediated mitochondrial membrane permeabilization both directly and indirectly. Mol Cell 17(4):525–535

33. Certo M, Moore Vdel G, Nishino M, Wei G, Korsmeyer S, Armstrong SA, Letai A (2006) Mitochondria primed by death signals determine cellular addiction to antiapoptotic BCL-2 family members. Cancer Cell 9(5):351–365

34. Simmons MJ, Fan G, Zong WX, Degenhardt K, White E, Gelinas C (2008) Bfl-1/A1 functions, similar to Mcl-1, as a selective tBid and Bak antagonist. Oncogene 27(10):1421–1428. doi:10.1038/sj.onc.1210771

35. Willis SN, Fletcher JI, Kaufmann T, van Delft MF, Chen L, Czabotar PE, Ierino H, Lee EF, Fairlie WD, Bouillet P, Strasser A, Kluck RM, Adams JM, Huang DC (2007) Apoptosis initiated when BH3 ligands engage multiple Bcl-2 homologs, not Bax or Bak. Science 315(5813):856–859

36. Liu X, Dai S, Zhu Y, Marrack P, Kappler JW (2003) The structure of a Bcl-xL/Bim fragment complex: implications for Bim function. Immunity 19(3):341–352

37. Kvansakul M, Wei AH, Fletcher JI, Willis SN, Chen L, Roberts AW, Huang DC, Colman PM (2010) Structural basis for apoptosis inhibition by Epstein-Barr virus BHRF1. PLoS Pathog 6(12):e1001236. doi:10.1371/journal.ppat.1001236

38. Lee EF, Czabotar PE, Smith BJ, Deshayes K, Zobel K, Colman PM, Fairlie WD (2007) Crystal structure of ABT-737 complexed with Bcl-xL: implications for selectivity of antagonists of the Bcl-2 family. Cell Death Differ 14(9):1711–1713. doi:10.1038/sj.cdd.4402178, 4402178 [pii]

39. Petros AM, Nettesheim DG, Wang Y, Olejniczak ET, Meadows RP, Mack J, Swift K, Matayoshi ED, Zhang H, Thompson CB, Fesik SW (2000) Rationale for Bcl-xL/Bad peptide complex formation from structure, mutagenesis, and biophysical studies. Protein Sci 9(12):2528–2534. doi:10.1110/ps.9.12.2528

40. Yan N, Gu L, Kokel D, Chai J, Li W, Han A, Chen L, Xue D, Shi Y (2004) Structural, biochemical, and functional analyses of CED-9 recognition by the proapoptotic proteins EGL-1 and CED-4. Mol Cell 15(6):999–1006

41. Denisov AY, Chen G, Sprules T, Moldoveanu T, Beauparlant P, Gehring K (2006) Structural model of the BCL-w-BID peptide complex and its interactions with phospholipid micelles. Biochemistry 45(7):2250–2256. doi:10.1021/bi052332s

42. Kvansakul M, van Delft MF, Lee EF, Gulbis JM, Fairlie WD, Huang DC, Colman PM (2007) A structural viral mimic of prosurvival Bcl-2: a pivotal role for sequestering proapoptotic Bax and Bak. Mol Cell 25(6):933–942

43. Gavathiotis E, Suzuki M, Davis ML, Pitter K, Bird GH, Katz SG, Tu HC, Kim H, Cheng EH, Tjandra N, Walensky LD (2008) BAX activation is initiated at a novel interaction site. Nature 455(7216):1076–1081. doi:10.1038/nature07396, nature07396 [pii]

44. Feng W, Huang S, Wu H, Zhang M (2007) Molecular basis of Bcl-xL's target recognition versatility revealed by the structure of Bcl-xL in complex with the BH3 domain of Beclin-1. J Mol Biol 372(1):223–235. doi:10.1016/j.jmb.2007.06.069

45. Herman MD, Nyman T, Welin M, Lehtio L, Flodin S, Tresaugues L, Kotenyova T, Flores A, Nordlund P (2008) Completing the family portrait of the anti-apoptotic Bcl-2 proteins: crystal structure of human Bfl-1 in complex with Bim. FEBS Lett 582(25-26):3590–3594. doi:10.1016/j.febslet.2008.09.028

46. Ku B, Woo JS, Liang C, Lee KH, Hong HS, E X, Kim KS, Jung JU, Oh BH (2008) Structural and biochemical bases for the inhibition of autophagy and apoptosis by viral BCL-2 of murine gamma-herpesvirus 68. PLoS Pathog 4(2):e25. doi:10.1371/journal.ppat.0040025

47. Smits C, Czabotar PE, Hinds MG, Day CL (2008) Structural plasticity underpins promiscuous binding of the prosurvival protein A1. Structure 16(5):818–829. doi:10.1016/j.str.2008.02.009

48. Day CL, Smits C, Fan FC, Lee EF, Fairlie WD, Hinds MG (2008) Structure of the BH3 domains from the p53-inducible BH3-only proteins Noxa and Puma in complex with Mcl-1. J Mol Biol 380(5):958–971. doi:10.1016/j.jmb.2008.05.071

49. Lee EF, Czabotar PE, van Delft MF, Michalak EM, Boyle MJ, Willis SN, Puthalakath H, Bouillet P, Colman PM, Huang DC, Fairlie WD (2008) A novel BH3 ligand that selectively targets Mcl-1 reveals that apoptosis can proceed without Mcl-1 degradation. J Cell Biol 180(2):341–355. doi:10.1083/jcb.200708096

50. Sinha S, Colbert CL, Becker N, Wei Y, Levine B (2008) Molecular basis of the regulation of Beclin 1-dependent autophagy by the gamma-herpesvirus 68 Bcl-2 homolog M11. Autophagy 4(8):989–997

51. Lee EF, Sadowsky JD, Smith BJ, Czabotar PE, Peterson-Kaufman KJ, Colman PM, Gellman SH, Fairlie WD (2009) High-resolution structural characterization of a helical alpha/beta-peptide foldamer bound to the anti-apoptotic protein Bcl-xL. Angew Chem 48(24):4318–4322. doi:10.1002/anie.200805761

52. Fire E, Gulla SV, Grant RA, Keating AE (2010) Mcl-1-Bim complexes accommodate surprising point mutations via minor structural changes. Protein Sci 19(3):507–519. doi:10.1002/pro.329

53. Liu Q, Moldoveanu T, Sprules T, Matta-Camacho E, Mansur-Azzam N, Gehring K (2010) Apoptotic regulation by MCL-1 through heterodimerization. J Biol Chem 285(25):19615–19624. doi:10.1074/jbc.M110.105452

54. Stewart ML, Fire E, Keating AE, Walensky LD (2010) The MCL-1 BH3 helix is an exclusive MCL-1 inhibitor and apoptosis sensitizer. Nat Chem Biol 6(8):595–601. doi:10.1038/nchembio.391

55. Dutta S, Gulla S, Chen TS, Fire E, Grant RA, Keating AE (2010) Determinants of BH3 binding specificity for Mcl-1 versus Bcl-xL. J Mol Biol 398(5):747–762. doi:10.1016/j.jmb.2010.03.058

56. Ku B, Liang C, Jung JU, Oh BH (2011) Evidence that inhibition of BAX activation by BCL-2 involves its tight and preferential interaction with

the BH3 domain of BAX. Cell Res 21(4):627–641. doi:10.1038/cr.2010.149

57. Czabotar PE, Lee EF, Thompson GV, Wardak AZ, Fairlie WD, Colman PM (2011) Mutation to Bax beyond the BH3 domain disrupts interactions with pro-survival proteins and promotes apoptosis. J Biol Chem 286(9):7123–7131. doi:10.1074/jbc.M110.161281

58. Lee EF, Clarke OB, Evangelista M, Feng Z, Speed TP, Tchoubrieva EB, Strasser A, Kalinna BH, Colman PM, Fairlie WD (2011) Discovery and molecular characterization of a Bcl-2-regulated cell death pathway in schistosomes. Proc Natl Acad Sci U S A 108(17):6999–7003. doi:10.1073/pnas.1100652108, 1100652108 [pii]

59. Ambrosi E, Capaldi S, Bovi M, Saccomani G, Perduca M, Monaco HL (2011) Structural changes in the BH3 domain of SOUL protein upon interaction with the anti-apoptotic protein Bcl-xL. Biochem J 438(2):291–301. doi:10.1042/BJ20110257

60. Lee EF, Smith BJ, Horne WS, Mayer KN, Evangelista M, Colman PM, Gellman SH, Fairlie WD (2011) Structural basis of Bcl-xL recognition by a BH3-mimetic alpha/beta-peptide generated by sequence-based design. Chembiochem 12(13):2025–2032. doi:10.1002/cbic.201100314

61. Ma J, Edlich F, Bermejo GA, Norris KL, Youle RJ, Tjandra N (2012) Structural mechanism of Bax inhibition by cytomegalovirus protein vMIA. Proc Natl Acad Sci U S A 109(51):20901–20906. doi:10.1073/pnas.1217094110

62. Boersma MD, Haase HS, Peterson-Kaufman KJ, Lee EF, Clarke OB, Colman PM, Smith BJ, Horne WS, Fairlie WD, Gellman SH (2012) Evaluation of diverse alpha/beta-backbone patterns for functional alpha-helix mimicry: analogues of the Bim BH3 domain. J Am Chem Soc 134(1):315–323. doi:10.1021/ja207148m

63. Smith BJ, Lee EF, Checco JW, Evangelista M, Gellman SH, Fairlie WD (2013) Structure-guided rational design of alpha/beta-peptide foldamers with high affinity for BCL-2 family prosurvival proteins. Chembiochem 14(13):1564–1572. doi:10.1002/cbic.201300351

64. Okamoto T, Zobel K, Fedorova A, Quan C, Yang H, Fairbrother WJ, Huang DC, Smith BJ, Deshayes K, Czabotar PE (2013) Stabilizing the pro-apoptotic BimBH3 helix (BimSAHB) does not necessarily enhance affinity or biological activity. ACS Chem Biol 8(2):297–302. doi:10.1021/cb3005403

65. Follis AV, Chipuk JE, Fisher JC, Yun MK, Grace CR, Nourse A, Baran K, Ou L, Min L, White SW, Green DR, Kriwacki RW (2013) PUMA binding induces partial unfolding within BCL-xL to disrupt p53 binding and promote apoptosis. Nat Chem Biol 9(3):163–168. doi:10.1038/nchembio.1166

66. Moldoveanu T, Grace CR, Llambi F, Nourse A, Fitzgerald P, Gehring K, Kriwacki RW, Green DR (2013) BID-induced structural changes in BAK promote apoptosis. Nat Struct Mol Biol 20(5):589–597. doi:10.1038/nsmb.2563, nsmb.2563

67. Friberg A, Vigil D, Zhao B, Daniels RN, Burke JP, Garcia-Barrantes PM, Camper D, Chauder BA, Lee T, Olejniczak ET, Fesik SW (2013) Discovery of potent myeloid cell leukemia 1 (Mcl-1) inhibitors using fragment-based methods and structure-based design. J Med Chem 56(1):15–30. doi:10.1021/jm301448p

68. Lee EF, Dewson G, Evangelista M, Pettikiriarachchi A, Gold GJ, Zhu H, Colman PM, Fairlie WD (2014) The functional differences between pro-survival and pro-apoptotic B cell lymphoma 2 (Bcl-2) proteins depend on structural differences in their Bcl-2 homology 3 (BH3) domains. J Biol Chem 289(52):36001–36017. doi:10.1074/jbc.M114.610758

69. Procko E, Berguig GY, Shen BW, Song Y, Frayo S, Convertine AJ, Margineantu D, Booth G, Correia BE, Cheng Y, Schief WR, Hockenbery DM, Press OW, Stoddard BL, Stayton PS, Baker D (2014) A computationally designed inhibitor of an Epstein-Barr viral Bcl-2 protein induces apoptosis in infected cells. Cell 157(7):1644–1656. doi:10.1016/j.cell.2014.04.034

70. Follis AV, Llambi F, Ou L, Baran K, Green DR, Kriwacki RW (2014) The DNA-binding domain mediates both nuclear and cytosolic functions of p53. Nat Struct Mol Biol 21(6):535–543. doi:10.1038/nsmb.2829

71. Marshall B, Puthalakath H, Caria S, Chugh S, Doerflinger M, Colman PM, Kvansakul M (2015) Variola virus F1L is a Bcl-2-like protein that unlike its vaccinia virus counterpart inhibits apoptosis independent of Bim. Cell Death Dis 6:e1680. doi:10.1038/cddis.2015.52

72. Rajan S, Choi M, Baek K, Yoon HS (2015) Bh3 induced conformational changes in Bcl-X revealed by crystal structure and comparative analysis. Proteins. doi:10.1002/prot.24816

73. Kim JS, Ku B, Woo TG, Oh AY, Jung YS, Soh YM, Yeom JH, Lee K, Park BJ, Oh BH, Ha NC (2015) Conversion of cell-survival activity of Akt into apoptotic death of cancer cells by two mutations on the BIM BH3 domain. Cell Death Dis 6:e1804. doi:10.1038/cddis.2015.118

74. Bruncko M, Oost TK, Belli BA, Ding H, Joseph MK, Kunzer A, Martineau D, McClellan

WJ, Mitten M, Ng SC, Nimmer PM, Oltersdorf T, Park CM, Petros AM, Shoemaker AR, Song X, Wang X, Wendt MD, Zhang H, Fesik SW, Rosenberg SH, Elmore SW (2007) Studies leading to potent, dual inhibitors of Bcl-2 and Bcl-xL. J Med Chem 50(4):641–662. doi:10.1021/jm061152t

75. Porter J, Payne A, de Candole B, Ford D, Hutchinson B, Trevitt G, Turner J, Edwards C, Watkins C, Whitcombe I, Davis J, Stubberfield C (2009) Tetrahydroisoquinoline amide substituted phenyl pyrazoles as selective Bcl-2 inhibitors. Bioorg Med Chem Lett 19(1):230–233. doi:10.1016/j.bmcl.2008.10.113

76. Sleebs BE, Czabotar PE, Fairbrother WJ, Fairlie WD, Flygare JA, Huang DC, Kersten WJ, Koehler MF, Lessene G, Lowes K, Parisot JP, Smith BJ, Smith ML, Souers AJ, Street IP, Yang H, Baell JB (2011) Quinazoline sulfonamides as dual binders of the proteins B-cell lymphoma 2 and B-cell lymphoma extra long with potent proapoptotic cell-based activity. J Med Chem 54(6):1914–1926. doi:10.1021/jm101596e

77. Perez HL, Banfi P, Bertrand J, Cai ZW, Grebinski JW, Kim K, Lippy J, Modugno M, Naglich J, Schmidt RJ, Tebben A, Vianello P, Wei DD, Zhang L, Galvani A, Lombardo LJ, Borzilleri RM (2012) Identification of a phenylacylsulfonamide series of dual Bcl-2/Bcl-xL antagonists. Bioorg Med Chem Lett 22(12):3946–3950. doi:10.1016/j.bmcl.2012.04.103

78. Schroeder GM, Wei D, Banfi P, Cai ZW, Lippy J, Menichincheri M, Modugno M, Naglich J, Penhallow B, Perez HL, Sack J, Schmidt RJ, Tebben A, Yan C, Zhang L, Galvani A, Lombardo LJ, Borzilleri RM (2012) Pyrazole and pyrimidine phenylacylsulfonamides as dual Bcl-2/Bcl-xL antagonists. Bioorg Med Chem Lett 22(12):3951–3956. doi:10.1016/j.bmcl.2012.04.106

79. Zhou H, Chen J, Meagher JL, Yang CY, Aguilar A, Liu L, Bai L, Cong X, Cai Q, Fang X, Stuckey JA, Wang S (2012) Design of Bcl-2 and Bcl-xL inhibitors with subnanomolar binding affinities based upon a new scaffold. J Med Chem 55(10):4664–4682. doi:10.1021/jm300178u

80. Wysoczanski P, Mart RJ, Loveridge EJ, Williams C, Whittaker SB, Crump MP, Allemann RK (2012) NMR solution structure of a photoswitchable apoptosis activating Bak peptide bound to Bcl-xL. J Am Chem Soc 134(18):7644–7647. doi:10.1021/ja302390a

81. Muppidi A, Doi K, Edwardraja S, Drake EJ, Gulick AM, Wang HG, Lin Q (2012) Rational design of proteolytically stable, cell-permeable peptide-based selective Mcl-1 inhibitors. J Am Chem Soc 134(36):14734–14737. doi:10.1021/ja306864v

82. Toure BB, Miller-Moslin K, Yusuff N, Perez L, Dore M, Joud C, Michael W, DiPietro L, van der Plas S, McEwan M, Lenoir F, Hoe M, Karki R, Springer C, Sullivan J, Levine K, Fiorilla C, Xie X, Kulathila R, Herlihy K, Porter D, Visser M (2013) The role of the acidity of N-heteroaryl sulfonamides as inhibitors of bcl-2 family protein-protein interactions. ACS Med Chem Lett 4(2):186–190. doi:10.1021/ml300321d

83. Souers AJ, Leverson JD, Boghaert ER, Ackler SL, Catron ND, Chen J, Dayton BD, Ding H, Enschede SH, Fairbrother WJ, Huang DC, Hymowitz SG, Jin S, Khaw SL, Kovar PJ, Lam LT, Lee J, Maecker HL, Marsh KC, Mason KD, Mitten MJ, Nimmer PM, Oleksijew A, Park CH, Park CM, Phillips DC, Roberts AW, Sampath D, Seymour JF, Smith ML, Sullivan GM, Tahir SK, Tse C, Wendt MD, Xiao Y, Xue JC, Zhang H, Humerickhouse RA, Rosenberg SH, Elmore SW (2013) ABT-199, a potent and selective BCL-2 inhibitor, achieves antitumor activity while sparing platelets. Nat Med 19(2):202–208. doi:10.1038/nm.3048, nm.3048 [pii]

84. Tanaka Y, Aikawa K, Nishida G, Homma M, Sogabe S, Igaki S, Hayano Y, Sameshima T, Miyahisa I, Kawamoto T, Tawada M, Imai Y, Inazuka M, Cho N, Imaeda Y, Ishikawa T (2013) Discovery of potent Mcl-1/Bcl-xL dual inhibitors by using a hybridization strategy based on structural analysis of target proteins. J Med Chem 56(23):9635–9645. doi:10.1021/jm401170c

85. Petros AM, Swann SL, Song D, Swinger K, Park C, Zhang H, Wendt MD, Kunzer AR, Souers AJ, Sun C (2014) Fragment-based discovery of potent inhibitors of the anti-apoptotic MCL-1 protein. Bioorg Med Chem Lett 24(6):1484–1488. doi:10.1016/j.bmcl.2014.02.010

86. Schilling J, Schoppe J, Sauer E, Pluckthun A (2014) Co-crystallization with conformation-specific designed ankyrin repeat proteins explains the conformational flexibility of BCL-W. J Mol Biol 426(12):2346–2362. doi:10.1016/j.jmb.2014.04.010

87. Koehler MF, Bergeron P, Choo EF, Lau K, Ndubaku C, Dudley D, Gibbons P, Sleebs BE, Rye CS, Nikolakopoulos G, Bui C, Kulasegaram S, Kersten WJ, Smith BJ, Czabotar PE, Colman PM, Huang DC, Baell JB, Watson KG, Hasvold L, Tao ZF, Wang L, Souers AJ, Elmore SW, Flygare JA, Fairbrother

WJ, Lessene G (2014) Structure-guided res-caffolding of selective antagonists of BCL-XL. ACS Med Chem Lett 5(6):662–667. doi:10.1021/ml500030p

88. Fang C, D'Souza B, Thompson CF, Clifton MC, Fairman JW, Fulroth B, Leed A, McCarren P, Wang L, Wang Y, Feau C, Kaushik VK, Palmer M, Wei G, Golub TR, Hubbard BK, Serrano-Wu MH (2014) Single diastereomer of a macrolactam core binds specifically to myeloid cell leukemia 1 (MCL1). ACS Med Chem Lett 5(12):1308–1312. doi:10.1021/ml500388q

89. Burke JP, Bian Z, Shaw S, Zhao B, Goodwin CM, Belmar J, Browning CF, Vigil D, Friberg A, Camper DV, Rossanese OW, Lee T, Olejniczak ET, Fesik SW (2015) Discovery of tricyclic indoles that potently inhibit Mcl-1 using fragment-based methods and structure-based design. J Med Chem 58(9):3794–3805. doi:10.1021/jm501984f

Chapter 17

Screening Strategies for TALEN-Mediated Gene Disruption

Boris Reljić and David A. Stroud

Abstract

Targeted gene disruption has rapidly become the tool of choice for the analysis of gene and protein function in routinely cultured mammalian cells. Three main technologies capable of irreversibly disrupting gene-expression exist: zinc-finger nucleases, transcription activator-like effector nucleases (TALENs), and the CRISPR/Cas9 system. The desired outcome of the use of any of these technologies is targeted insertions and/or deletions (indels) that result in either a nonsense frame shift or splicing error that disrupts protein expression. Many excellent do-it-yourself systems for TALEN construct assembly are now available at low or no cost to academic researchers. However, for new users, screening for successful gene disruption is still a hurdle. Here, we describe efficient and cost-effective strategies for the generation of gene-disrupted cell lines. Although the focus of this chapter is on the use of TALENs, these strategies can be applied to the use of all three technologies.

Key words TALEN, CRISPR, T7E1, Gene disruption, Gene editing, Indels, Screening

1 Introduction

Targeted gene disruption or genome editing has revolutionized our ability to assess protein function in mammalian cells. Unlike other technologies that modulate gene expression (e.g., RNAi), programmable nucleases result in complete and permanent loss of protein expression. Three main gene-editing technologies have emerged over recent years: zinc-finger nucleases, transcription activator-like effector nucleases (TALENs), and the CRISPR/Cas9 system [1–12]. These methodologies differ in both their mechanism of action and ease of construct assembly, but the desired result is the same: on-target double-stranded breaks (DSBs) at a desired locus, usually within the coding sequence of a gene of interest [13]. In most cell lines, DSBs are inaccurately repaired by nonhomologous end-joining (NHEJ), leading to insertions or deletions (indels) that disrupt protein expression. While it is feasible to exploit the other DSB repair mechanism, homology directed

Hamsa Puthalakath and Christine J. Hawkins (eds.), *Programmed Cell Death: Methods and Protocols*, Methods in Molecular Biology, vol. 1419, DOI 10.1007/978-1-4939-3581-9_17, © Springer Science+Business Media New York 2016

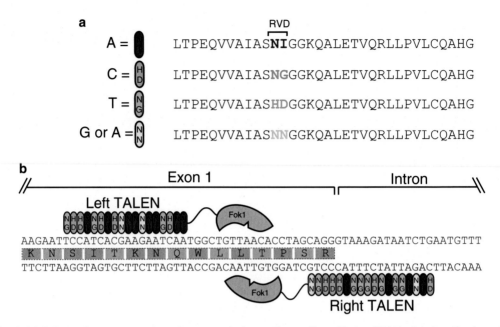

Fig. 1 (**a**) Schematic representation of a transcription activator-like effector (TALE) showing the four different repeat variable diresidues (RVDs). (**b**) Simplified illustration depicting a pair of TALENs targeting exon 1 of *GHITM*

repair (HDR) to make precise edits, this application of gene editing involves specific screening methodologies and thus will not be covered here.

TALENs consist of two domains, a DNA sequence-specific transcription activator-like effector (TALE) which is fused to the nuclease Fok1 to yield a TALEN [7–11]. TALEs can be further broken down to ~15–20 repeat variable diresidue domains (RVDs). There are four commonly used RVDs (NI, HD, NG, NN) named according to the two variable amino acids that define the domain's nucleotide specificity (Fig. 1a) [14, 15]. Fok1 is a nonspecific nuclease that is only active upon dimerization; therefore pairs of TALENs are constructed to promote dimerization over the desired DSB site (Fig. 1b). Construction of TALEN pairs is not trivial, and their efficient and cost-effective assembly is the major barrier to their use, especially when considering the low costs of entry of other systems such as CRISPR/Cas9. However, TALENs may have a specificity advantage over the CRISPR/Cas9 system due to their comparably large binding site (~60 nt vs. ~20 nt for CRISPRs; [13]) and thus may be more suitable for certain applications where low off-target activity is required. A number of companies design and assemble custom TALEN pairs or make available pre-existing TALEN pairs derived from large-scale libraries for a fee. However, most users choose to assemble TALENs in-house and several groups have developed well-documented systems for TALEN assembly and have made these available as kits through the non-profit plasmid repository Addgene (http://www.addgene.org).

TALEN design is a critical step and in our hands, target site selection is most efficiently achieved through web-based design tools (Table 1). While these tools assist the user in obeying TALEN design rules (e.g., rules specific to the TALE:DNA interaction), there are a number of additional considerations to be made specific to the gene being targeted (Fig. 2a). To ablate expression of a protein-coding gene, a common strategy is to target the translation initiation ATG. Care must be taken, however, as translation can alternately start from subsequent in-frame ATGs within the same exon [16]. Several alternate approaches are (1) target the most 3′ in-frame ATG within the first exon, (2) target an exon–intron boundary such that mRNA splicing will be disrupted, or (3) target a motif critical for the protein's biogenesis and function (e.g., a targeting signal or transmembrane domain). We have observed loss of protein function through all of these approaches [16–20]. Another consideration is the existence of different splice variants where translation is initiated from different coding exons (Fig. 2a, lower panel). In this case, we have had success targeting the first common coding exon [17, 20] subject to the considerations discussed above.

The repair of TALEN generated DSBs is a somewhat random event, with deletions (or less commonly insertions) ranging in size from a single to hundreds of base pairs. Each allele will often receive a different indel, and this is further complicated by the high frequency of aneuploidy in cultured cell lines. Characterization of all alleles present within the cell is therefore necessary to avoid non-protein function destroying mutations such as in-frame deletions. This is typically achieved through the generation of single cell-derived clonal populations in which all indels have been characterized. Typically researchers (and reviewers) demand two to three unique clones to control for clonal differences as well as potential off-target effects. At the commencement of a project we normally generate a clonal cell line in which all gene disruptions will be made (Fig. 2b). TALEN pairs are introduced into this line by transient transfection and single cells are derived through fluorescence activated cell sorting (FACS). To ensure efficient sorting of transfected cells, a fluorescent marker is co-transfected with the TALEN constructs using a limiting amount of DNA. In our hands, co-transfection of 1/10th the amount of a green fluorescent protein (GFP)-tagged construct relative to each TALEN construct is sufficient to allow FACS sorting of a majority of cells also harboring the TALEN constructs (data not shown). Following sorting, 10–20 single cell clones are expanded over the course of several weeks to cell numbers permissive of screening. If the TALENs are functional, at least ~10–50 % of clones should contain indels.

Table 1

List of the currently known TALEN targeting and design tools

Tool	Website URL	Usage	Kit specificity	Genome-wide off-target analysis.	TALENs scored/ ranked.	Detects restriction sites for screening.	Can specify target site class (e.g. UTR, 1st exon, splice site).	Outputs primers for screening and/ or sequencing.
ZiFiT [27]	http://zifit.partners.org/ZiFiT/	ZFN/TALEN/CRISPR	REAL/REAL-Fast/FLASH	no	no	no	no[a]	no
TALE-NT [28]	https://tale-nt cac.cornell.edu/	TALEN	Any	yes	no	yes	no	no
Mojo Hand [29]	http://www.ta.endesign.org/	TALEN/CRISPR	Any	no[b]	yes[c]	yes	no	no
E-TALEN [30]	http://www.e-talen.org/E-TALEN/	TALEN	Major kits only	yes	yes	yes	yes	no
SAPTA [31]	http://bao.rice.edu/	TALEN	Any	no[d]	yes	yes	no	no
CHOPCHOP [22]	https://chopchop.rc.fas.harvard.edu/	TALEN/CRISPR	Any	yes	yes	yes	yes	yes

[a]Gene sequence entered manually by user

[b]*Danio rerio* only

[c]Single search outputs and ranks both TALEN and CRISPR guides

[d]Companion site "PROGNOS" permits off-target analysis

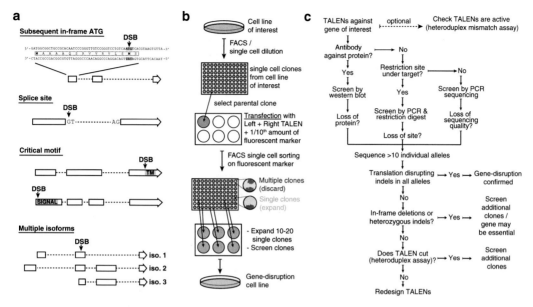

Fig. 2 (**a**) Example DSB targeting strategies for gene disruption. (**b**) Gene disruption workflow. A single-cell cloned cell line of interest is established as a parental line for gene disruption. The cell line is cotransfected with both TALEN pairs and a limiting amount (1/10th) of a cotransfection marker (e.g., cytosolic expressed GFP) to enable transfected single cells to be sorted by FACS into 96-well plates. Single cell colonies are expanded and screened for gene disruption. (**c**) Screening strategy decision tree

The choice of screening strategy largely depends on the reagents and equipment available to the researcher. Screening by Western blotting for loss of protein expression is the simplest and most robust approach, and the existence of a quality antibody usually directs our screening workflow (Fig. 2c). If an antibody is not available, we utilize one of two approaches that employ amplification of the targeted region by polymerase chain reaction (PCR): (1) mutation of a unique restriction site present at the target site, or (2) by sequencing the PCR product to detect indels. Once candidate clones are identified, sequencing of the individual alleles covering the target site in each clone is necessary to exclude non-translation disrupting as well as heterozygous indels.

2 Materials

The following protocols are intended for use with common lab cell lines cultured in Dulbecco's Modified Eagle Medium (DMEM) supplemented with 10 % fetal bovine serum (FBS). We have successfully used these protocols to generate gene disruptions in HEK293T cells [16, 19, 20], HeLa cells, mouse embryonic fibroblasts (MEFs) [18] and HCT116 cells [17] (*see* **Note 1**). All reagents for cell culture should be sterilized by autoclave, or prepared aseptically by passage through a 0.22 μm filter. Possession of basic tissue culture equipment (e.g., electronic pipette fillers, laminar flow hoods, CO$_2$ incubators) and consumables (e.g., pipette tips, 100 mm dishes, T75 flasks) is assumed, as is access to a staffed flow cytometry facility with 96-well plate single cell sorting capability. Reagents for screening should be prepared with autoclaved ultrapure water and reactions conducted in fresh, sterile plastic ware. For reagents commonly obtained from an international vendor, we have indicated the vendor's name to aid identification of the reagent. Substitutions can usually be made; however, the alternate vendor's product information sheets should be consulted to confirm parameters such as reagent concentration.

2.1 Maintenance of Cells in Culture, Transfection of TALENs, and Sorting of Transfected Cells

1. Sterile, phosphate buffered saline (PBS).

2. Incomplete DMEM: DMEM, high glucose (4.5 g/L glucose).

3. Transfection media: DMEM, high glucose (4.5 g/L glucose) supplemented with 10 % fetal bovine serum (FBS) (*see* **Note 2**).

4. Growth media: DMEM, high glucose (4.5 g/L glucose) supplemented with 10 % FBS and penicillin/streptomycin.

5. Freezing media: DMEM, high glucose (4.5 g/L glucose) supplemented with 20 % fetal bovine serum (FBS) and penicillin/streptomycin, 10 % Dimethyl sulfoxide (DMSO).

6. Cell sorting media: PBS supplemented with 10 % FBS and 1 mM ethylenediaminetetraacetic acid (EDTA).

7. 0.25 % Trypsin, phenol red.

8. Lipofectamine® 2000.

9. TALEN constructs: known concentration of plasmids encoding left- and right-TALEN constructs, suspended in sterile, ultrapure water.

10. Transfection marker, for example a known concentration of plasmid pEGFP-N1 (Clontech; *or see* **Note 3**) suspended in sterile, ultrapure water.

11. 70 μm cell strainer.

12. Multichannel pipette for 200 μl tips.

13. Preracked or sterilized 200 μl pipette tips.

14. 20 ml reservoir suitable for use with multichannel pipettes.

15. 96-well and 6-well tissue culture plates.

2.2 General Molecular Biology Reagents (Required for All Strategies)

1. SDS/proteinase K lysis buffer: 100 mM Tris–HCl (pH 8.8), 20 mM ammonium sulfate, 5 mM β-mercaptoethanol, 10 mM MgCl$_2$, 5 μM EDTA, 2 μM sodium dodecyl sulfate (SDS), 20 μg/ml proteinase K. Store in aliquots at −20 °C.

2. Robust proofreading DNA polymerase, e.g., Q5® High-Fidelity DNA Polymerase and programmable PCR thermocycler.

3. PCR purification kit, e.g., QIAquick PCR Purification Kit or Wizard® SV Gel or PCR Clean-Up System or equivalent.

4. Miniprep kit, e.g., QIAprep Spin Miniprep Kit or PureYield™ Plasmid Miniprep system or equivalent.

5. Sterile lysogeny broth (LB; 10 g/L tryptone, 5 g/L yeast extract, 10 g/L NaCl).

6. LB Agar plates supplemented with the appropriate antibiotic.

7. Competent *E. coli* cells for transformation (*see* **Note 4**).

8. pGEM®-4Z vector (Promega) (or *see* **Note 5**).

9. Restriction enzymes as needed, with compatible 10× buffers (*see* **Note 6**).

10. Oligonucleotides as needed (desalt purified by the manufacturer).

11. Alkaline Phosphatase, Calf Intestinal (10,000 U/ml).

12. T4 DNA Ligase (2,000,000 U/ml).

13. 2× QLB (Quick Ligation Buffer): 132 mM Tris–HCl (pH 7.6), 20 mM $MgCl_2$, 2 mM DTT, 2 mM ATP, 15 % PEG 6000.

14. DNA ladder, e.g., 1 kb DNA Ladder.

15. 6× DNA loading buffer: 30 % (v/v) glycerol, 20 mM Tris–HCl (pH 8.0), 60 mM EDTA, 0.5 % (w/v) SDS, 0.02 % (w/v) Orange G, 0.02 % (w/v) Xylene cyanol FF.

16. TAE buffer: 0.1 M acetic acid, 1 mM EDTA, 40 mM Tris–HCl (pH 8.0).

17. Agarose, low electroendosmosis (EEO).

18. Generic agarose gel casting, electrophoresis, and blue-light imaging system.

19. Blue-light compatible DNA stain, e.g., 1000× SYBR® Safe or equivalent (*see* **Note 7**).

2.3 Heteroduplex Mismatch Assay

1. T7 Endonuclease I (New England Biolabs), Guide-it™ Resolvase (Clontech), or CELII (sold as Surveyor™ Nuclease S by Transgenomic). Selection of appropriate reagent is discussed in Subheading 3.2.

2.4 SDS-PAGE and Western Blotting

1. Precast or self-made (*see* **Note 8**) SDS-PAGE and Western blotting system selected based on the requirements of your protein of interest.

2. 4× SDS sample buffer: 200 mM Tris–HCl (pH 6.8), 8 % (w/v) sodium dodecyl sulfate (SDS), 40 % (v/v) glycerol, 0.04 % (w/v) bromophenol blue. Can be stored in aliquots at −20 °C or at room temperature (*see* **Note 9**).

3. 1 M dithiothreitol (DTT). Stored in single-use aliquots at −20 °C.

4. Enhanced Chemiluminescence (ECL) reagent and compatible imaging system.

---

3 Methods

3.1 Primer Design and Basic PCR Amplification of the Target Site

All screening strategies converge with the sequencing of individual alleles (*see* Fig. 2c) therefore it is imperative that early in the project, oligonucleotide primers are designed that permit efficient amplification of the target region from genomic DNA. As this step requires cloning of the amplified products, we find it convenient to include restriction site overhangs—in our hands this does not affect the PCR efficiency or the PCR product's function in the other screening assays. Primers should be designed to yield a PCR product of ~500 bp, centered on the target site, with a T_m for the complementary region of 60–65 °C. We typically use the web-based "Primer-BLAST" tool [21] to search for specific primer pairs that yield no or few off-target products. The excellent TALEN design tool "CHOPCHOP" [22] is also capable of designing specific primers centered on the chosen target site. Restriction sites should be added to the primers such that they are unique (i.e., not present in the insert) and compatible with the vector chosen for downstream analysis. Overhang sequences should not impact on the primer T_m calculations. For synthesis, only desalt purification is required.

1. Harvest approximately 1×10^6 cells (equivalent to a confluent well of a 6-well dish) of the cell line of interest and transfer to a sterile 1.5 ml microcentrifuge tube. Centrifuge for 3 min at $300 \times g$.

2. Wash cells with PBS and centrifuge for 3 min at $300 \times g$.

3. Aspirate PBS and solubilize cells in 100 μl of SDS/proteinase K lysis buffer. Incubate for 1 h at 37 °C, followed by 10 min at 80 °C.

   Optional: Sonicate crude extract in a ultrasonic water bath for 5 min to reduce viscosity.

4. Centrifuge crude extract for 5 min at $16,000 \times g$ and transfer 90 μl of the supernatant to a new tube being careful not to disturb the pellet.

5. The amount of crude DNA extract to use in a PCR reaction needs to be titrated. Assemble the PCR reactions (*see* **Note 10**) and cycle as described in Table 2.

6. Cast a 2 % (w/v) agarose gel in TAE, supplemented with 1× SYBR Safe DNA stain.

Table 2
Protocol for generic PCR amplification using crude genomic DNA extract as template

| Reagent | Increasing amounts of crude genomic DNA extract (µl) | | | | | Cycle parameters | | |
	1	2	3	4	5	Step	Temperature (°C)	Time
5× reaction buffer	5	5	5	5	5	Denaturation	98	30 s
10 mM dNTPs[a]	0.5	0.5	0.5	0.5	0.5	35 cycles	98	10 s
10 µM Forward primer	1.25	1.25	1.25	1.25	1.25		55[b]	30 s
10 µM Reverse primer	1.25	1.25	1.25	1.25	1.25		72	30 s[c]
Crude genomic DNA extract[d]	0.25	0.5	1	1.5	2.5	Final extension	72	2 min
High-fidelity DNA polymerase	0.25	0.25	0.25	0.25	0.25	Hold	4	–
Nuclease-free water	16.5	16.25	15.75	15.25	14.25			

[a]Consisting of 2.5 mM each dNTP
[b]We find 55 °C to strike a good balance between specificity and efficiency for most amplicons
[c]Increase if product is >1000 bp.
[d]If excessively viscous, perform optional sonication step. Use of high amounts of crude DNA extract can inhibit PCR performance due to detergent present in the crude extract buffer

7. Combine 5 µl of each PCR reaction with 1 µl 6× DNA loading dye. Load onto agarose gel along with a suitable DNA ladder and perform electrophoresis at 150 V for ~15–30 min. Visualize using a blue-light transilluminator, or image using appropriate equipment.

3.2 Heteroduplex Mismatch Assay

The heteroduplex mismatch assay is a simple and time saving means to assess activity of TALEN pairs following their cotransfection but prior to single cell cloning and screening. Crude genomic DNA is extracted from cells, and the target region amplified by high-fidelity PCR. The PCR product, which will consist of a mixture of wild-type alleles from untransfected cells and mutant alleles from gene-disrupted cells, is melted and rehybridized generating mismatched heteroduplex DNA—mismatches resulting from as little as a single missing or mutated base are detected by the addition of a T7E1 resolvase or S1 endonuclease, both of which cleave mismatched DNA. The ratio between cleaved and uncleaved products can be compared by DNA gel electrophoresis (Fig. 3). There is no difference in experimental output between T7E1 resolvases and S1 endonucleases, although the former has been reported as being

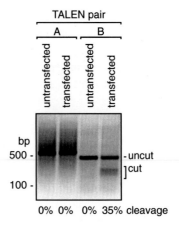

Fig. 3 Example results from a heteroduplex mismatch assay using nonfunctional (*A*) and functional (*B*) TALEN pairs. *Lanes 1* and *2* show the results following transfection of a nonfunctional TALEN pair. *Lanes 3* and *4* show the results from a functional TALEN pair with a cutting activity of 35 %, calculated by comparing the density of the cut bands with the density of both uncut and cut bands within a particular lane

more sensitive to low efficiency cutting, and the latter better for detecting single base changes [23]. In our hands, the Guide-it™ Resolvase (Clontech) has given the most robust results.

1. For each TALEN pair being tested, seed two wells of cells so as to achieve 70–90 % confluency on the day of transfection. Only one well will be transfected and the other will serve as an untransfected control. For HEK293T cells, we typically seed 500,000 cells per well of a 6-well dish the day prior to transfection.

2. In a sterile 1.5 ml microcentrifuge tube, combine 1.5 μg each TALEN construct and dilute with 500 μl incomplete DMEM (*see* **Note 11**).

3. Add 9 μl Lipofectamine® 2000, briefly vortex and incubate at room temperature for 5 min.

4. Aspirate media from one of the wells, carefully wash with PBS and overlay cells with 1.5 ml transfection media. Carefully add the transfection solution by pipette. Incubate cells overnight at 37 °C under 5 % CO_2.

5. Wash cells with PBS, add 500 μl 0.25 % trypsin and incubate for several minutes at 37 °C.

6. Detach cells with 500 μl growth media and transfer to sterile 1.5 ml microcentrifuge tube. Centrifuge for 3 min at $300 \times g$.

7. Prepare crude genomic DNA extract as described in Subheading 3.1.

8. Using the optimum amount of crude genomic DNA extract determined in Subheading 3.1, amplify the targeted region from both untransfected and transfected samples. Analyze 10 % of the reaction on a 2 % agarose gel. A clean, single PCR product is expected with an intensity of 1:1 with DNA ladder loaded such that each band represents 100 ng DNA (*see* **Note 12**).

9. In most cases, the PCR product can be used directly in the heteroduplex mismatch assay. Check the buffer composition used in the DNA polymerase buffer; the heteroduplex mismatch assay requires ≥ 1.5 mM $MgCl_2$ and ~50 mM salt (NaCl/KCl) at final concentration. If the buffer is incompatible or cannot be supplemented with additional $MgCl_2$, purify the PCR product with a commercial PCR purification kit. Elute the product in 10 mM Tris–HCl (pH 8.8), 1.5 mM $MgCl_2$, and 50 mM KCl aiming for a concentration of ~25–50 ng/μl.

10. Transfer 10 μl of the PCR product to a new tube. Perform a melt-hybridization in a programmable thermocycler (*see* **Note 13**) as follows: (1) 95 °C, 5 min, (2) ramp 95–85 °C at 2 °C/s, (3) ramp 85–25 °C at 0.1 °C/s, hold at 4 °C.

11. Add 1 μl of T7E1 (10 U), 1 μl Guide-it™ Resolvase, or 1 μl each Surveyor® Nuclease S and Enhancer S (*see* **Note 14**).

12. If using T7E1 or Guide-it™ Resolvase, incubate samples for 15 min at 37 °C. For Surveyor® incubate at 42 °C for 60 min.

13. Add 2 μl 6× DNA loading dye and run the entire reaction on a 2 % agarose gel.

14. Cutting efficiency can be quantified with appropriate software by comparing the density of the cut band(s) with the density of both uncut and cut band(s) within a particular lane.

3.3 Transfection of TALENs and Sorting of Transfected Cells

1. Parental cells should be seeded so as to achieve 70–90 % confluency at the time of transfection. For HEK293T cells, we typically seed 500,000 cells per well of a 6-well dish 24 h prior to transfection.

2. In a sterile 1.5 ml microcentrifuge tube, combine 1.5 μg each TALEN construct with 150 μg pEGFP-N1 (*see* **Note 15**) and dilute with 500 μl incomplete DMEM.

3. Add 1.5 μl Lipofectamine 2000 (*see* **Note 16**), briefly vortex and incubate at room temperature for 5 min.

4. Aspirate media from the cells to be transfected, carefully wash with PBS and overlay cells with 1.5 ml transfection media (*see* **Note 17**). Carefully add the transfection solution by pipette. Incubate cells overnight at 37 °C under 5 % CO_2.

5. Prepare 96-well plates for cell sorting by adding 200 µl growth media into each well using a multichannel pipette and buffer reservoir.

6. Aspirate transfection media and wash cells with PBS. Trypsinize using 200 µl 0.25 % trypsin for 5 min at 37 °C. Block trypsin using 2 ml cell sorting media and transfer cells through a 70 µm cell strainer to a tube suitable for FACS analysis.

7. Sort single cells into wells of the pre-prepared 96-well plate based on a medium level GFP fluorescence (*see* **Note 18**). Typically we sort a single plate based on recovery of ~20–50 single cell populations/plate, with a view to expanding and screening 12–24 clones. Recovery is highly dependent on both the cell type and proper use and calibration of the FACS, therefore additional 96-well plates may need to be sorted should recovery be less than expected.

3.4 Expansion of Candidate Gene Disruption Clones

1. Plates should be examined after 3–5 days for evidence of single cell populations. Mark wells containing single cell populations and exclude those containing more than one population—these should occur infrequently at the rate of 1–2 per 96-well plate.

   Optional: 7 days after sorting, change media using a multichannel pipette and two buffer reservoirs, one for waste and one for new media. Take care not to cross-contaminate clonal populations whilst using multichannel pipettes. Presterilized filter pipette tips are recommended.

2. Depending on the cell type, clonal populations will be clumped and overgrown 10–15 days post sorting. Mark 12–24 clonal populations for further screening (*see* **Note 19**).

3. Aspirate media from marked wells, and using single channel pipettes carefully wash with PBS. Add 20 µl of 0.25 % Trypsin and incubate for several minutes at 37 °C.

4. Detach cells with 180 µl of culture media. Using a pipette set to 100 µl, move half the cells from each well to a new well in a new 96-well plate prepared with 100 µl media per well. Supplement the original well with 100 µl media for freezing in Subheading 3.5. If performing SDS-PAGE and western blotting analysis in Subheading 3.6, a third plate should be assembled and the culture split to accommodate this. This is an opportune time to update well numbering for ease of culturing and to conserve reagents (as only one or two rows of the 96-well plate are required per gene disruption). Since the cells in the original plate will be subjected to cryo-storage, care should be taken to update the new numbering on this plate so a specific clone can later be revived.

5. Continue culturing clonal populations on the new plate until confluent. Repeat **step 3**, and using 180 μl media move the cells to a marked well of a 6-well plate. Supplement with 1.8 ml of growth media.

6. The clones growing in 6-well plates are used in the following sections, however, may be expanded further for additional analyses and/or freezing as per the needs of the researcher.

3.5 Freezing Candidate Clones in a 96-Well Plate

1. The original 96-well plate can be frozen as a backup, or for later analysis by carefully aspirating the media using a multichannel pipette.

2. Using the multichannel pipette, wash cells with PBS and add 20 μl 0.25 % trypsin. Incubate for several minutes at 37 °C.

3. Resuspend the cells in 180 μl freezing media.

4. Carefully (as the plate is not sealed) wrap the plate in paper towel and store at –80 °C (*see* **Note 20**).

To thaw cells: Swab with 70 % ethanol and place in a 37 °C incubator for 20 min. In a swinging centrifuge with microtiter plate adaptor, centrifuge for 5 min at $500 \times g$. Slowly aspirate 180 μl using a multichannel pipette and replace with 180 μl fresh culture media.

3.6 Screening by SDS-PAGE and Western Blot

The duplicate 96-well plate from Subheading 3.4 should reach 80–90 % confluency in about 3–7 days. Cells will be directly solubilized in the plate and analyzed by SDS-PAGE without normalization for protein concentration. While most populations will have similar cell numbers, there will no doubt be outliers. Despite such variability, we find this approach rapid and accurate enough for routine screening provided a loading control (e.g., actin, or Hsp70, *see* Fig. 4a) is used. *Aseptic technique is not required for the following steps.*

1. Note down on each well the approximate cell confluence as a factor of 100 (100 being the average density).

2. Aspirate media and wash cells by carefully adding 200 μl PBS to the side of each well, rocking the plate by hand and aspirating with the same tip.

3. Solubilize cells in 30 μl 1× SDS-PAGE sample buffer supplemented with 50 mM DTT by slowly pipetting up and down. *Optional:* Heat plate to 95 °C for 5 min with gentle shaking. Allow plate to cool down and centrifuge briefly.

4. Prepare a suitable SDS-PAGE gel (*see* **Note 8**) for electrophoresis and load 15 μl of each sample with a confluence factor of 100 as determined in **step 1**. Using this factor, load appropriate amounts of the remaining samples to a minimum of 5 μl and a maximum of 30 μl.

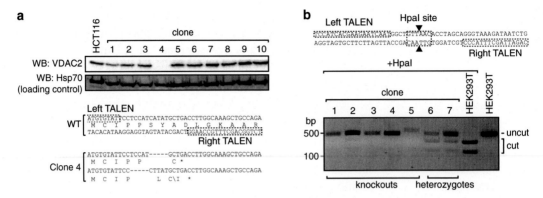

Fig. 4 (**a**) Example results of screening by SDS-PAGE and western blot. *Upper panel*, protein extracts from control (HCT116 parental line) and candidate clonal populations analyzed by SDS-PAGE and western blotting with antibodies against human VDAC2 and loading control Hsp70. *Lower panel*, targeted alleles present in clone 4 (*see* Subheading 3.9) (**b**) Example results following screening by loss of restriction sites. *Upper panel*, targeting strategy for gene disruption of the *GHITM* gene, showing targeting centered over the HpaI restriction site. *Lower panel*, the targeted region was amplified from control HEK293T cells and indicated clonal populations, and digested with the restriction enzyme HpaI. Gene disruption is evident from the loss of digested PCR products. Clones 6 and 7 are heterozygous for loss of the restriction site

5. Perform electrophoresis and western blotting as described by the manufacturer/literature for the gel and western blotting system being used.

3.7 Screening by Restriction Digestion

Screening by loss of restriction site is an effective means for identifying clones having undergone gene disruption. This strategy must be considered during the design process such that a unique (within the PCR of the target region) restriction site is located within the target site. In this assay, wild-type alleles (or those present in negative clones) will be cut by the restriction enzyme, however, alleles having undergone gene disruption will not be digested due to mutations in the restriction site (*see* Fig. 4b). This assay has the advantage of being the only strategy capable of revealing true heterozygotes DSB events (e.g., the TALENs only having generated DSBs on one allele) early in the screening process (Fig. 4b, compare clones 6 and 7 with clones 1–5).

1. Generate crude genomic DNA extracts (*see* Subheading 3.1) of clonal populations in the 6-well plates seeded in Subheading 3.4.

2. Using the optimum amount of crude genomic DNA extract determined in Subheading 3.1, amplify the targeted region from both untransfected and transfected samples. Analyze 10 % of the reaction on a 2 % agarose gel.

3. Purify the PCR product with a generic PCR purification kit, eluting in 22.5 μl nuclease free water.

4. To each sample, add 2.5 μl appropriate 10× restriction enzyme buffer and 0.2 μl (*see* **Note 21**) of the required restriction enzyme.

5. Incubate samples for 1 h at 37 °C or as is appropriate for the restriction enzyme being used.

6. Add 5 μl of 6× DNA loading buffer to each sample and resolve on a 2 % (w/v) agarose gel in TAE, supplemented with 1× SYBR Safe DNA stain. Visualize using a blue-light transilluminator, or image using appropriate equipment.

3.8 Sequencing the PCR Product as a Screening Strategy

In the absence of other available screening methodologies, we have been able to identify candidate clones with gene disruptions by direct sequencing of the PCR products amplified from individual clonal populations. This method relies on DSB repair rarely resulting in identical indels on all alleles (Fig. 5a). The mixed population of PCR products is sequenced by standard Sanger sequencing primed with one of the oligonucleotides used in its original amplification—the sequencing read will begin as normal but due to the mixture of alleles with different indels, the sequencing quality will deteriorate at the site of DSBs (*see* Fig. 5b, upper panel). Depending on the local costs associated with sequencing, this can be a reliable and cost effective means to isolate positive clones for further analysis.

1. Generate crude genomic DNA extracts (*see* Subheading 3.1) of clonal populations in the 6-well plates seeded in Subheading 3.4.

2. Using the optimum amount of crude genomic DNA extract determined in Subheading 3.1, amplify the targeted region

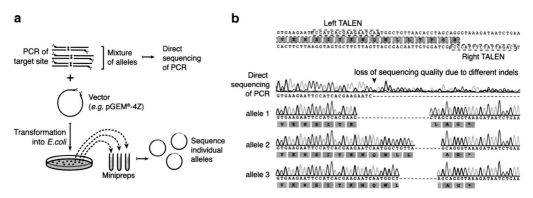

Reljić and Stroud Figure 5

Fig. 5 Screening by sequencing. (**a**) Schematic representation of the screening strategy. (**b**) *Upper sequencing chromatogram,* example result following direct sequencing of a mixed PCR product amplified from a single clone having undergone the above targeting strategy. *Lower chromatograms,* example results from sequencing individual alleles as described in (**a**). A total of 12 bacterial clones were sequenced, resulting in three unique indels

from both untransfected and transfected samples. Analyze 10 % of the reaction on a 2 % agarose gel.

3. Purify the PCR product with a generic PCR purification kit, eluting in a minimum volume of nuclease free water. Calculate the DNA concentration using its absorbance at 260 nm.

4. Using the forward primer designed in Subheading 3.1, assemble the sequencing reaction as per the requirements for sequencing a PCR product set by the local sequencing provider. A 10 µl reaction generally consists of 10–20 ng of DNA and 10 pmol primer diluted in nuclease free water.

3.9 Sequencing of Individual Alleles

The final step regardless of initial screening strategy is sequencing of the individual alleles present in a number of short-listed clones. The step is absolutely essential, as no other technique can fully exclude nontranslation disrupting indels such as in-frame deletions. The mixed population of PCR products representing all alleles (*see* **Note 22**) in a given clone is purified and the digested restriction sites introduced as overhangs to the primers (*see* Subheading 3.1) are used to ligate the mixture into a standard bacterial cloning vector. Individual alleles are isolated through transformation into competent *E. coli* and culturing of individual bacterial clones, each colony representing a unique allele. Plasmid DNA is isolated and the isolated alleles are sequenced using standard Sanger sequencing (*see* Fig. 5a). Given that cells in culture are typically diploid or triploid at a certain locus, the inevitable problem is deciding how many bacterial clones (and therefore representative alleles) to sequence. Our typical workflow involves the initial sequencing of eight clones. Sequences are aligned to the reference sequence (*see* Fig. 5b, lower panels) and duplicate reads are discarded. For example, when working with HEK293T cells we are usually satisfied upon detection of two or three unique alleles in a given clone.

1. Generate crude genomic DNA extracts (*see* Subheading 3.1) of clonal populations in the 6-well plates seeded in Subheading 3.4.

2. Using the optimum amount of crude genomic DNA extract determined in Subheading 3.1, amplify the targeted region from both untransfected and transfected samples. Analyze 10 % of the reaction on a 2 % agarose gel.

3. Purify the PCR product with a generic PCR purification kit, eluting in 25 µl of nuclease free water.

4. Assemble the following restriction enzyme digestions (*see* **Note 23**):

Clone specific PCR product(s)	Plasmid
21.5 µl purified PCR product	1 µg pGEM®-4Z or appropriate cloning vector
2.5 µl 10× buffer compatible with enzyme A and B	2.5 µl 10× buffer compatible with enzyme A and B
0.5 µl enzyme A	0.5 µl enzyme A
0.5 µl enzyme B	0.5 µl enzyme B
	Nuclease free water to 25 µl

5. Incubate the reactions for 2 h at 37 °C.

6. To the plasmid digest only, add 10 U (1 µl of 10,000 U/ml) alkaline phosphatase. Incubate both digests for an additional 30 min at 37 °C.

7. Purify digests with a generic PCR purification kit, eluting in 25 µl of nuclease free water.

8. For each clone, assemble the following ligation reaction (*see* **Note 24**):

Ligation/clone
2 µl digested/alkaline phosphatase treated plasmid
2 µl digested PCR product
1 µl T4 DNA ligase
5 µl 2× QLB buffer

9. Incubate ligations for 15 min at 37 °C.

10. In a sterile 1.5 ml microcentrifuge tube, combine 5 µl each ligation with 50 µl chemically competent *E. coli* (*see* **Note 25**) and incubate on ice for 10 min.

11. Heat shock at 42 °C for 90 s.

12. Immediately return transformation to ice and incubate for 2 min.

13. Add 500 µl LB and incubate for 60 min at 37 °C with shaking.

14. Plate ½ the suspension on LB Agar plates supplemented with the appropriate antibiotic for the cloning vector in use.

15. Incubate inverted plates at 37 °C overnight.

16. Using a sterile pipette tip, the following morning transfer eight colonies to pre-assembled 10 ml culture vials containing 5 ml LB supplemented with the appropriate antibiotic.

17. Incubate overnight at 37 °C with shaking.

18. Prepare miniprep scale plasmid DNA isolations using a generic miniprep kit. Calculate the DNA concentration using its absorbance at 260 nm.

19. Using the forward primer designed in Subheading 3.1, assemble sequencing reactions as per the requirements for sequencing a plasmid based template set by the local sequencing provider. A 10 μl reaction generally consists of 1 μg of DNA and 10 pmol primer diluted in nuclease free water.

4 Notes

1. While we have successfully made gene disruptions in HCT116 cells cultured in DMEM supplemented with 10 % FBS, optimal growth is achieved using McCoy's 5A (Modified) Medium (Life Technologies) supplemented with 10 % FBS and 1 mM sodium pyruvate.

2. Note omission of penicillin/streptomycin. As we have observed toxicity following high levels of TALEN expression, we avoid using commercial transfection enhancers such as Opti-MEM (Life Technologies) and directly replicating manufacturer's protocols as these have been optimized for high levels of expression. Cells used in the T7E1 assay will not be expanded therefore in this case the optimized protocols may be applied.

3. Can be replaced with any suitable marker for transfection according to the needs of the researcher and capabilities of the FACS; e.g., should cells already express a GFP tagged protein, another fluorophore should be utilized. Transient expression of the fluorophore should be nontoxic and should not overtly alter cellular function, yet should be bright enough for detection by FACS. Many empty vectors designed for C-terminal expression of GFP produce cytosolic localized protein driven by the strong CMV promoter, and these are highly suitable as transfection markers. Some TALEN constructs also express their own fluorophores in which case a transfection marker can be omitted.

4. These can be obtained commercially, however, as high transformation efficiency is not particularly important, competent cells may be made in-house using published protocols. We suggest using the excellent "one-step" method for generation of competent *E. coli* published by Chung et al. [24].

5. Can be replaced by any plasmid suitable for routine cloning using chemically competent *E. coli*.

6. Chosen as per the researcher's screening strategy and obtained from the preferred supplier.

7. May be substituted with ethidium bromide at 1 µg/ml final concentration (we typically use a 1000× working stock of 1 mg/ml). Note, ethidium bromide requires an ultraviolet transilluminator for visualization unlike other stains, which can be used with blue-light systems.

8. May be substituted with a self-made system. We commonly use the Tricine-SDS-PAGE and semidry electroblotting system developed by Hermann Schägger for which several detailed methods publications exist [25].

9. Alternatively, 200 mM DTT can be added to the 4× SDS sample buffer and aliquots for single-use stored at –20 °C.

10. Although in principle any high-fidelity DNA polymerase may be used, we have observed large differences in the quality and quantity of PCR product depending on polymerase. If feasible, multiple DNA polymerase systems should be tested to achieve optimal purity and amplification. For most downstream applications, a single strong band of the correct size will be required.

11. Cells will not be expanded and thus transfection toxicity will not be a problem, so reagents and protocols may be substituted with manufacturer recommended enhancers such as Opti-MEM® (Life Technologies) and more powerful transfection reagents such as Lipofectamine® LTX with PLUS™ Reagent (Life Technologies).

12. If a single PCR product cannot be obtained, gel purification of the correct band is possible. Elute the DNA in a buffer containing 10 mM Tris–HCl (pH 8.8), 1.5 mM $MgCl_2$, and 50 mM KCl at a concentration of ~25–50 ng/µl.

13. In the absence of a ramp function, melt-hybridization can be performed by incubation of the sample in a 95 °C water bath for 5 min, following which the water bath is turned off and allowed to cool to room temperature. Samples should be placed at 4 °C until use.

14. For the Surveyor assay, the manufacturer recommends to add additional $MgCl_2$ to a final concentration of 15 mM.

15. In order to sort only cells that have been transfected with the TALEN constructs, we co-transfect with a limiting amount (typically 1/10th each TALEN plasmid concentration) of pEGFP-N1. As discussed above, this may be replaced with any suitable fluorescent marker for cotransfection. Our experience is that the majority of cells with high level of GFP fluorescence also carry both TALEN plasmids.

16. We have optimized our transfection protocol to achieve low toxicity but good level of TALEN activity using the JDSx series of plasmids in HEK293T cells as described by Reyon et al. [26]. The amount of transfection reagent may need to be opti-

mized depending on the TALEN constructs used, as well as the parental cell types being studied.

17. Normal culture media lacking penicillin/streptomycin. We have found that omission of antibiotics results in higher numbers of single cell populations recovering after cell sorting.

18. We have found that gating on the highest level of fluorescence results in low numbers of single cell populations recovering after cell sorting. Best results are achieved by gating on average levels of fluorescence and excluding the highest quartile from single cell deposition.

19. While it is tempting to select slow or fast growing populations based on predicted gene disruption phenotypes, in our experience clonal populations with average growth rates tend to yield gene disruptions. This being the case, there are many examples where gene disruption has indeed impaired growth so our advice is to randomly select 12–24 clonal populations for an initial screen.

20. For a longer term solution, plates can be sealed using adhesive foil and stored inside a cryobox in liquid nitrogen vapor phase.

21. This amount of restriction enzyme is based on following assumptions, the enzyme is concentrated at 10 U/µl, and the PCR reaction yielded ~1 µg of DNA.

22. Although most lab cell lines are polyploid, we have found most loci in HEK293T and HeLa cell lines to be either diploid or triploid. We have observed up to five unique alleles in transformed MEFs, whilst the human colorectal cancer cell line HCT116 has been reliably diploid as reported.

23. Depending on the restriction enzymes chosen in Subheading 3.1, buffer compatibility may decree that two separate digests must be performed. If this is the case, following the initial digest using enzyme A purify the DNA with a generic PCR purification kit, eluting in 25 µl of nuclease free water, and repeat the reaction with enzyme B buffer conditions.

24. The insert:plasmid ratios indicated here have been optimized based on the average yield of a 25 µl PCR (25–50 ng/µl prior to purification), and both PCR and plasmid not having been gel extracted. Typically we ligate 150 fmol insert:50 fmol alkaline phosphatase treated plasmid, and achieve >80 % efficiency without using screening techniques such as colony PCR. Optimization may be required depending on experience and quality of reagents in use.

25. If using commercially prepared cells, refer to manufacturers for specific instructions regarding transformation.

Acknowledgments

We thank Mike Ryan, Michael Lazarou, Rochelle Tixeira, and Thanh N. Nguyen for reagents, advice, and discussions. This work was supported by the Australian National Health and Medical Research Council (NHMRC fellowship 1070916 to DAS) and the Australian Mitochondrial Disease Foundation (AMDF).

References

1. Cho SW, Kim S, Kim JM, Kim JS (2013) Targeted genome engineering in human cells with the Cas9 RNA-guided endonuclease. Nat Biotechnol 31(3):230–232

2. Cong L, Ran FA, Cox D, Lin S, Barretto R, Habib N, Hsu PD, Wu X, Jiang W, Marraffini LA, Zhang F (2013) Multiplex genome engineering using CRISPR/Cas systems. Science 339(6121):819 823

3. Hwang WY, Fu Y, Reyon D, Maeder ML, Tsai SQ, Sander JD, Peterson RT, Yeh JR, Joung JK (2013) Efficient genome editing in zebrafish using a CRISPR-Cas system. Nat Biotechnol 31(3):227–229

4. Jiang W, Bikard D, Cox D, Zhang F, Marraffini LA (2013) RNA-guided editing of bacterial genomes using CRISPR-Cas systems. Nat Biotechnol 31(3):233–239

5. Jinek M, East A, Cheng A, Lin S, Ma E, Doudna J (2013) RNA-programmed genome editing in human cells. eLife 2:e00471

6. Mali P, Yang L, Esvelt KM, Aach J, Guell M, DiCarlo JE, Norville JE, Church GM (2013) RNA-guided human genome engineering via Cas9. Science 339(6121):823–826

7. Christian M, Cermak T, Doyle EL, Schmidt C, Zhang F, Hummel A, Bogdanove AJ, Voytas DF (2010) Targeting DNA double-strand breaks with TAL effector nucleases. Genetics 186(2):757–761

8. Cermak T, Doyle EL, Christian M, Wang L, Zhang Y, Schmidt C, Baller JA, Somia NV, Bogdanove AJ, Voytas DF (2011) Efficient design and assembly of custom TALEN and other TAL effector-based constructs for DNA targeting. Nucleic Acids Res 39(12):e82

9. Li T, Huang S, Jiang WZ, Wright D, Spalding MH, Weeks DP, Yang B (2011) TAL nucleases (TALNs): hybrid proteins composed of TAL effectors and FokI DNA-cleavage domain. Nucleic Acids Res 39(1):359–372

10. Mahfouz MM, Li L, Shamimuzzaman M, Wibowo A, Fang X, Zhu JK (2011) De novo-engineered transcription activator-like effector (TALE) hybrid nuclease with novel DNA binding specificity creates double-strand breaks. Proc Natl Acad Sci U S A 108(6):2623–2628

11. Miller JC, Tan S, Qiao G, Barlow KA, Wang J, Xia DF, Meng X, Paschon DE, Leung E, Hinkley SJ, Dulay GP, Hua KL, Ankoudinova I, Cost GJ, Urnov FD, Zhang HS, Holmes MC, Zhang L, Gregory PD, Rebar EJ (2011) A TALE nuclease architecture for efficient genome editing. Nat Biotechnol 29(2):143–148

12. Urnov FD, Miller JC, Lee YL, Beausejour CM, Rock JM, Augustus S, Jamieson AC, Porteus MH, Gregory PD, Holmes MC (2005) Highly efficient endogenous human gene correction using designed zinc-finger nucleases. Nature 435(7042):646–651

13. Kim H, Kim JS (2014) A guide to genome engineering with programmable nucleases. Nat Rev Genet 15(5):321–334

14. Boch J, Scholze H, Schornack S, Landgraf A, Hahn S, Kay S, Lahaye T, Nickstadt A, Bonas U (2009) Breaking the code of DNA binding specificity of TAL-type III effectors. Science 326(5959):1509–1512

15. Moscou MJ, Bogdanove AJ (2009) A simple cipher governs DNA recognition by TAL effectors. Science 326(5959):1501

16. Stroud DA, Formosa LE, Wijeyeratne XW, Nguyen TN, Ryan MT (2013) Gene knockout using transcription activator-like effector nucleases (TALENs) reveals that human NDUFA9 protein is essential for stabilizing the junction between membrane and matrix arms of complex I. J Biol Chem 288(3):1685–1690

17. Ma SB, Nguyen TN, Tan I, Ninnis R, Iyer S, Stroud DA, Menard M, Kluck RM, Ryan MT, Dewson G (2014) Bax targets mitochondria by distinct mechanisms before or during apoptotic cell death: a requirement for VDAC2 or

Bak for efficient Bax apoptotic function. Cell Death Differ 21(12):1925–1935

18. Richter V, Palmer CS, Osellame LD, Singh AP, Elgass K, Stroud DA, Sesaki H, Kvansakul M, Ryan MT (2014) Structural and functional analysis of MiD51, a dynamin receptor required for mitochondrial fission. J Cell Biol 204(4):477–486

19. Formosa LE, Mimaki M, Frazier AE, McKenzie M, Stait TL, Thorburn DR, Stroud DA, Ryan MT (2015) Characterization of mitochondrial FOXRED1 in the assembly of respiratory chain complex I. Hum Mol Genet 24(10): 2952–2965

20. Stroud DA, Maher MJ, Lindau C, Vogtle FN, Frazier AE, Surgenor E, Mountford H, Singh AP, Bonas M, Oeljeklaus S, Warscheid B, Meisinger C, Thorburn DR, Ryan MT (2015) COA6 is a mitochondrial complex IV assembly factor critical for biogenesis of mtDNA-encoded COX2. Hum Mol Genet. doi:10.1093/hmg/ddv265

21. Ye J, Coulouris G, Zaretskaya I, Cutcutache I, Rozen S, Madden TL (2012) Primer-BLAST: a tool to design target-specific primers for polymerase chain reaction. BMC Bioinformatics 13:134

22. Montague TG, Cruz JM, Gagnon JA, Church GM, Valen E (2014) CHOPCHOP: a CRISPR/Cas9 and TALEN web tool for genome editing. Nucleic Acids Res 42(Web Server issue):W401–W407

23. Vouillot L, Thelie A, Pollet N (2015) Comparison of T7E1 and surveyor mismatch cleavage assays to detect mutations triggered by engineered nucleases. G3 (Bethesda) 5(3): 407–415

24. Chung CT, Niemela SL, Miller RH (1989) One-step preparation of competent Escherichia coli: transformation and storage of bacterial cells in the same solution. Proc Natl Acad Sci U S A 86(7):2172–2175

25. Schagger H (2006) Tricine-SDS-PAGE. Nat Protoc 1(1):16–22

26. Reyon D, Tsai SQ, Khayter C, Foden JA, Sander JD, Joung JK (2012) FLASH assembly of TALENs for high-throughput genome editing. Nat Biotechnol 30(5):460–465

27. Sander JD, Maeder ML, Reyon D, Voytas DF, Joung JK, Dobbs D (2010) ZiFiT (Zinc Finger Targeter): an updated zinc finger engineering tool. Nucleic Acids Res 38(Web Server issue):W462–W468

28. Doyle EL, Booher NJ, Standage DS, Voytas DF, Brendel VP, Vandyk JK, Bogdanove AJ (2012) TAL Effector-Nucleotide Targeter (TALE-NT) 2.0: tools for TAL effector design and target prediction. Nucleic Acids Res 40(Web Server issue):W117–W122

29. Neff KL, Argue DP, Ma AC, Lee HB, Clark KJ, Ekker SC (2013) Mojo Hand, a TALEN design tool for genome editing applications. BMC Bioinformatics 14:1. doi:10.1186/1471-2105-14-1

30. Heigwer F, Kerr G, Walther N, Glaeser K, Pelz O, Breinig M, Boutros M (2013) E-TALEN: a web tool to design TALENs for genome engineering. Nucleic Acids Res 41(20):e190

31. Lin Y, Fine EJ, Zheng Z, Antico CJ, Voit RA, Porteus MH, Cradick TJ, Bao G (2014) SAPTA: a new design tool for improving TALE nuclease activity. Nucleic Acids Res 42(6):e47

Chapter 18

Using CRISPR/Cas9 Technology for Manipulating Cell Death Regulators

Andrew J. Kueh and Marco J. Herold

Abstract

Clustered, regularly interspaced, short palindromic repeats (CRISPR)/Cas9 technology has been demonstrated to be a useful tool for generating targeted mutations in cell lines and mice. However, the use of CRISPR/Cas9 in a constitutively expressed manner can often result in low targeting efficiencies and lethality due to mutations in essential genes. Here, we describe the use of an inducible lentiviral vector platform, enabling rapid transduction and enrichment of CRISPR/Cas9 positive cells and high levels of targeted mutations upon induction.

Key words CRISPR/Cas9, Inducible, Lentiviral, Guide RNA

1 Introduction

CRISPR together with the Cas9 endonuclease originating from *Streptococcus pyogenes* have been extensively used for genome editing in a wide variety of eukaryotic and prokaryotic organisms [1–4]. The adaptation of the CRISPR/Cas9 system for genome editing features the use of a 20 nt single guide RNA (sgRNA) sequence to direct Cas9 endonuclease activity to any stretch of DNA sequence that has a protospacer adjacent motif (PAM) consisting of the nucleotides NGG, whereby N can be any of the DNA bases [5]. Cas9 endonuclease activity results in double stranded breaks, which stimulates the cellular DNA repair machinery to repair these breaks by the highly error prone non-homologous end-joining (NHEJ) pathway. The NHEJ repair process results in insertion/deletion (InDel) mutations, which can disrupt the open reading frame of a target gene and lead to its functional inactivation.

Here, we demonstrate the use of an inducible CRISPR/Cas9 lentiviral platform that can efficiently deliver Cas9 and sgRNA encoding vectors in a wide range of mouse and human cells [6]. In this system, sgRNAs are under the regulatory control of a

Hamsa Puthalakath and Christine J. Hawkins (eds.), *Programmed Cell Death: Methods and Protocols*, Methods in Molecular Biology, vol. 1419, DOI 10.1007/978-1-4939-3581-9_18, © Springer Science+Business Media New York 2016

tetracycline inducible H1 promoter. The exposure of cells to doxycycline (dox) relieves the Tet repressor from the H1 promoter, resulting in robust sgRNA expression in a temporally regulated manner. We also present a next-generation sequencing method to validate and characterize CRISPR/Cas9 mediated InDels in a high throughput and accurate approach. These techniques can be easily adapted to study apoptotic gene function in vitro and in vivo.

2   Materials

2.1   Vectors and sgRNA Cloning

1. pFgh1tUTG vector for dox-inducible expression of sgRNAs.
2. pFUCas9mCherry vector for constitutive expression of the Cas9 endonuclease.
3. Viral packaging vectors pMDL, pRSV-rev, and pVSV-G.
4. Oligonucleotide annealing buffer—Combine 2 μl of 1 M Tris–HCl (pH 7.5), 2 μl of 1 M MgCl$_2$, 0.5 μl of 3 M NaCl, and 9.5 μl of TE buffer (10 mM Tris–HCl, 1 mM EDTA, pH 8.0). This makes 14 μl of oligonucleotide annealing buffer, sufficient for one oligonucleotide annealing reaction.
5. T4 DNA ligation buffer.
6. Phenol/chloroform/isoamyl alcohol solution pH 7.8–8.2.
7. Chloroform/isoamyl alcohol solution.
8. 3 M NH$_4$OAc, pH 5.5.
9. 100 % ethanol.
10. 80 % ethanol.
11. Electrocompetent STBL4 bacterial cells.
12. 50–100 mg/ml Ampicillin stock solution.
13. Qiaprep Spin Miniprep Kit or equivalent.
14. DNAse-free water.
15. PNKinase.
16. 10× restriction enzyme buffer.
17. BsmB I restriction enzyme.
18. Bam H I restriction enzyme.
19. TE buffer: 10 mM Tris–HCl pH 8.0, 1 mM EDTA.
20. T4 DNA Ligase.
21. 20 mg/ml Glycogen.
22. Phospohorylated oligonucleotides.
23. Luria agar plates with ampicillin, 50–100 μg/ml final concentration.

2.2 Virus Production and Transduction

1. 293FT cells.

2. 10 % *D*ulbecco's *M*odified *E*agle *M*edia (DMEM), containing DMEM supplemented with 10 % fetal calf serum (FCS), 100 U/ml penicillin, 100 U/ml streptomycin, and 2 mM L-alanyl-L-glutamine.

3. HBS buffer: 0.28 M NaCl, 0.05 M HEPES, 1.5 mM Na_2PO_4. Adjust pH to 6.95–7.05 with NaOH.

4. Polybrene.

5. 0.5 M $CaCl_2$.

6. 10 cm tissue culture plates.

7. 0.45 μm syringe filters.

8. 6-well tissue culture plates.

9. 10 mg/ml doxycycline stock concentration (10,000×).

10. LSR IIW or FACSCalibur flow cytometer or equivalent.

2.3 Validation of Knock Out Cell Lines

1. Standard Western blotting reagents, including polyacrylamide gels, electrophoresis tanks, transfer membranes, and ECL detection reagents.

2. Next-generation sequencing reagents, including MiSeq reagent kit v2, AMPure XP DNA clean-up, and size selection beads or equivalent.

3. MiSeq Forward indexing oligonucleotide: 5'CAAGCAGAAGACGGCATACGAGATCCGGTCTC GGCATTCCTGCTGAACCGCTCTTCCGATCTNNNNN NNNGTGACCTATGAACTCAGGAGTC)3'.

4. MiSeq Reverse indexing oligonucleotide: 5'AATGATACGGCGACCACCGAGATCTACACTCTT TCCCTACACGACGCTCTTCCGATCTNNNNN-NNNCTGAGACTTGCACATCGCAGC3'

   The sequence NNNNNNNN is where unique 8 bp indexes are inserted into forward and reverse indexing oligonucleotides.

5. GoTaq Green mix polymerase.

6. DNAse-free water.

7. Forward primer: **5'GTGACCTATGAACT-CAGGAGTC***CGAGGCTGCTTTTCTTCG* 3' (bold: overhang; italics: gene specific).

8. Reverse primer: **5'CTGAGACTTGCACATCGC-AGC***AACTCGTCCTCCTCCTCCTC* 3' (bold: overhang; italics: gene specific).

9. Magnetic rack.

2.4 Functional Assays

1. 5.0 mg/ml ionomcycin stock (1000×).

2. 50 μg/ml Etoposide stock (1000×).

3. Dead Cell Apoptosis Kit with Annexin V Alexa Fluor® 488 and Propidium Iodide (PI) or equivalent.

3   Methods

3.1   Designing sgRNAs

1. Use the Optimized CRISPR Design website, crispr.mit.edu for sgRNA design. Select the genome of the organism to be targeted (e.g., mm9 for the mouse genome). Enter a DNA sequence, up to 250 nucleotides long, of the gene to be targeted by Cas9. A list of candidate sgRNAs will be generated, ordered by specificity and efficiency. The higher the sgRNA score, the greater the specificity with fewer off-targets. Select sgRNAs with a score greater than 75 and select at least two sgRNAs per target gene (*see* **Note 1**).

2. Order selected sgRNA sequences with a 5′ "TCCC" 4 bp overhang for the complementary sequence and a 5′ "AAAC" 4 bp overhang for the reverse complementary sequence. For instance, an sgRNA with sequence "GGCAACTATGGCTTCCACCT" should be ordered as (Forward: 5′**TCCC**GGCAACTATGGCTTCCACCT 3′) along with (Reverse: 5′ **AAAC**AGGTGGAAGCCATAGTTGCC 3′) (*see* **Note 2**).

3.2   Cloning of sgRNAs into the dox-Inducible pFgh1tUTG Lentiviral Vector

1. Resuspend oligonucleotides in DNAse-free water to a concentration of 100 µM. Anneal oligonucleotides, by adding 3 µl of forward sgRNA oligonucleotide (containing 4 bp overhang) and 3 µl of reverse sgRNA oligonucleotide (containing 4 bp overhang) to 14 µl of oligo annealing buffer in a PCR tube. Heat PCR tube to 96 °C for 5 min and place on ice immediately.

2. Phosphorylate annealed oligonucleotides by adding 5 µl of annealed oligonucleotides to 12 µl DNAse-free water, 2 µl of T4 DNA ligation buffer and 1 µl of PNKinase. Incubate mixture for 20 min at 37 °C followed by heat inactivation for 10 min at 70 °C. Add 1 µl of phosphorylated oligonucleotides to 99 µl of DNAse-free water to produce a 1:100 diluted oligonucleotide solution (*see* **Note 3**).

3. Generate the overhangs for inserting the annealed oligos into the pFgh1tUTG vector, by cutting the vector with BsmB I. Add 10 µl of 10× restriction enzyme buffer to 5 µg of pFgh1tUTG vector, 4 µl (10 U/µl) of BsmB I restriction enzyme and adjust reaction volume to 100 µl with DNAse-free water.

4. Incubate the mixture from **step 3** above at 55 °C for 4 h followed by 80 °C for 20 min to heat inactivate the restriction enzyme. Add 1 µl (50 ng) of digested pFgh1tUTG to 19 µl of DNAse-free water to produce a 1:20 diluted pFgh1tUTG solution.

5. Ligate sgRNA oligonucleotides into the cut pFgh1tUTG vector, by adding 3 µl of diluted pFgh1tUTG solution from **step 4** above to 3 µl of diluted phosphorylated oligonucleotide solution from **step 2**, 2 µl of 10× ligation buffer, 0.8 µl DNA ligase, and 11.2 µl DNAse-free water. Leave overnight at room temperature.

6. Next day add 180 µl of TE buffer to the 20 µl ligation reaction, followed by 200 µl of phenol/chloroform/isoamyl alcohol solution and vortex well. Centrifuge at 13,000 RPM ($17,900 \times g$) in a bench top centrifuge for 5 min.

7. Transfer 200 µl of the top aqueous phase into a new microfuge tube, add 200 µl of chloroform/isoamyl alcohol solution, and vortex well. Centrifuge at 13,000 RPM ($17,900 \times g$) in a bench top centrifuge for 5 min.

8. Transfer 200 µl of the top aqueous phase from **step 7** above into a new microfuge tube, add 20 µl of 3 M NH$_4$OAc, 1 µl (20 µg) glycogen, and vortex well.

9. Add 500 µl 100 % ethanol, vortex well, and incubate solution at −20 °C overnight.

10. Next day centrifuge at 13,000 RPM ($17,900 \times g$) in a bench top centrifuge for 15 min at 4 °C. Discard supernatant by carefully aspirating.

11. Wash the DNA pellet by adding 500 µl of 80 % ethanol and vortexing briefly. Centrifuge at 13,000 RPM ($17,900 \times g$) in a bench top centrifuge for 15 min at 4 °C and discard supernatant and allow pellet to air dry for 15 min at room temperature.

12. Resuspend purified DNA pellet in 5 µl of DNAse-free water and use all of the resuspended DNA to transform electrocompetent STBL4 bacterial cells.

13. Plate out transformed bacterial cells on ampicillin (100 µg/ml) supplemented L-Broth agar and incubate plates at 37 °C overnight in an incubator.

14. After incubation, select six to eight colonies per sgRNA and complete a miniprep protocol for each colony according to manufacturer's instructions.

15. Digest minipreps from **step 10** with Bsmb I and BamH1. If the sgRNA sequence has been successfully ligated, this double digestion will linearize the plasmid. Release of a 1.2 kb fragment indicates unsuccessful cloning (*see* Fig. 1).

16. Perform Sanger sequencing for each miniprep using sequencing primer CAGACATACAAACTAAAGAAT for the sequencing reaction and run samples on a DNA analyzer (Applied Biosystems, ABI 3730XL). Verify that the sgRNA sequence of interest is correct.

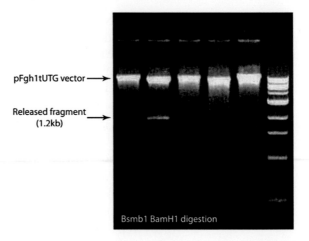

Fig. 1 Diagnostic digest for successful sgRNA cloning into the pFgh1tUTG vector. The successful ligation of the sgRNA into the pFgh1tUTG vector removes the Bsmb1 cloning site, resulting in a single linearized band when double digested with Bsmb1 and BamH1. If the ligation was unsuccessful, Bsmb1 and BamH1 double digestion of the empty pFgh1tUTG vector will release a 1.2 kb fragment (*lane 2*)

3.3 Virus Production and Transduction of Cells

1. Culture 3×10^6 293FT cells in 10 ml of 10 % DMEM media on a 10 cm dish at 37 °C in an incubator overnight.

2. The following day, aspirate media and replace with fresh 10 % DMEM.

3. Make lentiviral packaging cocktail, by adding 5 µg of pMDL vector, 2.5 µg of pRSV-REV vector, 3 µg of pVSV-G vector, 10 µg of pFgh1tUTG vector and/or pFUCas9mCherry vector, 250 µl of 0.5 M CaCl₂, 500 µl of HBS, and 250 µl of DNAse-free water. Vortex the solution for 10 s and incubate for 10 min at room temperature (*see* **Note 4**).

4. Transfect 293FT cells with lentiviral packaging constructs, by adding lentiviral packaging cocktail from **step 3** above, drop wise over 293FT cells while swirling the dish constantly. Incubate cells overnight at 37 °C in an incubator (viral particles will form in the media overnight).

5. Seed cells to be transduced in a 6-well plate such that they reach 40–50 % confluence the following day (*see* **Note 5**).

6. On the following day, collect the virus-containing supernatant from **step 4** above and filter sterilize using a 0.45 µm syringe filter to remove any contaminating 293FT cells.

7. Add fresh 10 % DMEM media to the 293FT cells and incubate overnight at 37 °C for collecting a second batch of viral complexes.

8. Supplement 3 ml of filtered viral supernatant from first batch (**step 6**) with 8 µg/ml polybrene (*see* **Note 6**). Add this onto target cells cultured in 6-well plates.

9. Spin the plate at 2200 rpm ($\sim$500$\times g$) for 2 h at 32 °C. Place cells at 37 °C overnight in an incubator.

10. The following day, remove virus-containing media from the 6-well plate and reinfect cells with a second batch of viral particles as described in **step 8**. 293FT cells can then be discarded.

11. Replace virus-containing media from 6-well plates with fresh media. Passage cells when confluent.

3.4   Flow Cytometric Analysis, dox Induction and Sorting of Cells

1. The pFgh1tUTG sgRNA vector and pFUCas9mCherry vector feature a GFP and a mCherry selection marker, respectively, allowing efficient identification of transduced cells by flow cytometry. Following viral transduction and expansion, assess the cell lines by flow cytometry using GFP and mCherry detection parameters.

2. If GFP and mCherry double positive cells represent less than 50 % of the cell population, sort GFP and mCherry double positive cells using a cell sorter and reassess GFP and mCherry double positivity after cells have expanded.

3. Repeat cell sorting if necessary until GFP and mCherry double positive cells represent more than 85 % of the cell population (*see* Fig. 2).

4. Induce expression of sgRNAs in cell lines, by adding dox in cell culture media to a final concentration of 1 μg/ml. Culture cells in dox supplemented media for 3 days before replacing culture media with fresh dox-free media. By this stage, sgRNAs would have complexed with constitutively expressed Cas9 protein and generated InDels in target gene sequences.

5. The cell lines can now be expanded and analyzed as a bulk population or further sorted (gate on GFP mCherry double positive cells) in 96-well plates to generate clonal cell populations (*see* **Note 7**).

3.5   Validation of Knock Out Cell Lines

1. If an antibody is available to detect the gene product of interest, perform Western immunoblotting analysis on material derived from expanded single-cell clonal cell lines or from bulk cell populations (*see* **Note 8**).

2. To validate the precise genome editing outcome induced by CRISPR/Cas9, perform next-generation sequencing of cell lines (*see* **Note 9** and Fig. 3). Start by generating gene-specific oligonucleotides (with 60 °C annealing temperatures) that flank the sgRNA used, producing an amplicon approximately 300 bp long. Extend gene-specific oligonucleotides with the overhangs GTGACCTATGAACTCAGGAGTC for the forward oligonucleotide and CTGAGACTTGCACATCGCAGC for the reverse oligonucleotide. Position each overhang at the

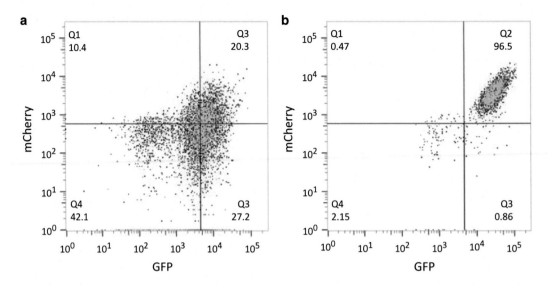

Fig. 2 Flow cytometric analysis and enrichment of pFgh1tUTG and pFUCas9mCherry double transduced cells. (**a**) Flow cytometric analysis of HeLa cells 1 week after transduction with pFgh1tUTG and pFUCas9mCherry vectors. Double positive cells transduced with both vectors were enriched by flow cytometric cell sorting, gating on the GFP and mCherry high population (Q2). (**b**) Flow cytometric analysis of HeLa cells 1 week after sorting and enrichment

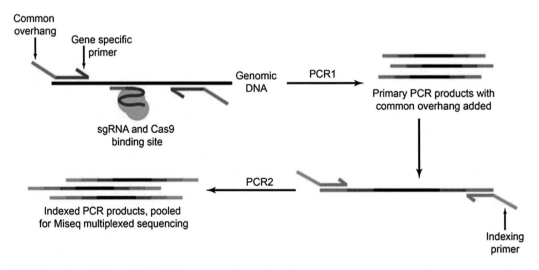

Fig. 3 Overview of the next-generation sequencing protocol

5′ end of each oligonucleotide. For instance, the following overhangs (green) are positioned at the 5′ position of the gene-specific oligonucleotides (red):

Forward primer: 5′ **GTGACCTATGAACTCAGGAGTC**-CGAGGCTGCTTTTCTTCG 3′

Reverse primer: 5′ **CTGAGACTTGCACATCGCAGC**AAC-TCGTCCTCCTCCTCCTC 3′.

3. Perform the first PCR reaction by combining 1 μl of DNA (100 ng/μl) with 10 μl of GoTaq Green mix polymerase (2× concentration, Promega), 0.5 μl of 10 μM forward oligonucleotide, 0.5 μl of 10 μM reverse oligonucleotide, and 8 μl of DNAse-free water. Run the PCR reaction for 18 cycles (95 °C 2 min, 60 °C 30 s, 72 °C 30 s) (*see* **Note 10**).

4. To purify amplicons from the first PCR reaction, add 20 μl of Agencourt AmPure XP beads (Beckman Coulter) to each PCR reaction. Incubate for 10 min at RT to allow beads to bind DNA amplicons. Place PCR plates on a magnetic rack and leave until magnetic beads have been fully attracted to the magnet. Aspirate and discard supernatant from each well. Add 150 μl of 80 % ethanol to each well and gently pipette to wash magnetic beads. Remove ethanol and repeat ethanol wash one more time. Remove ethanol and allow magnetic beads to air dry at room temperature (*see* **Note 11**).

5. To elute DNA, remove PCR plate from magnetic rack, add 30 μl of DNAse-free water to each well and pipette to mix. Once magnetic beads appear homogeneous, place PCR plate back onto a magnetic rack and allow beads to accumulate toward the magnet. Transfer 10 μl of clear eluted DNA solution to a fresh PCR plate and add 10 μl of GoTaq Green mix (2× concentration, Promega), 0.5 μl of 10 μM forward indexing oligonucleotide, and 0.5 μl of 10 μM reverse indexing oligonucleotide. Run PCR reactions for 24 cycles (95 °C 2 min, 60 °C 30 s, 72 °C 30 s).

6. Combine all PCR reactions in a clean trough and mix well by pipetting. Pipette a 100 μl aliquot in a 1.5 ml tube and add 100 μl of magnetic beads. Perform purification of amplicons as described in **step 4**. Resuspend air-dried magnetic beads in 100 μl of DNAse-free water. Quantify DNA concentration and molarity using a Qubit Fluorometer (Life technologies) and 2200 Tapestation (Agilent technologies), respectively.

7. Dilute dual indexed library pool to a 2 nM concentration and proceed with library preparation using the Miseq Reagent Kit V2 (Illumina) according to manufacturer's instructions. Perform sequencing with a 281 cycle forward read followed by a 44 cycle second index read.

8. Conduct functional assays—for an example of a functional assay (*see* **Note 12** and Fig. 4).

4   Notes

1. Beware of sgRNAs that have a high degree of complementarity with predicted off-targets in gene exons as they can cause unintended frameshift mutations. The 12 nucleotides of the sgRNA

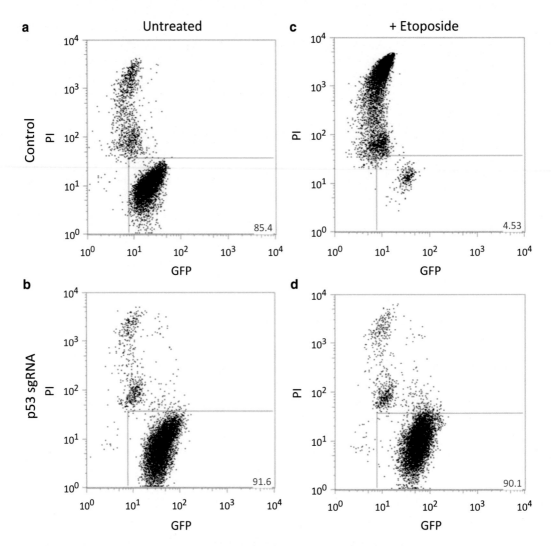

Fig. 4 PI cell viability assay of control and p53 knockout Eμ-myc murine lymphoma cell lines following 0.05 μg/ml etoposide treatment for 24 h. (**a** and **b**) Untreated control and p53 knockout Eμ-myc murine lymphoma cell lines displaying high levels of PI negative viable cells. (**c**) Control cell lines display a significant increase in PI positive dead cells in response to etoposide treatment compared to (**d**) p53 knockout Eμ-myc murine lymphoma cell lines which display resistance to etoposide treatment

adjacent to the PAM site form the seed region of the sgRNA. In particular, mismatches in this seed region reduce the chance of Cas9 activity at potential off-targets.

Ensure that sgRNAs target essential exons used by all protein coding splice variants of the target gene. Targeting a proximal exon increases the chance of frameshift mutations caused by InDels. Gene sequences coding for functional domains such as catalytic domains are also particularly good targets and increase the chance of nonfunctional gene products being produced due to CRISPR/Cas9 induced InDels. Multiple sgRNAs can also be used in tandem to delete entire exons. Ordering at least two sgRNAs per gene target allows validation of phenotypes observed and reduces the likelihood that cellular phenotypes are due to off-targets. An alternative sgRNA design site is E-CRISP (http://www.e-crisp.org/E-CRISP/). CCTop (http://crispr.cos.uni-heidelberg.de/index.html) also offers an alternative sgRNA off-target prediction tool.

2. When ordering sgRNA sequences, ensure that the PAM site is *not* included.

3. If experiencing difficulty with cloning, try diluting oligonucleotide solution 1:1000 and 1:10,000. Use these dilutions in parallel to the 1:100 diluted solution during the sgRNA ligation step.

4. Transducing cells with the pFUCas9mCherry vector alone first will yield a stock of Cas9 positive cells that can then be transduced with any sgRNA of interest. However, if time is limited, the pFUCas9mCherry vector and the pFh1tUTG sgRNA vector can be simultaneously transduced.

5. Depending on cell type, target cells can also be cultured in 24-well or 48-well plates.

6. Polybrene increases viral transduction efficiencies but may be highly cytotoxic to certain cell types. Protamine sulfate can be used as an alternative to polybrene.

7. If mutation of the target gene causes a lethal cellular phenotype, analyze the bulk cell population immediately after dox induction. A clonal population can also be derived prior to dox treatment.

8. For negative controls, include samples that were (a) not transduced with viral vectors and (b) transduced with viral vectors but not treated with dox. Depending on the cell lines used, gene of interest and efficiency of transduction, it may be necessary to screen 5–20 single-cell clonal lines to isolate sufficient knockout cell lines. At least two knockout single-cell clonal lines per sgRNA should be used to perform functional assays.

9. Gene-specific primers featuring a common overhang sequence flank the predicted Cas9 cutting site in the first PCR reaction. Primary PCR products from this reaction are added to a secondary PCR reaction along with primers featuring indexing barcodes. The final indexed PCR products are then pooled for multiplexed sequencing on the Illumina Miseq next-generation sequencing platform.

10. A proof-reading polymerase can be used in this PCR reaction if required. Performing the PCR reactions in a 96-well plate format allows the use of multichannel pipettes and speeds up the completion of the protocol.

11. The magnetic beads will assume a flaky appearance when completely dry.

12. If proapoptotic genes have been targeted and mutated, this can be functionally validated using several cell death assays. For instance, the loss of Bim can be validated by the gain of resistance to 5 μg/ml ionomycin treatment. Similarly, the loss of p53 can be validated by the gain of resistance to 0.05 μg/ml etoposide treatment. Quantify cell viability by flow cytometry using Annexin V-PI staining.

Acknowledgement

We thank B Aubrey for contributing images (Fig. 4) and L Tai for technical assistance. This work was supported by the National Health and Medical Research Council, Australia (program grant 1016701) and project grant APP1049720 (to MJH). This work was made possible through Victorian State Government Operational Infrastructure Support and Australian Government National Health and Medical Research Council Independent Research Institutes Infrastructure Support Scheme.

References

1. Sander JD, Joung JK (2014) CRISPR-Cas systems for editing, regulating and targeting genomes. Nat Biotechnol 32(4):347–355. doi:10.1038/nbt.2842

2. Doudna JA, Charpentier E (2014) Genome editing. The new frontier of genome engineering with CRISPR-Cas9. Science 346(6213):1258096. doi:10.1126/science.1258096

3. Hsu PD, Lander ES, Zhang F (2014) Development and applications of CRISPR-Cas9 for genome engineering. Cell 157(6):1262–1278. doi:10.1016/j.cell.2014.05.010

4. Mali P, Esvelt KM, Church GM (2013) Cas9 as a versatile tool for engineering biology. Nat Methods 10(10):957–963. doi:10.1038/nmeth.2649

5. Jinek M, Chylinski K, Fonfara I, Hauer M, Doudna JA, Charpentier E (2012) A programmable dual-RNA-guided DNA endonuclease in adaptive bacterial immunity. Science 337(6096):816–821. doi:10.1126/science.1225829

6. Aubrey BJ, Kelly GL, Kueh AJ, Brennan MS, O'Connor L, Milla L, Wilcox S, Tai L, Strasser A, Herold MJ (2015) An inducible lentiviral guide RNA platform enables the identification of tumor-essential genes and tumor-promoting mutations in vivo. Cell Rep 10(8):1422–1432. doi:10.1016/j.celrep.2015.02.002

Chapter 19

Lentiviral Vectors to Analyze Cell Death Regulators

Ueli Nachbur and Gabriela Brumatti

Abstract

Ectopic expression of proteins involved in cell death pathways is an important tool to analyze their role during apoptosis or other forms of cell death. Lentiviral vectors offer the advantages of high rate of transduction and stable integration of donor DNA into the genome of the host cell, leading to reproducible and relevant readouts compared to classical overexpression by transfection of naked plasmid DNA.

Here, we describe the production and application of lentiviral vectors to express cell death proteins in eukaryotic cells. A packaging cell line, usually HEK293T cells, is transfected with viral packaging plasmids and your gene of interest, which is flanked by long terminal repeat sites with an internal ribosome entry site in the 5′UTR (*Un translated region*). Virions are harvested from the supernatant and can be directly used to transduce target cells. Varied selection markers as well as a variety of promoters that regulate expression of the gene of interest make this system attractive for a wide range of application in many cell lines or in whole organisms.

Key words Lentivirus, Transduction, Inducible gene expression

1 Introduction

The analysis of single proteins, or mutants thereof, involved in signaling pathways is a powerful method to investigate molecular mechanisms of cell death. Overexpression using classical transfection of naked DNA is inefficient and often results in unspecific bystander effects, which are not reflective of the mechanisms that occur under endogenous conditions. Furthermore, transfection efficiency varies greatly between cell types. A particular problem in cell death research is that uncontrolled overexpression of effector proteins, such as caspases or proapoptotic members of the Bcl2 family, leads to cell death in the absence of a death stimulus [1–4]. Expression of such effectors therefore have to be tightly regulated and stable integration of donor DNA into the host genome, in combination with controlled gene expression using inducible promoters, offers the possibility to study cell death proteins under more physiological conditions.

Hamsa Puthalakath and Christine J. Hawkins (eds.), *Programmed Cell Death: Methods and Protocols*, Methods in Molecular Biology, vol. 1419, DOI 10.1007/978-1-4939-3581-9_19, © Springer Science+Business Media New York 2016

Retroviruses are a family of enveloped RNA viruses that replicate in their host cell through reverse transcription. Lentiviruses are a genus of the Retroviridae family and have the advantage over other retroviruses to infect nondividing cells. Lentiviruses integrate their DNA into the host cell and are able to deliver relatively large amounts of RNA compared to other retrovirus family members, making them ideal tools for in vitro research. Furthermore, for research purposes and safety, lentiviruses are rendered replication incompetent and, thus, are unable to replicate in the host cell.

Here, we describe a protocol using the second generation of lentiviral plasmids as deposited with Addgene by Weinberg and Trono labs. To produce lentiviral particles, packaging cells are transfected with three mammalian expression plasmids: a plasmid encoding the gene of interest; a plasmid containing the viral envelope protein; and a plasmid encoding the viral components Gag, Pol, Rev, and Tat (*see* Fig. 1). In contrast to the second generation system, the third generation uses two packaging plasmids and does not use the Tat protein. The third generation system therefore offers increased biosafety, but it is more cumbersome as it involves transfection of a total of four plasmids in the packaging cell line. Due to better transduction efficiency in our hands, we favor working with the second generation system.

The packaging cell line is an easy-to-transfect cell type, which is able to produce proteins in large amounts. HEK293T cells (HEK293 containing the SV40 Large T antigen allowing for plasmid replication and enhanced protein expression) are a popular choice as they are easy to grow in culture and produce large amounts of virions. The packaging plasmid, **pCMV delta R8.2** encodes essential virion proteins Gag (matrix, nucleoprotein), Pol (transcriptase), Rev (reverse transcriptase), and Tat (transactivator of transcription). The Envelope plasmid, **pCMV VSV-G** [5], encodes for VSV-G, the glycoprotein G of the Vesicular stomatitis virus. VSV-G is an envelope glycoprotein and pseudotypes the lentiviral particles to allow transduction to all mammalian cell types. The gene of interest is transfected in a separate **transfer plasmid**, which harbors a *l*ong *t*erminal *r*epeat (**LTR**) sequence allowing integration into the host cell genome, an internal ribosome entry site (**IRES**) in the 5′UTR as well as a Ψ (**Psi**) sequence, which serves as a packaging sequence for the viral RNA genome. The viral genes on the packaging and envelope plasmids do not have a Ψ sequence or LTRs, therefore RNA transcripts are not packed into the virions and consequently target cells will not produce further lentiviruses. In addition to the packaging, envelope and transfer plasmid we recommend cotransfection with a **Marker plasmid** (e.g. **pEGFP-N1**) to monitor successful transfection of the packaging cell line.

Upon transfection with the lentiviral expression plasmids, the packaging cell line produces viral proteins as well as mRNA of the transfer plasmid, which is packed into the virions due to its Ψ sequence.

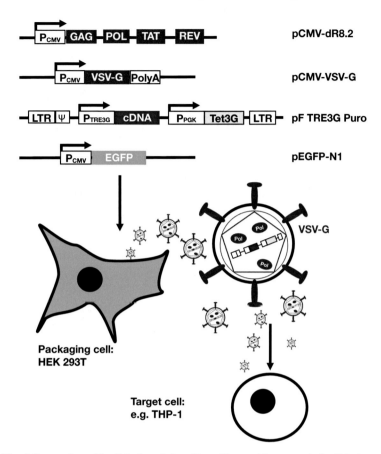

Fig. 1 Generation of lentiviral particles. Plasmids used to generate lentiviral particles: The virus packaging plasmid (pCMV-dR8.2), the envelope plasmid (pCMV-VSV-G), the transfer plasmid (pF TRE3G Puro) containing the cDNA of choice and a puromycin resistance marker, and the marker plasmid (pEGFP-N1) are transfected into HEK 293T cells in a ratio of 3:1:5:1. The packaging cells (HEK 293T) will express GFP and produce lentiviral particles containing the transfer DNA and reverse polymerase to infect the target cells. The envelope protein VSV-G ensures infection in all mammalian cells. Alternatively, plasmid containg the envelope protein ENV can be used for the infection of rodent cells only, providing a safer approach and tool for transduction of cells. The transfer DNA will be integrated in the host cell genome upon infection

Along with the lentiviral RNA, the packaging cell equips the virions with viral reverse transcriptase to transcribe the virion RNA genome into DNA, once it has been delivered to the host cell.

Upon infection of the target cell, the LTR sequences in the virion genome allow the reversely transcribed DNA to integrate into the host genome. The site of insertion is unpredictable and was considered to occur randomly. However, recent studies show that the DNA integrates preferentially in active sites of the genome, which are more accessible for integration [6]. Due to the unpredictable nature of insertion into the host genome, the level of gene expression between different cells can vary considerably.

Therefore in many cases it is required that single cell clones are selected with moderate expression of the gene of interest. Selection of individual clones is particularly encouraged if the gene of interest is driven by an inducible promoter. If the gene has been inserted in a hyperactive region of the genome or downstream of an active promoter, leaky expression (i.e., expression in the absence of the inducing agent) could ensue. Therefore, isolation of clonal lines may be warranted.

The choice of using either inducible or cell type-specific promoters to study your gene of interest is a major advantage of the lentiviral transduction method. In cell death research, inducible Doxycycline or Tamoxifen-inducible promoters are often used, as expression of the gene of interest can be regulated by titrating these inducing agents.

Here, we describe virus production in 293T cells and the transduction of both an adherent cell line, e.g., mouse embryonic fibroblasts (MEFs), and a suspension cell line, human monocytic THP-1 cells. The lentivirus produced contains the protein of choice as well as a puromycin selection marker. In adherent cells, selection of single clones can be achieved by seeding a small number of cells in normal tissue culture plates. For cells in solution, single cell colonies are generated using limiting dilution in 96-well plates or by culturing cells in 0.3 % agar.

2 Materials

2.1 Plasmids

1. Lentiviral packaging plasmid: pCMV delta R8.2 (Addgene, deposited by the Trono lab).

2. Lentiviral envelope plasmid: pCMV-VSV-G (Addgene, deposited by the Weinberg lab).

3. Transfer plasmid: Any plasmid which includes LTRs, IRES, a Ψ sequence, and your gene of interest. We routinely use pF TRE3G Puro, a derivate of the pRetroX vector (Clonetech) [7] (*see* **Note 1**).

4. Marker plasmid: pEGFP-N1 (Clonetech) or similar.

2.2 Cell Culture and Transfection

1. Packaging cell line: Human embryonic kidney (HEK) 293T cells, sourced from ATCC.

2. Target cell lines: Here, we use human monocyte cell line THP-1 (ATCC) and mouse embryonic fibroblast MEFs immortalized with SV40 large T antigen [8], generated as described before [9] (*see* **Note 2**).

3. Cell culture medium: 293T and MEFs are cultured in *D*ulbecco's *M*odified *E*agle *M*edium (DMEM) supplemented with 10 % fetal calf serum, 2 mM l-glutamine, and (optional)

antibiotic solution containing 100 IU/ml Penicillin and 100 µg/ml Streptomycin. THP-1 cells are cultured in RPMI-1640 medium supplemented with 10 % fetal calf serum, 2 mM l-glutamine, and antibiotic solution. All cell lines are cultured at 37 °C in a humified incubator at 10 % CO_2.

4. *P*hosphate *b*uffered *s*aline (PBS).

5. Trypsin-EDTA.

6. Transfection reagent: This protocol uses the calcium phosphate coprecipitation protocol for transfection of viral vectors. The solutions required are: (1) 2× HBSS (50 mM HEPES, pH 7.05, 280 mM NaCl, 1.5 mM Na_2HPO_4); (2) 2.5 M $CaCl_2$. Both solutions must be sterile-filtered and stored at –20 °C.

7. Sterile water.

8. Vortex.

9. Hemocytometer for counting cells.

2.3 Lentivirus Production and Harvest

1. 10 cm tissue culture plates.

2. 20 ml syringe.

3. 0.4 µm syringe filter.

4. 5 ml Cryotubes for storage of virion containing supernatants.

5. Fluorescent microscope.

2.4 Transduction of Target Cells

1. 6-well tissue culture plates.

2. Polybrene (*see* **Note 3**).

3. Temperature regulated bench top centrifuge with inserts for tissue culture plates.

4. 10 ml Falcon tubes.

2.5 Selection of Infected Cells

1. 6-well cell culture plates.

2. Selection marker: Here, we use 10 mg/ml Puromycin Dihydrochloride in 20 mM HEPES.

2.6 Selection of Single Cell Clones

1. Two water baths, set to 37 °C and 50 °C, respectively.

2. 10 cm tissue culture dishes and 48-well tissue culture plates (for adherent cells); 6-well tissue culture plates or 96-well U-bottom cell culture plates (for suspension cells).

3. Trypsin solution.

4. 3 % agar in sterile water.

5. Dissection microscope for picking of single clones.

2.7 Testing of Single Cell Clones (Inducible Gene Expression)

1. 1 mg/ml Doxycycline stock (1000×).

2. 24-well plates (for nonadherent cells) or 6-well plates (for adherent cells).

3. NP40 buffer (150 mM sodium chloride 1.0 % NP-40 (Triton X-100 can be substituted for NP-40) 50 mM Tris pH 8.0) with protease inhibitors (*see* **Note 4**).

4. RIPA buffer (150 mM sodium chloride 1.0 % NP-40 or Triton X-100 0.5 % sodium deoxycholate 0.1 % SDS (sodium dodecyl sulfate) 50 mM Tris, pH 8.0) containing protease inhibitors (*see* **Note 4**).

5. Western blot reagents and apparatus.

6. Antibodies for detecting your protein of interest.

3   Methods

Carry out all procedures in sterile conditions in a tissue culture cabinet. Cell culture media are warmed to 37 °C in a water bath (*see* **Notes 5** and **6**).

3.1   Lentiviral Production

1. The day before transfection, seed 1×10^6 HEK 293T cells in a 10 cm tissue culture plate for each transfection. Let adhere for a minimum of 16 h.

2. The following day, aspirate cell culture medium and replace with 9 ml fresh medium per well.

3. Label two 1.5 ml microfuge tubes for each transfection:

Tube A: 64 µl 2 M $CaCl_2$	Tube B: 500 µl 2×HBSS
3 µg pCMV delta R8.2	
1 µg pCMV-VSV-G	
5 µg pFTRE3G	
1 µg pF-EGFP	
Top up to 500 µl with sterile water	

4. Add contents of tube A to tube B in a drop wise manner while gently vortexing intermittently.

5. Incubate at room temperature for 30 min. A small precipitate may form.

6. Using a 1000 µl pipette, add the transfection mix drop wise to the cells in a 10 cm dish.

7. Incubate cells over night at 37 °C in a humified CO_2 incubator.

8. The next day, check under a fluorescent microscope for transfection efficiency (GFP positive cells).

9. Carefully aspirate supernatant. Add 6 ml of fresh medium to cells. Incubate for 24 h for maximum virion production.

10. After 24 h, recover the virion containing supernatant using a 20 ml syringe and remove cells and cellular debris using a 0.4 μm syringe top filter.

11. The virion containing supernatant can either be used immediately or stored at –80 °C for up to 6 months. We recommend to store the virion supernatant in small aliquots (1–2 ml) to avoid multiple freeze thawing.

3.2 Transduction of the Target Cell Line

3.2.1 Using Adherent Cells, e.g., MEFs

1. The day before infection, seed 5×10^5 MEFs into each well of a 6-well plate. Culture for 14–16 h at 37 °C in a humified CO_2 incubator.

2. The following day, add polybrene at 4 μg/ml to the virion containing supernatant (*see* **Note 7**).

3. Remove medium from MEFs and replace with 2 ml virion containing supernatant per well.

4. Spin the plates in a cell culture centrifuge at $1000 \times g$ for 45 min at 30 °C. Keep cells at room temperature for minimum 30 min after spin. Alternatively, MEF cells may be infected by culturing cells with virion supernatant for 2 h in the cell culture cabinet, no spin required (*see* **Note 8**).

5. Incubate the plates at 37 °C in a humified incubator with 10 % CO_2 for 6–14 h.

6. Aspirate and discard the supernatant. Replace with 2 ml fresh medium per well. *Critical*: Some cell lines do not tolerate polybrene for more than 6 h!

7. Split the cells when they reach 90 % confluence.

3.2.2 Using Suspended Cells, e.g., THP-1 Cells

1. Seed 2.5×10^5 cells in a volume of 1 ml into each well of a 6-well plate.

2. Add polybrene at 4 μg/ml to the virion containing supernatant. Add 1 ml of virion containing supernatant to 1 ml of cell suspension in each well.

3. Spin the plates in a cell culture centrifuge at $2000 \times g$ for 90 min at 30 °C. Keep cells for a minimum 30 min in the cell culture cabinet after spin.

4. Transfer cells to a 37 °C in a humified CO_2 incubator (10 %) and leave for 6–14 h.

5. We recommend a double infection for suspension cells: Spin cells at $300 \times g$ for 5 min at 30 °C, aspirate, and discard supernatant. Add 1 ml of fresh culture media and 1 ml of virion supernatant with polybrene. Repeat procedure described in **step 3**.

6. Incubate the plates for another 6–14 h at 37 °C in a humified incubator with 10 % CO_2. Collect cells from all wells in a 10 ml Falcon tube. Spin at $300 \times g$ for 5 min at room temperature.

Discard supernatant (to remove polybrene) and resuspend the pellet in 3 ml fresh medium.

7. Split the cells when they reach a density of 2×10^6 cells/ml.

3.3 Selection of Infected Cells

To test for infection efficiency, transduced cells can be selected with a marker that is included in the transfer plasmid. This can be either, as described here, resistance to a antibiotic such as puromycin or geneticin (G418), or a fluorescent marker. If a fluorescent marker is used, infected cells can be identified using a fluorescent microscope and cells can be sorted by Flow cytometry. Here, we describe a protocol using puromycin as the selection agent.

3.3.1 Using Adherent Cells, e.g., MEFs

1. Seed four wells of a 6-well plate with 2.5×10^5 transduced cells per well, let adhere for 14–16 h. As a control, seed four wells of nontransduced cells. If it is unknown what concentration of antibiotic efficiently kills the cell line used, it is recommended that a titration treatment of the cells to be carried out with the selected antibiotic. The optimal concentration kills 100 % of nontransduced cells in 3–4 days.

2. The next day, remove supernatant and add 2 ml fresh medium to each well.

3. Add puromycin at increasing concentrations (1 µg/ml, 2 µg/ml, 4 µg) to the cells. Leave one well of transduced and nontransduced cells as control.

4. After 48–72 h, check for cell death using microscopy. For numerical assessment, a FACS based cell death assay can be performed.

5. Choose the concentration where you observe near 100 % cell death in nontransduced cells.

6. Transduced cells from the corresponding concentration can be further cultured as a polyclonal population.

7. To maintain selection pressure on the cells, keep the cells culturing using 0.5–2 µg/ml of puromycin in the cell culture medium for the first week.

3.3.2 Using Suspended Cells, e.g., THP-1 Cells

1. Seed 1×10^6 cells in at least four wells of a 6-well plate in a volume of 2 ml/well. Also seed four wells of nontransduced cells as control.

2. Add puromycin at increasing concentrations (1 µg/ml, 2 µg/ml, 4 µg/ml) to the cells. Leave one well of transduced and nontransduced cells as control.

3. The next day, check for cell death using FACS.

4. Choose the concentration in which you observe near 100 % cell death in nontransduced cells.

5. Transduced cells from the corresponding concentration can be further cultured as a polyclonal population.

6. To selection pressure on the cells, keep the cells culturing using 0.5–2 µg/ml puromycin in the cell culture medium for the first week.

3.4 Isolation of Single Cell Clones

Selection of single cell clones is not essential, but can be favorable for consistent results. It is essential, that **at least three independent clones are selected and analyzed**. In particular when using inducible promoters, we encourage selection of single cell clones with minimal expression in the uninduced stage.

3.4.1 Using Adherent Cells, e.g., MEFs

1. Trypsinize and harvest the transduced, polyclonal cell population. Centrifuge at $300 \times g$ for 5 min at room temperature. Resuspend in 1 ml medium and count the cells using a hemocytometer.

2. Seed 1000, 100, or 10 cells each in a 10 cm plate. Let colonies form over the next few days by incubating at 37 °C in a humified incubator with 10 % CO_2. Check daily for colony formation.

3. Once the colonies are distinguishable, aspirate supernatant. Wash the cells with 10 ml PBS. Aspirate PBS and add 5 ml Trypsin solution.

4. Prepare ten wells of a 48-well plate with 200 µl medium per well.

5. Check the cells under a microscope. Once the cells start to lift from the plate, use a 200 µl pipette and sterile pipette tips to aspirate cells from individual clones. Hold the plate at an angle and start picking colonies from bottom end of the plate to top to avoid mixing cells. *Critical*: Colonies have to be picked as quickly as possible or they dry out! Resuspend each clone in one well of the prepared 48-well plate.

6. Expand the clones and test basal expression and induction of your gene of interest.

3.4.2 Using Suspended Cells, e.g., THP-1 Cells

For suspension cells, single clones can be isolated using either limiting dilution or culture in soft agar (*see* **Note 9**).

Limiting dilution

1. Harvest and count cells (from Subheading 2.6).

2. Dilute cells to 100 cells/10 ml and 200 cells/10 ml in fresh medium.

3. Seed 200 µl of the low density cell dilutions (100 cells/10 ml) in a U-bottom 96-well plate. Incubate at 37 °C in a humified incubator with 10 % CO_2. Check daily for cell growth.

4. After 1 week, determine how many wells are harboring cells. Select the cell dilution in which less than 50 % of the wells have a cell clone. Most of these clones are likely to originate from a single cell.

5. Prepare ten wells of a 48-well plate with 200 µl medium per well.

6. Select ten clones and transfer them to the 48-well plate.

7. Expand the clones and test basal expression and induction of your gene of interest.

Colony formation in soft agar

1. Prepare 50 ml of culture media and keep it in a 37 °C water bath.

2. Setup a water bath to 50 °C to keep agar solution.

3. Slowly heat a 3 % agar solution in a microwave. Stopping every 30 s to avoid spilling. Make sure all the agar is dissolved and solution is smooth and liquid. Equilibrate it in the 50 °C water bath until used.

4. Harvest cells (from Subheading 2.6), spin at $300 \times g$ for 5 min at room. Resuspend cells in RPMI/20 % FCS and count.

5. Dilute cells in media with 20 % FCS to a concentration of 2000 cells/ml.

6. Seed cells at 1000 (500 μl), 250 (125 μl), and 62 (30 μl) cells/well.

7. Add 5 ml of agar to 50 ml of warm RPMI to obtain a 0.3 % agar solution.

8. Quickly add 3 ml of the agar mix per well, swirling to mix cells and media. Avoid creating bubbles.

9. Let the agar set in the plate in the tissue culture hood with the lid open to avoid condensation on the lid.

10. Once set, place 7 ml of sterile water in the gaps between the wells of the plate. This maintains the plate humid and avoids agar from drying out.

11. Place plates back into the incubator and incubate for 10–15 days.

12. To culture single colonies, add 100 μl of fresh culture media containing 10 % FCS and necessary supplements in a U-bottom 96-well plate.

13. Collect single colonies with a 100 μl tip attached to a pipette to reduce the risk of contamination. Place tip in the 100 μl in a 96-well plate, pipetting gently up and down.

14. Place the 96-well plate in the incubator and culture for 5–7 days. Once colonies grow, expand to larger culture dishes for further analysis.

3.5 Testing of Single Cell Clones (Inducible Gene Expression)

1. For nonadherent cells, seed each clone in at least two wells of a 24-well plate (250,000 cells; 0.5 ml medium/well). Also seed nontransduced cells as a control.

2. Add 5–20 ng/ml doxycycline in one of the wells per clone; leave the other well untreated (*see* **Note 10**).

3. 4 h after addition of doxycycline, lyse the cells in 100 µl lysis buffer of choice (*see* **Note 11**).

4. Test expression of your protein of choice by Western blot analysis.

5. For adherent cells, these same steps are followed except 6-well plates are used instead of 24-well plates.

4 Notes

1. Transfer plasmids may contain markers themselves. They may have fluorescent markers such as GFP or Texas-Red or protein-tags such as HA or FLAG. The expression of those markers may be used to identify cells that have been successfully infected. Infected cells can be identified by Western blotting or flow cytometry and selection may be done by single cell sorting. HA and FLAG tags may also be useful for immunoprecipitation assays. Those markers may be dependent or independent of expression of the gene of interest.

2. Lentiviruses are very effective in infecting primary cells and this protocol is therefore highly suitable for immortalization of primary cells. Examples are immortalization of primary MEFs with the SV 40 large T antigen [9] or the generation of factor dependent myeloid cells from fetal liver progenitor cells using HoxB8 or HoxA9 [10, 11].

3. Addition of the cationic polymer polybrene has been shown to increase virion absorption in target cells [12]. Additionally we recommend centrifugation of supernatants onto cultured cells (spin infection) to maximize transduction rates.

4. For cell lysis, the buffer of choice may depend on whether the protein is cytosolic or nuclear. For cytosolic protein extraction, NP40 buffer (or equivalent) containing protease inhibitors and for nuclear protein extraction, RIPA buffer may be used.

5. All materials are to be kept sterile and only cell culture grade reagents and plastic ware are recommended. Media and solutions are stored at 4 °C and warmed to 37 °C before usage. Avoid freeze thawing of chemicals and reagents.

6. Appropriate waste management for any solutions and plastic ware containing, or in contact with lentiviral particles needs to be in place. In particular, solutions containing virions need to be inactivated using 4 % hypochlorite solution. All work is carried out in PC2 facilities or according to local regulations.

7. Thaw virion supernatant in a 37 °C water bath immediately before use.

8. Lentiviruses are temperature sensitive and should be used for infections at the optimal temperature of 30–32 °C. Higher

temperature may inactivate the virion and decreases efficiency of infection.

9. The limiting dilution method is relatively simple, but it is difficult to identify whether the clones originate from a single cell or from a small number of cells. Colony forming assay in soft agar is more elaborate but identifies single cell clones with higher accuracy.

10. If the gene of interest induces cell death, a cell death inhibitor (e.g., Q-VD-OPh for caspase-dependent apoptosis) can be added to the cell to avoid loss of protein due to cell death.

11. If the protein of interest is tagged with a fluorescent marker, protein expression should be tested by flow cytometry.

References

1. Boyd JM, Gallo GJ, Elangovan B, Houghton AB, Malstrom S, Avery BJ, Ebb RG, Subramanian T, Chittenden T, Lutz RJ et al (1995) Bik, a novel death-inducing protein shares a distinct sequence motif with Bcl-2 family proteins and interacts with viral and cellular survival-promoting proteins. Oncogene 11(9):1921–1928

2. Wang K, Yin XM, Chao DT, Milliman CL, Korsmeyer SJ (1996) BID: a novel BH3 domain-only death agonist. Genes Dev 10(22):2859–2869

3. Inohara N, Ding L, Chen S, Nunez G (1997) harakiri, a novel regulator of cell death, encodes a protein that activates apoptosis and interacts selectively with survival-promoting proteins Bcl-2 and Bcl-X(L). EMBO J 16(7):1686–1694

4. Zong WX, Lindsten T, Ross AJ, MacGregor GR, Thompson CB (2001) BH3-only proteins that bind pro-survival Bcl-2 family members fail to induce apoptosis in the absence of Bax and Bak. Genes Dev 15(12):1481–1486

5. Stewart SA, Dykxhoorn DM, Palliser D, Mizuno H, Yu EY, An DS, Sabatini DM, Chen IS, Hahn WC, Sharp PA, Weinberg RA, Novina CD (2003) Lentivirus-delivered stable gene silencing by RNAi in primary cells. RNA 9(4):493–501

6. Marini B, Kertesz-Farkas A, Ali H, Lucic B, Lisek K, Manganaro L, Pongor S, Luzzati R, Recchia A, Mavilio F, Giacca M, Lusic M (2015) Nuclear architecture dictates HIV-1 integration site selection. Nature 521(7551):227–231

7. Moujalled DM, Cook WD, Murphy JM, Vaux DL (2014) Necroptosis induced by RIPK3 requires MLKL but not Drp1. Cell Death Dis 5:e1086

8. Tevethia MJ (1984) Immortalization of primary mouse embryo fibroblasts with SV40 virions, viral DNA, and a subgenomic DNA fragment in a quantitative assay. Virology 137(2):414–421

9. Vince JE, Wong WW, Khan N, Feltham R, Chau D, Ahmed AU, Benetatos CA, Chunduru SK, Condon SM, McKinlay M, Brink R, Leverkus M, Tergaonkar V, Schneider P, Callus BA, Koentgen F, Vaux DL, Silke J (2007) IAP antagonists target cIAP1 to induce TNFalpha-dependent apoptosis. Cell 131(4):682–693

10. Brumatti G, Salmanidis M, Kok CH, Bilardi RA, Sandow JJ, Silke N, Mason K, Visser J, Jabbour AM, Glaser SP, Okamoto T, Bouillet P, D'Andrea RJ, Ekert PG (2013) HoxA9 regulated Bcl-2 expression mediates survival of myeloid progenitors and the severity of HoxA9-dependent leukemia. Oncotarget 4(11):1933–1947

11. Salmanidis M, Brumatti G, Narayan N, Green BD, van den Bergen JA, Sandow JJ, Bert AG, Silke N, Sladic R, Puthalakath H, Rohrbeck L, Okamoto T, Bouillet P, Herold MJ, Goodall GJ, Jabbour AM, Ekert PG (2013) Hoxb8 regulates expression of microRNAs to control cell death and differentiation. Cell Death Differ 20(10):1370–1380

12. Davis HE, Morgan JR, Yarmush ML (2002) Polybrene increases retrovirus gene transfer efficiency by enhancing receptor-independent virus adsorption on target cell membranes. Biophys Chem 97(2–3):159–172

Chapter 20

Proteomic Profiling of Cell Death: Stable Isotope Labeling and Mass Spectrometry Analysis

Andrew I. Webb

Abstract

Proteins directly control almost all cellular processes and researchers in many biological areas routinely use mass spectrometry for the characterization of proteins. Amongst a growing list of available quantitative proteomic techniques, Stable Isotope Labeling by Amino acids in Culture (SILAC) remains one of the most simple, accurate, and robust techniques for cultured cellular systems. SILAC enables strict quantitative peptide measurements, thus removing false positives and facilitates large-scale kinetics of entire proteomes. In this, chapter we describe an optimized labeling strategy and experimental design for SILAC workflows for characterizing the components downstream of cell death stimuli.

Key words Mass spectrometry, Proteomics, SILAC, Proteome quantitation

1 Introduction

Programmed cell death is an essential cellular mechanism for regulating normal physiological processes and is crucial during development and in the maintenance of a healthy immune system [1]. In addition to its role in the controlled removal of cells, the selective induction of apoptosis in diseases such as cancer has become an important focus. Thus it is imperative to identify the pathways and components that are involved in the cell death pathways and to characterize their role under different conditions and stimuli. The family of Cys-dependant Asp-specific proteases called caspases drive the key mechanisms involved in programmed cell death. These caspases cleave C-terminally of aspartate and are made up of 11 distinct functional genes in the human proteome. Upon activation, these caspases initiate a downstream cascade of activation, deactivating, and translocation events on substrate proteins [2].

Several methods have been introduced to identify substrates cleaved in a caspases-dependant manner, including those that can identify the exact location of cleavage sites. Mass spectrometry based methods can be divided into those that aim to determine the

Hamsa Puthalakath and Christine J. Hawkins (eds.), *Programmed Cell Death: Methods and Protocols*, Methods in Molecular Biology, vol. 1419, DOI 10.1007/978-1-4939-3581-9_20, © Springer Science+Business Media New York 2016

peptides at around the specific cleavage site, and those that are applied at the global proteome level for the identification of substrates and related biological effects (both up- and downstream of caspase activation). The former methods mostly entail blocking or modifying all preexisting N-termini and depleting the subsequent de novo generated N-termini by the protease (which acts a handle for covalent attachment). Both positive and negative selection methods have been developed and have proven highly effective at mapping substrate cleavage sites [3–10]. A significant advantage of these peptide-based methods is that they identify the exact site of protease cleavage. However, as it is limited to a single peptide for protein identification, they are generally limited in the number of identifications possible (i.e., peptide parameters may not be optimal for MS identification). Additionally, contextual information about the substrate protein is not readily detectable (abundance changes and posttranslational modifications).

In contrast, global approaches applied at the whole proteome level aim to quantitate as many proteins as possible (without necessarily determining the site of cleavage) and do not bias MS analysis toward caspase substrates. Historically, two-dimensional gel electrophoresis has been used to differentiate proteome differences during cell death [11–13]; however, these techniques are limited in throughput, sensitivity, and reproducibility. Recent developments in Ultra-high Pressure nanoflow Liquid Chromatography (UPLC), mass spectrometry, and experimental workflows have dramatically improved the depth of sequencing and run-to-run reproducibility and have circumvented the methods mentioned above. In particular, Stable Isotope Labeling with Amino acids in Culture (SILAC) facilitates highly accurate peptide-based ratio information for every peptide identified. Comprehensive proteome coverage and accurate quantitation allows for subtle (>1.5-fold) quantitative measurements of individual proteins following triggers of cell death from the entire global proteome of a cell. This extremely high level of specificity in such comprehensive data sets is crucial for teasing out subtle nuances of death cellular signaling.

SILAC introduces a mass difference between two proteomes facilitating a reference for relative quantitation. As the two proteomes are only distinguishable by the isotope used (12C/13C, 14N/15N, and 1H/2H), they are not subject to variations in both sample processing and between LC/MS runs and are generally considered much more accurate than label-free strategies. Stable isotope methods can be subdivided into two classes: (1) Metabolic—that utilize biological incorporation of the isotopes into cells (typically the essential amino acids Arginine and Lysine are used). (2) Chemical—that utilize covalent attachment of a reagent to introduce a mass tag. This protocol will be limited to metabolic incorporation and any culture system where the amino acid source is defined can be labeled with SILAC.

In this protocol, we describe the adaption and testing of cells in SILAC media for efficient incorporation (which is particularly important when working with new cell lines for the first time). SILAC and conventional sample analysis differ only in the preparation of the media, adaption of cells, and mixing of the protein lysates prior to sample processing and MS.

Here, we also describe the use of UPLC coupled to nano-electrospray ionization mass spectrometry (nano-LCMS) for SILAC sample data acquisition (see **Note 1**). As researchers will have access to a variety of nano-LCMS systems from different vendors, we will describe the general principles and minimum requirements for SILAC sample analysis. For an introduction to MS peptide sequencing and proteomics, please refer to refs. 14–16. For data analysis, we will briefly detail the important steps using MaxQuant as an example data analysis workflow (that can be used with Thermo Orbitraps, Bruker QTOFs, and Sciex TripleTOFs).

Experimental design—SILAC involves incorporating stable isotope containing amino acids during cellular protein synthesis and typically involves Arginine and Lysine containing a combination of substituted 13C and 15N atoms in the amino acid molecule. Two populations of cells are grown in separate medium formulations in (1) in light media (containing the natural isotope abundance) and (2) heavy medium containing the SILAC amino acids chosen (see Fig. 1). The aim is to completely replace the labeled amino acid in the proteome (typically take five to eight cell passages labeled media). As SILAC depends on MS for the readout, even a small percentage of unlabelled amino acid in the labeled population can contribute to the unlabelled signal, thereby introducing quantification errors into the data. In practice, through the process of cell division and protein degradation, proteomes are generally rapidly labeled. At least 97 % incorporation should be seen before beginning an experiment (and confirmed for each new cell line before an experiment is attempted). When lysates from light and heavy labeled samples are mixed together, processed, and analyzed with MS, they are differentiated by the residue specific mass corresponding to the labeled amino acid residues in the peptide. As the quantitative information is encoded in the SILAC residue, they must be selected specifically for the experiment (i.e., for trypsin it is recommended to label Arg and Lys, as it cleaves at these basic residues, leaving charged C-termini that helps facilitate MS sequencing of all digested peptides). The area under the curve of the light and heavy labeled peptides provides the quantitative information for comparison of there relative abundance.

The following protocols describe the steps required for efficient and robust generation of samples for highly accurate quantitative comparison. The labeling conditions and samples preparation has been optimized over a wide range of human and murine cell lines and provides significant quantitative coverage of the proteome

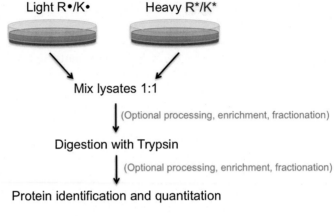

Light R•/K•   Heavy R*/K*

Mix lysates 1:1

(Optional processing, enrichment, fractionation)

Digestion with Trypsin

(Optional processing, enrichment, fractionation)

Protein identification and quantitation

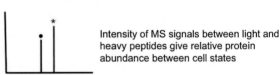

Intensity of MS signals between light and heavy peptides give relative protein abundance between cell states

Fig. 1 Experimental design flow chart for SILAC. Cells are prepared in natural (light) amino acids and "heavy" SILAC amino acids. Cells incorporate the heavy amino acids after five to eight cell doublings and generally have no effect on morphology or growth rates. When light and heavy cell populations are mixed, they remain distinguishable by MS by the encoded isotopic mass differences. Protein abundances are determined from median relative MS peptide signal intensities. SILAC provides highly accurate relative quantification without any chemical derivatization or manipulation

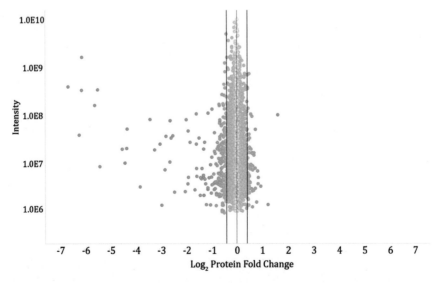

Fig. 2 A typical dataset from a SILAC experiment. Changes in MEF cell protein levels during expression of a necroptosis-inducing mutant of MLKL. Log2 protein expression ratios (WT-MLKL versus Mutant-MLKL) for >4200 mouse proteins (x axis) plotted against the protein intensity (summed peptide intensities per protein) (y axis)

(>5000 proteins quantifiable with optimized recent high-end UPLC and MS instruments). An example of the expected results can be seen in Fig. 2. Induced expression of activating mutant of MLKL was used to stimulate programmed necrosis (necroptosis) in a murine embryonic fibroblasts (MEFs) [17] and SILAC experiment was designed to identify downstream effects in this relatively uncharacterized death signaling pathway. In this example, 61 proteins (negative fold change from >4200 proteins) were significantly unregulated after a 3-h induction of the mutant MLKL protein.

2 Materials

Organic solvents are HPLC grade and reagents of the highest grade available are recommended.

2.1 Reagents

1. Cell line of choice (U937 cells used in the example shown).
2. Cell culture medium (DMEM or RPMI SILAC media—i.e., commercial cell culture medium without arginine, lysine in this example).
3. Dialyzed fetal bovine serum (FBS).
4. Glutamine.
5. L-Arginine monohydrochloride (l-Arg).
6. L-Lysine hydrochloride (L-Lys).
7. SILAC amino acids: L-arginine-13C6 monohydrochloride or L-arginine-13C615N4 hydrochloride and L-lysine-13C615N2 hydrochloride or L-Lysine-4,4,5,5-D4 hydrochloride (see **Note 2**).
8. Proteomics grade modified trypsin.
9. Urea.
10. SDS.
11. Tris–HCl.
12. Dithiothreitol (DTT).
13. Iodoacetamide.
14. Ammonium bicarbonate.
15. Trifluoroacetic acid (TFA).
16. Formic acid (FA).
17. Acetonitrile (ACN).
18. Sartorius Vivacon 500 30k MWCO filter units.

2.2 Equipment

1. Mass spectrometer with nano-electrospray source (best results from high resolution instrument capable of resolving >30,000 resolution—Orbitraps or QTOFs).

2. Nanoflow HPLC (best results from ultra-high pressure instruments—Waters NanoAcquity or Thermo UHPLC).

3. Protein and peptide identification software tools (*see* **Note 3**).

4. Quantitation software (*see* **Note 4**).

5. Bench-top microcentrifuge (>16,000×*g*, with cooling).

6. Oven (for 37 °C incubation for trypsin digestion).

7. Vacuum evaporator centrifuge.

8. Waters NanoAcquity trapping column (150 μm ID 5 μm Symmetry×20 mm)—or equivalent.

9. Waters NanoAcquity analytical column (75 μm ID 1.7 μm BEH×250 mm)—or equivalent (*see* **Note 5**).

10. Filter-aided sample preparation microfuge tubes (FASP).

2.3 Buffers and Reagent Preparation

1. Phosphate buffered saline (PBS).

2. Amino acid stock solutions: Prepare concentrated 0.1 ml stock solutions by dissolving amino acids in PBS or FBS free culture medium. Arginine (0.798 mM), lysine (0.398 mM) are prepared as 500 times concentration stocks for use in DMEM (RPMI Arg 0.925 mM and Lys 0.274 mM). Filter amino acid solutions through a 0.22-μm syringe filter and store at −20 °C for up to 12 months.

3. Stable isotope-labeled amino acid stock solutions are prepared in the same manner but the increased molecular weight of the amino acids bearing 13C or 15N should be taken into account for equimolar amounts in both light and heavy media.

Filter Aided Sample Preparation Buffers

1. Lysis buffer: 100 mM Tris–HCl, pH 8.0, 100 mM DTT, 4 % SDS.

2. Wash buffer 1: 100 mM Tris–HCl, pH 8.0, 8 M Urea.

3. Wash buffer 2: 50 mM Ammonium bicarbonate.

4. Digestion buffer: 50 mM Ammonium bicarbonate and trypsin at 1:100 trypsin:protein ratio.

3 Methods

3.1 Preparation of SILAC Media (Triplex Labeling Optional)

1. Measure out 45 ml media in a 50 ml tube and add 5 ml dialyzed fetal calf serum.

2. Add 0.1 ml Arginine and Lysine stock solutions to tubes labeled as follows:

Arg 0—for light label (to Light 50 ml Tube).

Arg 6—for medium label (to Medium 50 ml Tube).

Arg 10—for heavy label (to Heavy 50 ml Tube) (for triplex only).

Lys 0—for light label (to Light 50 ml Tube).

Lys 4—medium label (to Medium 50 ml Tube).

Lys 8—for heavy label (to Heavy 50 ml Tube) (for triplex only). (*see* **Note 6**).

3. Add Glutamine and antibiotics as required.

3.2 Adaptation of Cells from Normal to SILAC Media

1. Passage cells in a 6-well plate, growing in normal medium to 80 % confluency and seed 10–15 % (or appropriate for the specific cell line) of the original cells into two culture dishes, each containing light and heavy SILAC medium, respectively.

2. Change or subculture medium (using either light or heavy SILAC medium) every 2–3 days.

3. From the seventh passage, keep a small number of heavy labeled cells (2e5 cells) to check for SILAC incorporation (for this sample proceed to Subheading 3.3) (*see* **Note 7**).

4. For experiment samples, expand the last passage (typically at passage 8 is more than sufficient for most cell lines) into the required number of cells (5e5–2e6 for Shotgun proteomics).

3.3 Sample Preparation of Cell Lysates

1. Wash cells 2× in PBS and lyse cells with Lysis buffer (30 μl per 1e6 cells).

2. Vortex the lysate for 1 min.

3. To the 30 μl of Lysis cell mixture add 170 μl 8 M urea wash buffer and pellet the debris by centrifuging for 10 min at 16,000–20,000×g in a bench top centrifuge at 18 °C.

4. Collect the supernatant in a Sartorious vivcon 500 filter unit, taking care to avoid DNA or the cell pellet (*see* **Note 8**).

5. Spin filter unit at 14,000×*g* for 15 min and add 100 μl iodoacetamide to a final concentration of 25 mM to alkylate cysteines. Gently vortex to mix and incubate in the dark for 20 min. Then spin at 14,000×*g* for 15 min at 18 °C.

6. Aspirate and discard flow through.

7. Add 200 μl 8 M Urea wash buffer and spin at 14,000×*g* for 15 min at 18 °C. Repeat two more times.

8. Aspirate and discard flow through.

9. Add 200 μl 50 mM ammonium bicarbonate and spin at 14,000×*g* 15 min at 18 °C.

10. Aspirate and discard flow through.

11. Repeat **step 9** two more times.

12. On the final wash, leave the remaining flow through to prevent unwanted drying of the filter membrane.

13. Add Trypsin at an enzyme: substrate ratio of 1:100 and incubate at 37 °C overnight.

14. Transfer filter unit to a FASP microfuge tube and spin at 14,000×*g* at 18 °C for 8 min.

15. Add 40 μl 50 mM ammonium bicarbonate and spin at 14,000×*g* at 18 °C for 8 min. Repeat once more.

16. Acidify pooled flow through to a final concentration of 1 % Formic acid.

17. Concentrate flow through using Vacuum evaporator centrifuge until dry.

18. Proceed to nanoscale LC–MS to identify proteins and peptides in a shotgun analysis.

3.4 Mass Spectrometry, Peptide Identification, and Protein Quantitation

1. Inject approximately 1–2 μg of peptide into a column for nanoflow LC–MS analysis. Typical gradient lengths of 2–4 h per sample offer the highest yield of identifications per analysis time. By increasing the number of replicates and utilizing the "match-between runs" feature of the MaxQuant we observed significantly more identifications.

2. From the acquired data, identify peptides and proteins using MaxQuant search software [18] making sure to include the modified masses of SILAC amino acids to the search parameters (*see* **Note 9**).

3. Find the ratio of summed signal intensities (area under the curve) from the light and heavy peptide extracted ion chromatograms to give the relative peptide abundance ratio between the two cell states (found in the Peptides.txt output if using Maxquant).

4. Obtain peptide ratios for all validated peptides in a protein and average these to give the average protein ratio (found in the Poteins.txt output if using Maxquant).

5. Statistical analysis of replicate samples can be performed using a variety of software packages (R, Matlab or Perseus or equivalent). Typically a minimum of three biological replicates is performed and a *t*-test performed on the ratios of all identified peptides per protein group.

4 Notes

1. The setup and operation protocol is outside the scope of this protocol and would be left to the host instrument facility.

2. Either combination of l-Arg and L-Lys are amenable for duplex experiments or all can be used in triplex labeling experiments.

3. Maxquant used in this example—Many alternatives available including Mascot, MS+GF+, SpectrumMill, XTandem, SEQUEST, or equivalent.

4. MaxQuant is used in this example. Alternatives include MSQuant, SpectrumMill, Proteome discoverer, OpenMS, or other mass spectrometer instrument vendors' software capable of handling SILAC data.

5. Longer direct injection columns can provide increased peak capacity and identifications at the expense of long injection times and longer analysis times.

6. Can be stored at 4 °C for up to 2 months.

7. You must perform this labeling check if this is the first time SILAC is used with this cell stock to avoid incomplete incorporation and potential errors in quantification. For phagocytic cell lines, Arginase conversion of heavy Arginine to heavy Proline should also be monitored at this step (and if present it can generally be rectified by doubling the free Proline concentration in the media).

8. DNA may not pellet completely and will appear as a gel-like clump, which is easily removed when aspirating with a pipette.

9. For up to date explanations of the software refer to the MaxQuant webpage—http://141.61.102.17/maxquant_doku/doku. php?id=start and the MaxQuant summer school tutorial videos—https://www.youtube.com/channel/UCKYzYT m1cnmc0CFAMhxDO8w0

References

1. Strasser A, O'Connor L, Dixit VM (2000) Apoptosis signaling. Annu Rev Biochem 69(1): 217–245

2. Chowdhury I, Tharakan B, Bhat GK (2008) Caspases—an update. Comp Biochem Physiol B Biochem Mol Biol 151(1):10–27

3. Gevaert K, Goethals M, Martens L, Van Damme J, Staes A, Thomas GR, Vandekerckhove J (2003) Exploring proteomes and analyzing protein processing by mass spectrometric identification of sorted N-terminal peptides. Nat Biotechnol 21:566–569

4. Van Damme P, Martens L, Van Damme J, Hugelier K, Staes A, Vandekerckhove J, Gevaert K (2005) Caspase-specific and nonspecific in vivo protein processing during Fas-induced apoptosis. Nat Methods 2:771–777

5. Staes A, Van Damme P, Helsens K, Demol H, Vandekerckhove J, Gevaert K (2008) Improved recovery of proteome-informative, protein N-terminal peptides by combined fractional diagonal chromatography (COFRADIC). Proteomics 8:1362–1370

6. Mahrus S, Trinidad JC, Barkan DT, Sali A, Burlingame AL, Wells JA (2008) Global sequencing of proteolytic cleavage sites in apoptosis by specific labeling of protein N termini. Cell 134:866–876

7. Impens F, Colaert N, Helsens K, Ghesquiere B, Timmerman E, De Bock PJ, Chain BM, Vandekerckhove J, Gevaert K (2010) A quantitative proteomics design for systematic identification of protease cleavage events. Mol Cell Proteomics 9:2327–2333

8. Impens F, Colaert N, Helsens K, Plasman K, Van Damme P, Vandekerckhove J, Gevaert K (2010) MS-driven protease substrate degradomics. Proteomics 10:1284–1296

9. Kleifeld O, Doucet A, auf dem Keller U, Prudova A, Schilling O, Kainthan RK, Starr AE, Foster LJ,

Kizhakkedathu JN, Overall CM (2010) Isotopic labeling of terminal amines in complex samples identifies protein N-termini and protease cleavage products. Nat Biotechnol 28:281–288

10. Kleifeld O, Doucet A, Prudova A, auf dem Keller U, Gioia M, Kizhakkedathu JN, Overall CM (2011) Identifying and quantifying proteolytic events and the natural N terminome by terminal amine isotopic labeling of substrates. Nat Protoc 6:1578–1611

11. Thiede B, Dimmler C, Siejak F, Rudel T (2001) Predominant identification of RNA-binding proteins in Fas-induced apoptosis by proteome analysis. J Biol Chem 276:26044–26050

12. Thiede B, Siejak F, Dimmler C, Rudel T (2002) Prediction of translocation and cleavage of heterogeneous ribonuclear proteins and Rho guanine nucleotide dissociation inhibitor 2 during apoptosis by subcellular proteome analysis. Proteomics 2:996–1006

13. Agard NJ, Wells JA (2009) Methods for the proteomic identification of protease substrates. Curr Opin Chem Biol 13:503–509

14. Aebersold R, Mann M (2003) Mass spectrometry-based proteomics. Nature 422(6928):198–207

15. Steen H, Mann M (2004) The ABC's (and XYZ's) of peptide sequencing. Nat Rev Mol Cell Biol 5(9):699–711

16. Medzihradszky KF (2005) Peptide sequence analysis. Methods Enzymol 402:209–244

17. Hildebrand JM, Tanzer MC, Lucet IS, Young SN, Spall SK, Sharma P, Pierotti C, Garnier JM, Dobson RC, Webb AI, Tripaydonis A, Babon JJ, Mulcair MD, Scanlon MJ, Alexander WS, Wilks AF, Czabotar PE, Lessene G, Murphy JM, Silke J (2014) Activation of the pseudokinase MLKL unleashes the four-helix bundle domain to induce membrane localization and necroptotic cell death. Proc Natl Acad Sci 111(42): 15072–15077

18. Cox J, Matic I, Hilger M, Nagaraj N, Selbach M, Olsen JV, Mann M (2009) A practical guide to the MaxQuant computational platform for SILAC-based quantitative proteomics. Nat Protoc 4(5):698–705

INDEX